6.95
Pub
377 p
1968

LABORATORY
GUIDE

for Biological Science

WILLIAM T. KEETON

MICHAEL W. DABNEY

ROBERT E. ZOLLINHOFER

CORNELL UNIVERSITY

ILLUSTRATED BY *Paula DiSanto Bensadoun* AND *Monica Howland*

W · W · NORTON & COMPANY · INC · NEW YORK

ISBN 0 393 09823 0
PRINTED IN THE UNITED STATES OF AMERICA
3 4 5 6 7 8 9 0

CONTENTS

NOTE TO THE INSTRUCTOR

Despite ever-widening use of excellent Biological Sciences Curriculum Study programs in the high schools, many colleges and universities find that although entering students are eager to learn and understand science, they have very limited laboratory experience. This book is written for such students and for others with a limited background in biology, and is intended to help instructors provide students with a meaningful and satisfying laboratory experience. Although some of the experiments in this book are suitable for use with honors-level students or others with superior laboratory background, we recommend instead that each school develop an honors program to suit the special needs of its own students.

Although this guide is adaptable to any integrated laboratory program and will properly complement most introductory biology textbooks, its topical organization and evolutionary undertone parallel the style and approach of W. T. Keeton's *Biological Science*.

Laboratory Guide for Biological Science represents an effort to integrate observational and experimental biology. In most of its exercises, there is sufficient student direction to free the instructor for individualized help; many questions — some of which are unanswerable and are included primarily to stimulate discussion — build a dialogue with the student and continually invite his active participation in thinking about biological problems. In writing this book, we have aimed (1) to give students the opportunity to see and handle materials, alive whenever possible, they might not otherwise see; (2) to help them develop a feeling for the design and limitations of experiments, and to offer them some of the excitement of independent discovery; (3) to help them gain an appreciation for the diversity of life; (4) to help them develop some proficiency with commonly used laboratory equipment; and (5) to foster the gradual development of self-sufficiency in the conduct of laboratory experiments and observations, in the handling of organisms and other laboratory materials, and in the analysis of data.

Each of the thirty topics includes enough material to fill four hours of laboratory work in either one four-hour or two two-hour sessions. However, the book is adaptable to two-and-a-half hour or three-hour sessions meeting only once each week. It should be emphasized that it is *essential* to delete materials for a three-hour laboratory session, such as that conducted at Cornell and at many other universities and colleges. The organization of each topic lends itself easily to the omission or addition of experiments.

To keep preparation time to a minimum, we have written a complete Instructor's Guide. It includes full directions for preparing each part of each topic, suggested deletions and rearrangements, logistics for experiments extending beyond a single laboratory period, and other suggestions intended to make the job of ordering and preparing as trouble-free as possible. No supplementary source material should be required for the preparator. Finally, there is a capsule explanation of the coverage within each topic, designed to help the instructor decide which topics to include and which to exclude from the laboratory sequence. A copy of the Instructor's Guide accompanies each examination copy of the *Laboratory Guide for Biological Science*, and additional copies may be obtained upon request from the publisher.

This book assumes the presence of simple materials and inexpensive apparatus readily available in most college laboratories or easily within modest budgets. Live materials are a major expense, and we have recommended them liberally. If the topics are followed closely, a few additional slides and other pieces of equipment will probably be necessary. One item, a closed Plexiglas manometer system, is used in three exercises, and can be made in most college shops; it will also be available from one biological supply company. An option (Appendix 4) is offered to the single topic in which the manometer experiments are the only ones performed (Topic 3, Enzyme Activity), and the other experiments requiring the manometer can be omitted or similiar exercises can be substituted.

Instructors concerned about systematic coverage of plants and animals should note that much information about organisms is conveyed in experiments and observations throughout the book. Moreover, in each topic the student classifies each organism he has not previously encountered, by using a listing in Appendix 2, and enters that organism's generic name in a "phylogenetic tree." As the student's experience with organisms expands, he will begin to see and understand relationships between the major groups. There are five topics devoted to systematic coverage of plants and invertebrate animals; vertebrate animals are used so extensively in other topics that no special exercise on them is included.

The authors welcome comments and suggestions for improvement from users of this book, students and faculty alike.

Foremost among contributors to this book are the artists, Paula DiSanto Bensadoun and Monica Howland. Mrs. Bensadoun skillfully prepared the halftones direct from our dissections, and she made several other figures. Mrs. Howland drafted the other line drawings, tables, and graphs, often from our very rough sketches. Both biologists, they have contributed immeasurably to the teaching value of this book, and the authors are very grateful for their help.

Steven C. Carlin of the Graduate Department of Biochemistry at Brandeis University critically read and edited many parts of the manuscript. He prepared Topic 30 (Microbial Ecology) and Appendix 7 (Microbiological Technique). Figures for these sections and FIGURE 4-1 were made from his sketches. The authors are indebted to him for tireless help and criticism.

Credit is due to Dr. T. H. Wilson of Harvard University for Topic 11 (Active Transport), an adaptation of exercises developed in conjunction with Dr. G. Wiseman of the University of Sheffield, England. Dr. Barbara Stay of the University of Iowa is due credit for ideas that have stimulated the development of many parts of this book; she also collected the electron micrographs that are reproduced in Topic 1 (The Structure of Cells).

The Photography Department of the Carolina Biological Supply Company generously provided many of the photographic illustrations. We are indebted to Mr. T. H. Mackintosh and to William West, both of that department, for their assistance. Special thanks are due Dr. Raymond O. Flagg, Head of Botany, who gave willingly of his time to discuss the availability of biological materials. The company President, Dr. T. E. Powell, Jr., and Dr. Kenneth Perkins, Head of Zoology, both helped us in many ways.

Ward's Natural Science Establishment, Triarch Incorporated, the Biological Sciences Curriculum Study, the U. S. Forest Products Laboratory, Bausch & Lomb, and the AO Instrument Company all provided illustrations and other forms of assistance. We gratefully acknowledge their generous help.

Richard B. Root, Carol McFadden, Kenneth Hall, Herbert W. Israel, Ronald Howard, Susan Gaufin, Mary Basl, and a host of other Cornell colleagues read parts of the manuscript or provided illustrations, and made many valuable suggestions. Sister Ann Martin of Seton Hill College read Topic 18 and suggested improvements; Arthur C. Benke of the University of Georgia read Topic 29 and made similar suggestions. Many Cornell students have served as "guinea pigs" for new experiments.

Mrs. Bertha J. Blaker typed more than half the manuscript and we are grateful for her cheerfulness in meeting our deadlines for this mammoth task. Mrs. Walter Biesemeyer and her Mountain House in Keene, New York, provided the very favorable environment in which substantial parts of the manuscript were written.

It is a special pleasure to acknowledge the patience, help, and encouragement extended by the staff of W. W. Norton & Company, including Kitty D. Wright, who copyedited our manuscript, and Margaret F. Plympton, who designed the book.

Finally, we should like to thank patient wives and numerous friends without whose support, encouragement, and forbearance this book would never have been written.

Ithaca, N.Y.
March, 1968

W.T.K.
M.W.D.
R.E.Z.

NOTE TO THE STUDENT

While the science of two centuries ago was little more than a plaything of the wealthy, we now live in a civilization dominated by science and its products. Whether such dominance is desirable or not, whether scientific activity is destined to increase in importance or to be swallowed up in the rebirth of some other human concern, you have a responsibility to try to understand the goals and methods of science that have engendered this knowledge explosion. Your insights into the nature of science and its meaning for you as an individual in a complex society and civilization will develop slowly. Though this course will start you on the road toward developing such insights, you will not automatically possess them all when you have finished the course. One of the most important places in which to begin understanding science is the laboratory, since that is where the real business of science proceeds.

Growing to understand science is a little like getting to know a person: someone else can give you a description, but the words always fall short of personal contact. It is the same way with science; if you are to know it well, you must deal with it intimately. To offer you a series of opportunities for this contact, to help you begin to formulate a meaningful understanding of scientific endeavor, is the fundamental goal of this book.

The purpose of your laboratory experience, then, is to clarify the meaning and methods of science by giving you the opportunity to become a part-time scientist. It will not teach you to become a professional biologist, though it can help to give you background that would be important if you entered a career in the natural sciences. With your effort, this book and your instructor together can help you to learn to think effectively about biological problems; they will afford you first-hand contact with a wide variety of living organisms and a wide variety of biological ideas; finally, they will encourage you to do some biological exploring yourself, and occasionally to carry out experiments of your own design.

Attention to the following suggestions will save you much time and trouble in the future.

1. Always *preread* the laboratory work, and come to your laboratory session not only with a clear idea of what you are to do and in what order you will do it, but also prepared to ask questions about anything you do not understand. Prereading will help you know what to bring to the laboratory and will clarify your instructor's supplementary directions. During the sessions, you will not have time to waste on an initial reading of the instructions, especially in laboratories where experiments are to be performed.

2. One of the most important features of scientific inquiry is exchange of information. A scientist publishes his results to make them known to others. You should encourage the same kind of exchange with your neighbors and with your instructor. Remember that your laboratory instructor, whether graduate student or professor, is a professional biologist. While this does not mean that he (or anyone else) has all the answers, your contacts with him and the degree to which you profit from his experiences and insights form a very important part of your laboratory experience.

3. This book and your instructor will pose many questions to encourage discussion and invite your active thinking about biological problems. Learn also to become a questioner yourself. Some of the questions we raise will be unanswerable, either because they have not been tested, or because by their very nature they are not susceptible to experimental test. Never allow this to discourage you; rather, let it emphasize the open-ended and undogmatic nature of scientific inquiry.

4. Always know exactly what you are doing in the laboratory and why you are doing it. Nothing could be worse than to drift through a laboratory session in a state of heady confusion. While later laboratory sessions, readings, and lectures will doubtless clarify what has come before, you should always clear up problems promptly as you encounter them. If you feel you are losing perspective during a laboratory session, ask your instructor for help.

5. Consideration for the many others who use your laboratory facilities requires that all cooperate in leaving the laboratory room as neat as possible. If possible, leave the laboratory neater than you found it. Never wait for your instructor to ask you to clean up; take the initiative to clear your own area. Make sure that microscope lenses are perfectly clean, and that the other equipment you use each week is properly returned to its storage place. Offer to help the instructor when you see him straightening up other parts of the laboratory.

6. Occasionally, demonstrations may be set up on one of the tables in the laboratory room. You are always responsible for the content of such demonstrations.

7. Bring your text to the laboratory. References to *Biological Science* (Second Edition) and *Elements of Biological Science* are listed at the beginning of each topic. If your class is using another text, your instructor will supply page references to it. Familiarize yourself with the location of additional reference books, and make liberal use of them.

8. Biology has a language of its own, which is a means of succinct description, but which can be troublesome at times. Terms you should know are indicated throughout this book in boldface; some students find it helpful to maintain a list of such terms. Most of the scientific words you will encounter are constructed of Greek and Latin roots; if you learn these roots and employ the words frequently, you will soon be comfortable even with unfamiliar scientific words in entirely new contexts.

9. *Most important of all*, remember that this book is not meant to limit your laboratory experiences. It is only a guide; where it stops, you should pick up the ball. Experiment and observe generously, not just with what is required but with anything and everything that interests you.

We sincerely hope that your contacts with science will stimulate more than a superficial curiosity about what it means to be a living organism yourself; we hope they will help you learn to distinguish between the kinds of questions you can approach as a scientist and those you cannot; we hope that they will begin to build in you a sense of the potency and limitations of science, and that they will convey some of the excitement of independent discovery that spurs men to embrace science for their lifework; finally, we trust that these contacts will enhance your ability to function meaningfully in a science-oriented society. Whether or not you should eventually choose to find a career in some natural science, this course should help lay the groundwork for an intelligently critical attitude toward science in general and biological science in particular.

You are about to embark on experiences that can be among the most enriching of your entire college career. On occasions when the going seems rough, remember that even if you become a professional scientist yourself, you may never renew these wide-ranging opportunities for close contact with so many different parts of biology. We therefore urge you to make the very most of the weeks ahead.

THE STRUCTURE OF CELLS

Biological Science, pp. 45-51, 63-92.
Elements of Biological Science, pp. 35-38, 46-66.

Advances in all branches of science often have been brought about by the development of instruments capable of extending the normal range of human sensory equipment; instrumentation has thus made possible the fruitful exploration of parts of the natural environment to which the human senses are not normally attuned. The light microscope is one such tool, and its development wrought a major revolution in biological thought by opening up the world of cells.

A light microscope is really only a sophisticated arrangement of magnifying lenses, constructed for convenient observation of small objects; simple microscopes were in use long before the principle was applied to the study of living materials.

In this laboratory, you will be introduced to a variety of living materials — mostly cells — and to the means for studying these materials with microscopy. It is supposed that you are already familiar with the basic parts and operation of the microscope; if you are not, or if your instructor directs a review, turn to the Appendix to this Topic on page 12, and study it carefully before you proceed.

PART I. MAKING WET MOUNTS

Many of the specimens you examine will be mounted in water on glass slides. Such a **wet mount** is prepared by placing a drop of material to be examined on a slide or, if the material is dry, by placing it directly on the slide and adding a drop of water. The mount is then covered with a **cover slip,** a thin piece of glass or plastic that helps prevent the drop from drying and flattens the preparation to avoid the uneven refraction that would occur if light passed through the top of a drop of fluid.

A. Human epidermal cells

Gently scrape the inside of your cheek with a clean, flat toothpick, and mount the scrapings on a clean, dry slide. Let the scrapings dry, and then add a drop of tap water. Holding a cover slip (with fingers or forceps) by its edges to avoid fingerprinting it, place one side of the cover slip at one edge of the drop. Now gently lower the cover slip over the rest of the drop; if this is done slowly, you will find that water flows evenly beneath the

Note:

Page or chapter references for the material covered in each laboratory will be given at the beginning of each topic. They refer to W. T. Keeton, *Biological Science*, 2nd ed. (New York: Norton, 1972) and W. T. Keeton, *Elements of Biological Science*, (New York: Norton, 1973).

glass. (With a little practice, you will learn that the amount of fluid required to mount a specimen depends upon the size and thickness of the specimen; if too much water is added, the cover slip may "float" about on the fluid, and if too little water is used, part of the area under the cover slip may remain dry. You can add more fluid with a dropper at the edge of the cover slip, and withdraw excess fluid by soaking it up with a bit of filter paper or toweling.) Examine the slide under low power, remembering to decrease the light intensity if necessary.

You have removed some of the **epidermal cells** that form a protective surface inside the mouth. Like similar cells all over the body, these are constantly being worn off and replaced by new cells of the same type. These cells usually come off the cheek in masses, so look for groups of cells; when you have found one, study its margins under high power to see the individual cells. Locate the **nucleus**, a small spherical body in the center of the cell, and the **cytoplasm** (a term that refers collectively to all of the cell's contents apart from the nucleus). The **plasma membrane** bounding the cell is too thin to see with light microscopy (refer to the electron micrographs at the end of this Topic), although you will see the boundary where the cytoplasm ends and the mounting fluid begins. The structure of the cytoplasm is an extremely complicated arrangement of membranes and organelles, but very little of this fine structure can be seen with a light compound microscope.

B. Staining

It is often helpful to increase the contrast of certain cell structures so that they may be seen more clearly. Chemicals that dye parts of cells for this purpose are called stains. A great variety of stains is available; some color all parts of cells more or less indiscriminately, while others act more specifically on particular structures or chemical compounds within the cell.

Obtain some more cheek cells, place them directly on a clean, dry slide, and allow the scrapings to dry. Place the slide on a paper towel at your desk and add a drop or two of methylene blue stain. After two minutes, rinse off the stain with water; this is best accomplished by dripping a little water from your fingers or a dropper over the slide while you hold it above the sink. Dry the bottom of the slide, add a drop of water and a cover slip, and observe. What feature of the cheek cell stains most prominently?

Stains of this type usually kill the cells, but there are a few **vital stains** that may be used with living materials.

PART II.
DEPTH OF FOCUS
A. Colored threads

Obtain a slide with three or four colored threads mounted together. Under low power, find a point where several of the threads can be seen together in the same field. Then, with the fine adjustment, slowly focus up and down. Notice that as you do so, different parts of the threads, and different threads, become distinct; when one is in focus, the others, above and below, are somewhat blurred. By continuous fine focusing, up and down through the specimen, you can perceive the depth dimension that is not evident when the focus is resting at one point. By focusing up and down, determine the order of threads, top to bottom, on your slide.

Top_____
_____
_____
_____

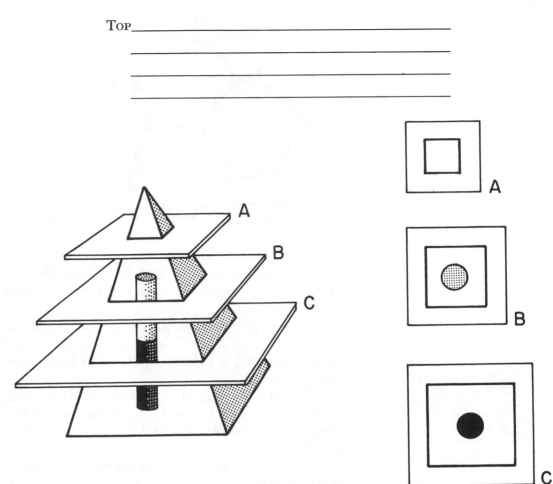

Fig. 1-1. Depth of focus. Sections A, B, and C illustrate how sections taken at different levels through a specimen reveal different structures.

Examine the threads under high power, and notice that you can see considerably less depth than under low power. Indeed, you may not be able to distinguish one whole thread clearly under high power. As you increase the magnification, you can always expect the perceivable depth of the specimen to decrease.

The vertical distance that will remain in focus at any particular point is called the **depth of focus** or depth of field; it is a constant value for each of the objectives. By focusing up and down through specimens that are thicker than the depth of focus (see Figure 1-1), you can get around this limitation in depth perception. Changing the focus in this manner is a little like making successive thin slices of the object and then viewing them one by one. Indeed, the structure of some small objects is actually studied by making and observing serially a large number of such slices. Expert microscopists keep themselves continuously aware of depth by making constant changes in the fine adjustment.

B. Onion epidermis Cut an onion into quarters or eighths, and with forceps strip a small piece of the thin epidermal layer from the inside of one of the thick, modified leaves of the onion bulb. Place it gently on a clean glass slide, and add a

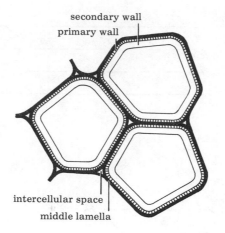

secondary wall
primary wall

intercellular space
middle lamella

Fig. 1-2. Three adjacent plant cells, showing cell walls.

drop of tap water. If necessary, straighten out the epidermis with forceps. Then, taking care to avoid trapping air bubbles, cover with a cover slip. Observe the preparation under low power, adjusting the light intensity until it is favorable.

The onion epidermis is a **tissue** (an aggregation of cells similar in structure and function) made up of many cells. Use the fine adjustment to focus on one cell at various depths under high power, and observe the following features:

1. **Cell wall,** a rigid but porous structure composed of the polysaccharide **cellulose** and several other substances; it is secreted by the plasma membrane. Cell walls (FIGURE 1-2) give plant cells a rigidity that is absent in animal cells.

2. Layer of cytoplasm, containing many small granules but otherwise nearly transparent; it is located just inside the cell wall and plasma membrane, which is not visible.

3. **Cell vacuole,** a large fluid-filled sac, occupying most of the central part of every cell and pressing against the cytoplasm; it sometimes contains a red pigment, and is often crossed by thin cytoplasmic strands.

4. Spherical nucleus, located in the cytoplasm along one of the walls; it contains several smaller bodies, the **nucleoli** (singular, nucleolus). Chromosomes are not visible in the nucleus of a cell that is not dividing. Check the nucleus in several cells. Is it always in the same position? Can you see any evidence that the nucleus is connected directly to parts of the cytoplasm other than that in which it rests directly?

Examine several nuclei to see how many nucleoli each contains. You will need to use the fine focus under high power to count the number of nucleoli. Is the number constant from cell to cell?

Direct your attention to the layer of cytoplasm just inside the cell wall. Would you describe it as thick or thin relative to other parts of the cell?

Read Appendix 1, on drawing, and make a large drawing of the onion cell.

C. Paramecium

Examination of the protozoan *Paramecium* will serve as a critical test of your ability to observe objects with the microscope and to perceive depth of focus.

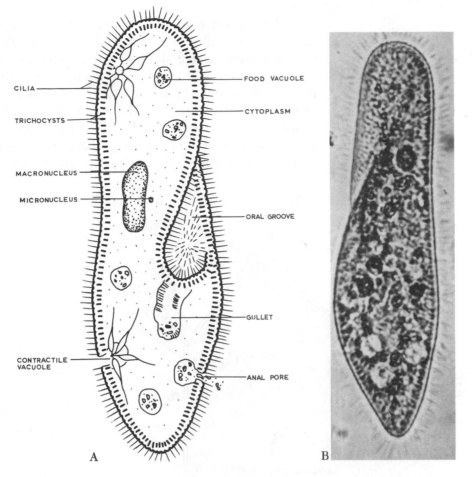

CILIA

TRICHOCYSTS

MACRONUCLEUS

MICRONUCLEUS

CONTRACTILE
VACUOLE

FOOD VACUOLE

CYTOPLASM

ORAL GROOVE

GULLET

ANAL PORE

A   B

FIG. 1-3. *Paramecium.* (A: *Courtesy Carolina Biological Supply Company.*)

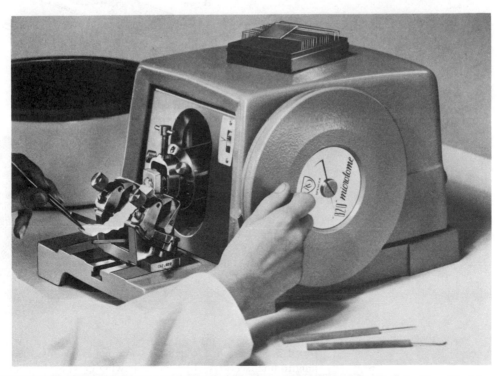

FIG. 1-4. Microtome in use. (*Courtesy AO Instrument Company, Buffalo, New York*)

Make a wet mount of *Paramecium* by mixing a drop of the available culture with a drop of methyl cellulose on a slide. (Methyl cellulose is a viscous substance that slows the animal's swimming.) Mix thoroughly with a toothpick, and add a cover slip. Examine the slide under low power, and locate several animals. Using Figure 1-3, study the animal's structure, and check off the structural features on the figure one by one as you locate them. It will be necessary to use the fine focus continuously, to examine a large number of animals, and to change magnification; an animal in one position may not reveal the same structures as an animal positioned in another way.

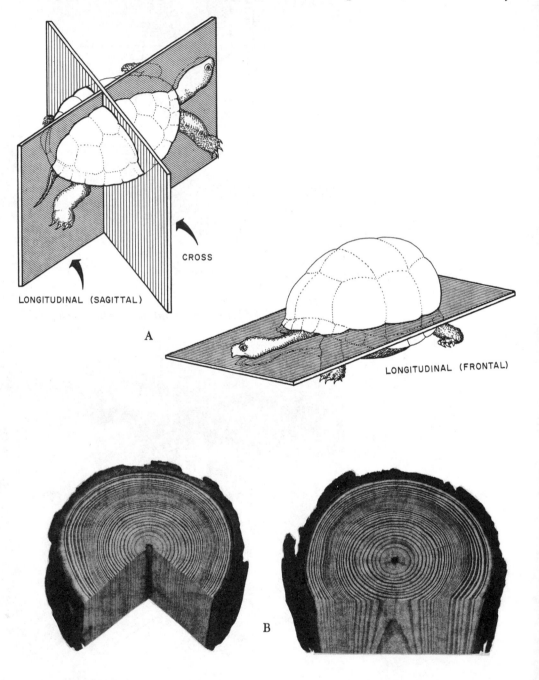

Fig. 1-5. Sections. (A) Turtle showing planes of cross, (frontal) longitudinal, and (sagittal) longitudinal sections. (B) Wood, showing cross and radial sections (left photo), cross and tangential sections (right photo). (*B: Courtesy Carolina Biological Supply Company.*)

**PART III.
SECTIONS**

Unless an object is unusually small and thin (e.g., *Paramecium*), it is necessary to cut it into thin, transparent slices and to mount the slices on slides, before the object can be viewed with the microscope. Such slices are called **sections** and are prepared on a special instrument called a **microtome** (FIGURE 1-4). Sections are named according to the plane in which they have been cut, as follows (see FIGURE 1-5):

cross section (c., c.s., or x.s.) · — cut at right angles to the longitudinal axis of an object

longitudinal section (l.s.) — cut parallel to the longitudinal axis of an object

median section (med. or m.) — cut along the middle of an object

Similarly, a *radial* section would be a cut made along a radius of a circular object, and a *tangential* section would be a cut parallel to a tangent to a circular object.

Since a single section clearly shows only a part of the structure of an object, it is essential that the viewer be aware of the plane in which the cut was taken. This information is usually marked (often abbreviated as above) on the slide label, along with other identifying information, such as the generic name of the organism and the manner in which the slide was prepared (e.g., "whole mount," "section," "squash," etc.).

Your instructor may provide examples of different types of sections.

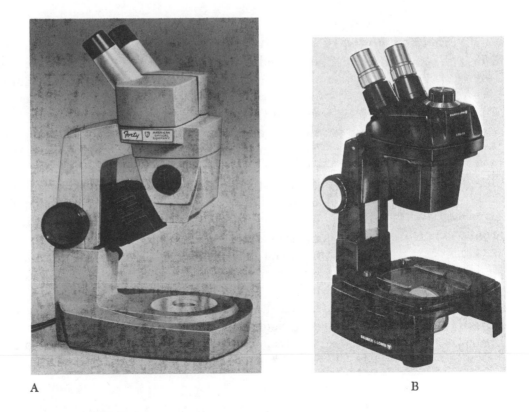

A

B

FIG. 1-6. Dissecting microscopes. (A: *Courtesy AO Instrument Company, Buffalo, New York. B: Courtesy Bausch & Lomb.*)

PART IV. DISSECTING MICROSCOPE
A. Structure and operation

The other microscope you will be using in this course is the binocular dissecting microscope. As the term "binocular" implies, this microscope has a pair of oculars, one for each eye; most dissecting microscopes also have a double objective system, arranged in such a way as to render a stereoscopic (three-dimensional) image. FIGURE 1-6 illustrates two commonly used dissecting microscopes.

Notice that the distance between the oculars can be adjusted. Pull them apart or push them together until you have them adjusted to the correct position for your eyes. What is the magnification of the oculars? (If your microscope has an ocular on which the focus can be adjusted, look through the *nonadjustable* ocular first — with your other eye closed — and obtain a clear focus on some object in the usual manner. Then open the other eye, close the first, look through the adjustable ocular and rotate the milled cuff around it until you obtain a clear image.)

Since the magnification of the dissecting microscope is lower than that of your compound microscope, the working distance is greater; hence much larger objects can be examined. Some dissecting microscopes have objectives of several different magnifications mounted on a revolving turret; if your instrument is of this type, determine the power of each objective. Other microscopes may have a rotating dial in place of a turret, or an adjustment knob atop the body which, when turned, permits continuous change in the magnification throughout a range marked on the knob. This latter type is called a "zoom" microscope. Whatever the arrangement on your instrument, determine how the magnification may be changed. What is the lowest possible magnification? The highest?

B. Practice with the dissecting microscope

1. With forceps, obtain a small moss plant from the clump at the supply desk, and place it on a dry slide, without cover slip.

If your microscope is equipped with a substage mirror and a transparent stage, adjust the illuminator to discover whether you get a clearer view of the moss with or without illumination from beneath. Light passing through an object and then into the microscope is **transmitted** light, while light going from the illuminator directly to the object (and reflected by the object up into the microscope) is **reflected** light. One of the decisions a microscopist must make is whether to use transmitted or reflected light. Which form have you already used with the compound microscope?

Whether you use reflected or transmitted light depends, of course, on the size and opacity of your specimen. If you decide to use reflected light, it is well to remember that the illumination must usually be more intense, since more of the light is lost by scattering before it reaches the microscope lenses.

If the stage of your dissecting microscope has a black side and a white side, determine which is better for examination of the moss.

Examine the moss carefully, turning it to a variety of positions. How would you describe the gross structure of a moss plant?

2. Net a guppy from the aquarium in the laboratory, and place it in a small covered petri dish of aquarium water. Examine the structure, gill move-

ments, and swimming movements of the fish under the dissecting microscope.

PART V. CHOICE OF THE PROPER INSTRUMENT

The usefulness of microscopy depends a good deal upon the wisdom, imagination, and good judgment of the person using the instrument. He must decide when to use a compound microscope, when a dissecting microscope, and when other types of magnification (if available). He must decide what magnification to use when the magnification is adjustable.

Look at the moss with the compound microscope (without cover slip) and with a hand lens, if one is available. Which of these instruments do you find the most useful for an object like the moss, which is opaque, three-dimensional, and large enough to be seen with the naked eye? Which instrument would you use to determine the cellular structure of a single moss leaf?

Perform the following observations, deciding in each case which microscope is most appropriate.

1. With a razor blade, make very thin sections — transparent if possible — of potato tuber, and examine the cells that compose the tuber. The oval bodies in these cells are starch storage organelles, called **leucoplasts.** Stain a section with iodine (dropped at the edge of the cover slip), and note the resulting color in the leucoplasts; this color is a positive test for starch.

2. Examine a drop of blood, obtained from your finger with a lancet and diluted on a slide with two more drops of 0.9% salt solution.

3. Make cross and longitudinal sections of the plant stems provided by your instructor, and examine them for general appearance and cellular structure.

4. Examine the organisms in pond water provided by your instructor.

PART VI. FIELD SIZE

Estimate the size of the visual field of your dissecting microscope and the low power of your compound microscope by examining a millimeter ruler. If you make subdivisions with a razor blade between the millimeter markings, you can also roughly estimate the size of your high-power field.

LOW COMPOUND:

HIGH COMPOUND:

DISSECTING LOWEST:

DISSECTING HIGHEST:

PART VII. OTHER TYPES OF MICROSCOPY

One can arrange a light microscope for use with polarized light, dark field, fluorescence, phase contrast, and a number of other types of microscopy that render some cell components very clearly. Some of these may be on demonstration, with explanatory material beside them.

As you already know, a great deal has been discovered about the internal "fine" structure of cells. The limitations of light as a refractable

Fig. 1-7. Electron microscope. (*Courtesy Perkin-Elmer Corporation.*)

medium and the practical problems of manufacturing very tiny lenses make it impossible to attain magnifications higher than about 2000X with the light compound microscope. Most of our knowledge of cellular fine structure has therefore been derived from **electron microscopy**, a technique similar in principle to light microscopy, but which employs electrons instead of light, and magnets in place of glass lenses. Electron beams have a much shorter wavelength than visible light, and refracting them with magnets can yield resolutions several hundred to several thousand times higher than conventional light microscopy.

An electron microscope is illustrated in Figure 1-7, and Figure 1-8 shows an electron micrograph of a common type of cell surrounded by smaller plates of the various common organelles. Review Appendix 1, on drawing, and carefully label the cell parts in Figure 1-8.

Examine other microscope equipment on demonstration.

QUESTIONS FOR FURTHER THOUGHT

1. What is meant by the term parfocal?

2. What is meant by depth of focus? With which objective of your compound microscope would depth of focus be greatest?

3. What means, other than the iris diaphragm, could you employ to control the light entering a compound microscope?

4. By what means, besides changing objectives, could you alter the total magnification of a microscope?

5. Why is it necessary to center a detail in the low-power field before switching to a higher power?

6. Is the cell wall of a plant cell alive? Where does it come from?

7. Suppose you baked a Ping-Pong ball into the center of a loaf of bread. Would *any* cross section reveal the ball? Would a median longitudinal section reveal the ball? What would be the appearance of the Ping-Pong ball in section?

8. Briefly describe a test for starch.

9. Define or describe and explain: stain, primary image, ocular.

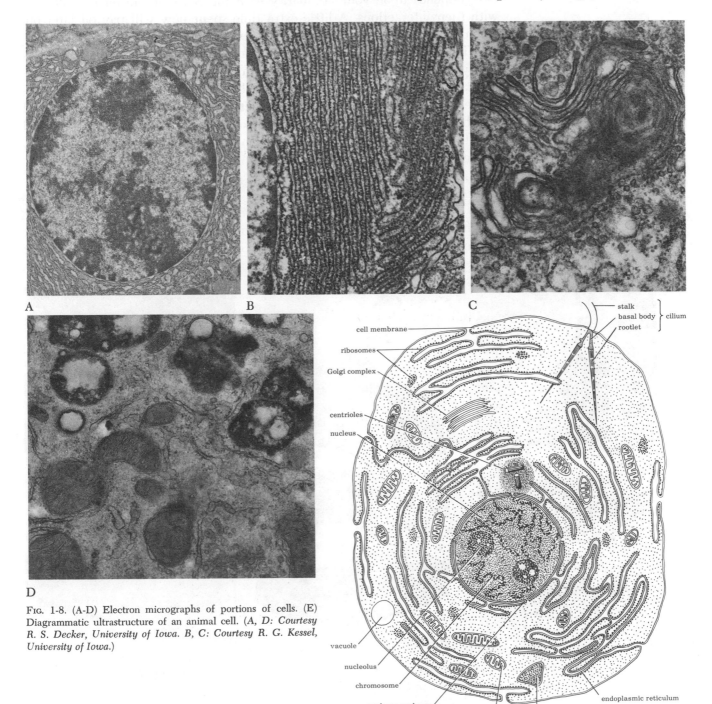

Fig. 1-8. (A-D) Electron micrographs of portions of cells. (E) Diagrammatic ultrastructure of an animal cell. (A, D: *Courtesy R. S. Decker, University of Iowa. B, C: Courtesy R. G. Kessel, University of Iowa.*)

If you have never used a microscope before, or if you need a review, you should study this appendix in detail. Proper, comfortable use of the instrument demands practice, but the practice afforded you in the first few exercises depends upon your familiarity with the parts of the microscope and with their interactions. A little extra time spent now will pay off later on, when you are asked to do much more exacting microscopy.

The most important thing to remember about using a microscope is that you should be physically and mentally comfortable as you use it. Concentration and hard work are indispensable parts of learning any new technique, but if your eyes or back are strained, if the microscope field is dull and blurry and you do not know how to correct it, or if you are confused about what you are examining, stop at once to find out what the problem is. If necessary, seek help to correct it.

PART I.
PARTS OF THE
MICROSCOPE

A compound microscope is a delicate and expensive instrument, and must be treated gently. Familiarize yourself with the parts and operation of your

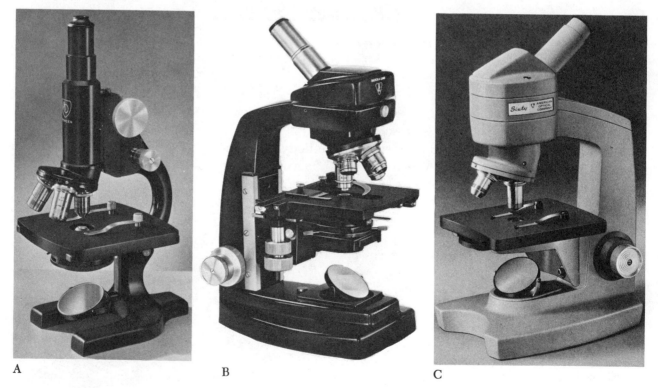

A   B   C

FIG. A-1. Compound microscopes. (*A, C: Courtesy AO Instrument Company, Buffalo, New York. B: Courtesy Bausch & Lomb.*)

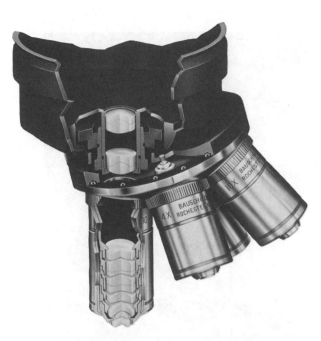

Fig. A-2. Cutaway of nose-piece assembly and objective, showing the arrangement of optical elements. (*Courtesy Bausch & Lomb.*)

instrument, referring when necessary to Figure A-1.

1. Remove your assigned compound microscope from its cabinet by the **arm.** Holding it upright, and supporting the **base** with your free hand, take the instrument to your desk and put it down gently, arm toward you. (A microscope should always be carried in this fashion, with *both hands.*)

2. The arm supports a vertical **body tube,** which in turn supports the magnifying elements of the microscope. At the top of the tube is the **ocular,** or eyepiece, one of the magnifying elements. It is usually removable. The other magnifying elements are screwed into a revolving **nosepiece** at the bottom of the body tube; they are called **objectives** (FIGURES A-1 and A-2). Together, the ocular and an objective constitute the magnifying system of your microscope.

3. Underneath the body tube and nosepiece is a flat plate, the **stage,** upon which objects to be examined are placed. Directly beneath the stage opening you will notice a system of lenses, the **condenser** (it may be absent in some microscopes), which serves to concentrate light reflected up through it by the **mirror** below. On most instruments, one side of the mirror is concave and the other flat. Your instructor will tell you which side is best to use with your instrument and light source.

4. Light plays an extremely important role in the operation of a compound microscope. As you can see by reference to FIGURE A-3, light is reflected up through the stage opening, passes through the specimen, and thence into the body tube, ultimately forming an **image** on the retina, the light-sensitive portion of the operator's eye. The critical importance of light makes necessary its careful adjustment, and several controls are available for this purpose. The **iris diaphragm** may be found attached below the condenser, or if the condenser is absent, just below the stage. To see the diaphragm clearly, remove the ocular (so that it does not fall out) and gently turn your microscope over. Look up through the stage opening, find the **iris diaphragm control,** a small handle at one side of the diaphragm, and

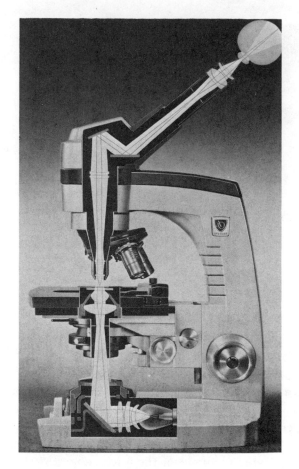

FIG. A-3. Path of light through a compound microscope. (*Courtesy AO Instrument Company, Buffalo, New York.*)

push it back and forth, noting how the size of the iris opening changes to regulate the amount of light passing through it. This control is one of the most important on your instrument. (*Note:* In place of a diaphragm, some microscopes may have a revolving plate with holes of different sizes to admit varying amounts of light.) Open the diaphragm fully and reposition your microscope; replace the ocular.

5. Just as with a magnifying glass, viewing an object with the microscope requires that the lens be a certain specified distance from the object. That distance is a property of the lens system, and is constant for any particular objective; it is called the **working distance.** At the working distance from an object, an objective is in focus. Changes and adjustments in the focus, which are necessary when an object is first placed on the microscope, and essential in order to perceive depth, are accomplished by means of **coarse** and **fine adjustment** knobs, usually located on the arm. Try turning each of these knobs, noting carefully how each affects the position of the objectives. The coarse adjustment is used to obtain an approximate focus, and the fine adjustment to obtain an exact and clear focus.

PART II. MAGNIFICATION

The magnification of most objectives and oculars is engraved upon them. On the ocular, the marking can be found on the top edge or on the smooth cylinder that fits inside the body tube; on the objectives, magnification is engraved on the side of the cylinder. The marking "10X" means that the element so marked yields an image ten times larger in each dimension than the object being viewed.

Remember that in using a microscope of this kind, you are employing *two* sets of magnifiers. On low power (10X), for example, the objective forms an image (inside the body tube) that is ten times larger than the viewed object; the 10X ocular then magnifies this "primary" image another ten times. The image that finally reaches your eye has been enlarged to 100X the size of the object. The *total magnification* for any combination of objective and ocular can be computed simply as the *product* of the magnifications of each element. Prepare a table listing the total magnification that is possible with each of the various combinations of oculars and objectives that is possible for the microscopes you are using.

In addition to using the terms "low" and "high" as common names of the most frequently used objectives, the objectives may be designated by magnifying power, working distance, or numerical aperture (a number which describes the resolving power* of the lens). Compare the values listed in the table below with those on the objectives of your microscope.

COMMON NAME	MAGNIFICATION	WORKING DISTANCE	NUMERICAL APERTURE
Scanning Lens	2.5-4	25-55 mm	About 0.10
Low Power	10	5-10 mm	About 0.25
High Power	40-50	0.15-0.60 mm	0.65-1.00
Oil Immersion	90-100	0.05-0.15 mm	About 1.25

←—— INCREASED RESOLVING POWER

PART III. BASIC STEPS IN USE OF THE MICROSCOPE
A. Adjusting the light

Place the microscope on a firm base, arm toward you. Open the iris diaphragm or its equivalent fully. Place a 3 x 5-inch index card or piece of white paper over the stage opening. Position the light source about six inches away from the mirror and adjust the mirror until you see a bright spot of light on the paper. When you see that spot, the illuminator and mirror are properly adjusted to render the brightest possible illumination, and it will be unnecessary to make any further change in the light except to adjust intensity with the iris diaphragm.

B. Focusing

On the paper you used to fix the light, write a lower-case letter "e" about the size of a type character in this book. Put the paper back on the stage

* Explained in the last section of this Appendix.

so that the letter lies right in the middle of the stage opening. Now turn the nosepiece at the bottom of the body tube until the *lowest* power objective clicks into position. Then, *watching from the side of the microscope,* lower the objective until it is about one-quarter inch from the paper. Look into the ocular and focus slowly upward. As you turn the coarse adjustment, the objective moves upward; when it reaches the correct distance from the paper (what is this distance called?), the letter will be in focus. (*Note:* If you failed to get the letter into focus, it may not have been centered in the stage opening, or you may have "overshot" the working distance by focusing upward too rapidly. Try again, making certain that you *never* move the objective back down toward the slide *without first* looking from the side of the instrument.)

1. What is the appearance of the "e" under low power? Is it right side up or upside down?

2. When you move the paper to the right, while looking into the microscope, which way does the image appear to move? Which way does the image move when you move the paper toward you? Away from you?

These phenomena are called **inversions;** because of the microscope optics, the object and its movements are inverted. Practice moving the paper about, or writing and observing different letters, until you are used to the inversions.

C. Changing magnification

When you want to see more detail, you must switch to a higher power objective. Before doing so, center the detail you wish to examine right in the middle of the lower power field. Then rotate the nosepiece until the next higher power objective clicks into place. Then use the *fine focus* to make the object clear. Unless your specimen is extremely thick, or your microscope is not equipped with *parfocal* objectives (see note below), it is usually unnecessary to make more than a minor adjustment in the fine focus, or to raise the body tube before you change objectives. Do make certain, however, that the nose of the higher powered objective does not contact the specimen.

1. How much area of the "e" can you see under the higher power relative to what you saw under the lower power?

2. How did the brightness of the light change when you changed objectives? If you did not see any change, go back to the lower power and make the change again. If it were necessary to increase the light intensity after switching objectives (as it often is), by what means would you do so?

(*Note:* Most microscopes have their objectives machined so that when all are firmly screwed into place, it is unnecessary to make a major change in the focus when magnification is changed. Objectives so aligned are spoken of as being **parfocal** to one another. If your instrument was in sharp focus under the lower power, it should not be necessary to make more than a small change in the fine adjustment to obtain a clear focus after switching objectives.)

Troubleshooting

1. If you "lost" the object when you changed magnification, your ob-

jectives may not be parfocal, or you may have failed properly to center the object intended for examination. If the objectives are not parfocal, simply refocus (according to B above) whenever you change objectives. If you did not center properly, try moving the object slightly, or change back to the lower power and then re-center.

2. If the image you get is unclear, the slide, mirror, or magnifying elements may require cleaning. Follow the directions in the next section for proper cleaning.

3. If the image you obtain seems to be in focus, and cleaning does not help to alleviate a dull or "washed out" appearance, you may not be using the correct amount of light. Use the iris diaphragm to adjust the light intensity.

PART IV. IMPORTANT CAUTIONS

These few points of operation and maintenance will save your microscope — and your eyes — from much unnecessary wear and tear. Learn them *now*.

1. *Never* move the body tube downward (except, of course, for minor adjustment focusing) while looking into the microscope. Any time a major downward change in focus is necessary, watch from the side, so that you are always aware of the position of the body tube. Otherwise, you are apt to ram slides with objectives, causing serious damage to slides, which are expendable, and to objective lenses, which are not. Some modern microscopes are equipped with stops to prevent this ramming damage no matter how careless the student operator, but you will probably have occasion to use microscopes without this feature, so form the habit of *always* looking from the side as you focus downward.

2. You can avoid a great deal of eyestrain by keeping both eyes *open* while you are looking through the microscope. You will soon learn to disregard the image from the "off" eye by concentrating attention on the microscope field.

3. Dirt, especially eyelash oil and dust, is apt to accumulate on the surface lens of your ocular. Since many other students use your instrument, make it a habit to clean the ocular each time you use the instrument. Use *only* lens paper made for the purpose, *never* any other material. It is a good practice to clean the objectives as well, since soil, even a small amount that may not be visible to you, seriously interferes with image formation by a compound microscope. If, after making sure that the magnifying elements, mirror, slide, and cover slip are completely clean, you continue to have difficulty obtaining a clear image, ask your instructor to help. (*Note:* It is possible to tell whether dirt is on the ocular by rotating it while looking into the microscope; if it is on the ocular, the dirt moves.)

4. Keep the stage of the microscope *dry* at all times. Fluids on the metal parts will cause corrosion.

PART V. RESOLUTION

A more accurate way of saying that cells are normally invisible to us would be to state that the human eye cannot normally **resolve** points within the limits of dimensions of most cells. The print on this page is easy for you to resolve at usual reading distances, but if you move the page away, it becomes increasingly difficult to discern specific letters with certainty, particularly if they resemble one another. Your perception of letters and words is dependent upon the ability of your eye to separate different points in the pattern of ink that forms each letter. A microscope lens system is capable of greater resolution than is your eye; its properties in this respect are called its **resolving power.** The resolving power of a lens is described by a number called **numerical aperture,** calculated from certain physical properties of the lens and the angles at which light enters and leaves it. The numerical aperture is often engraved on the objective.

At normal reading distances and illumination, the top curve and cross-piece in the small letter "e" are easily resolved by your eye: that is, you perceive them as distinct lines, and hence recognize an "e." At great distances, the two lines are not seen as distinct, but blend instead into one line; the letter "e" then begins to look like the letter "c." At great enough distances it becomes impossible to distinguish the two letters. This is because the retina, or light-sensitive part of your eye, contains a limited number of sensing cells. When images from two actually separate points fall on the same sensing cell, the two points are perceived as one. The farther away the two points from your eye, the closer together they come on the retina.

Biological Science, pp. 51-63, 238-241,
382-383, 736-738, 767-768.
Elements of Biological Science, pp. 38-46,
150-152, 492.

TOPIC 2 THE BEHAVIOR OF LIVING CELLS

As you have already seen, cells are complex units with highly specialized internal structures that fulfill a variety of important functions. Mitochondria, for example, function as tiny factories that convert the energy contained in foodstuffs into a usable form, by degrading the relatively large food molecules in a stepwise fashion that allows the energy to be trapped in small packets as ATP, the "currency" in nature's "energy bank." A host of other organelles function to store and to synthesize compounds the cell requires, and to integrate its activities. Finally, the semipermeable membrane bounding each cell confers upon it the ability to maintain, by selective acquisition of raw materials and elimination of wastes, an essentially constant chemical environment in which all the parts can function smoothly. The living cell is thus a functional as well as a structural unit, and this laboratory is designed to help you think about it from that point of view. In addition, you will be introduced to some of the questions that can be asked about cell function, and see the manner in which they may arise. We will examine today two simple aspects of cell behavior, and simple physical models for each of them: cytoplasmic streaming, and the problem of water regulation.

PART I. LOCOMOTORY BEHAVIOR IN AMOEBA

One important outlet for chemical energy is movement. Movement in most animals results from the action of muscles on some form of skeleton, but motion is an important capacity of all cells, and this capacity frequently shows up as a phenomenon called **cytoplasmic streaming.** Some organisms, including two that you will examine today, depend upon movements of the cytoplasm for locomotion, but many cells of multicellular organisms — perhaps all cells — utilize the capacity for movement in other ways. *Paramecium*, the protozoan you saw in the last laboratory, exhibits movements of the cytoplasm which serve to move food vacuoles and thus, very probably, to expedite the diffusion of food to all parts of the organism. *Nitella*, a plant you will examine in the second section of this unit, utilizes streaming in much the same way. Many cells engulf small food particles by forming small evaginations of the cytoplasm and plasma membrane around the material to be taken in.

Although one can clearly see changes in the physical structure of the cytoplasm in streaming regions of some cells, the actual mechanism of streaming is poorly understood. It probably involves a change in shape of protein molecules in the cytoplasm, but how the changes in molecular shape are related to the release of ATP, and how streaming can be triggered in certain regions of a cell while other regions remain quiescent, are problems we still do not understand.

A. *Living* Amoeba   Take a clean slide to your instructor, and obtain a drop or two of culture with several living *Amoeba*. Locate the animals under the dissecting microscope, noting the general size, changes in shape, and method of locomotion. Locomotion of this type, common to many kinds of cells, is called **amoeboid movement.**

Now add a cover slip; you should support the cover slip either with small pieces of broken cover slip, or with vaseline. The latter is preferable, since it serves the double function of raising the cover slip (so that the specimens are not crushed) and protecting the preparation from drying out.

VASELINE SUPPORTED   UNSUPPORTED

FIG. 2-1. Vaseline-supported and unsupported cover slips. Vaseline is a good means of supporting the cover slip when a relatively thick or fragile specimen is to be studied. It may be applied either to the slide or to the cover slip with a toothpick, or with a syringe into which the jelly has been drawn after melting.

FIGURE 2-1 illustrates the difference between a wet mount supported with vaseline and an unsupported mount. Under low and high powers, identify:

1. Cytoplasm, containing many granules.
2. Nucleus, a slightly ovoid structure near the center of the cell.
3. Food vacuoles, small, dark, food-filled areas in the cytoplasm.
4. Contractile vacuole, a clear, spherical area, filled with fluid and often carried near the part of the animal farthest from that actually moving forward in a particular direction.

Remember that the plasma membrane is too thin to see, although you will be able to see the boundary between the cytoplasm and the mounting medium.

It may be necessary to decrease the light intensity before you will be able to see these structures clearly, especially under low power. High power will show more detail; watch especially the contractile vacuole, which will be seen to pulsate at regular intervals if it is closely observed. What is the function of this behavior of the contractile vacuole?

Watch carefully as the *Amoeba* moves along the glass surface. You will note that the shape of the cell can change constantly, and the granular cytoplasm inside does not remain still, but is commonly seen in a streaming motion. Observe the streaming under high power, as the animal flows along with outward bulgings of the cell called pseudopodia (singular, pseudopod). In the flow of cytoplasm associated with the formation of a pseudopod, you should be able to distinguish

5. A clear, outer layer of cytoplasm.
6. Two inner, granular layers of cytoplasm (the outer composed of granules fixed in position, and the inner of granules in motion).

If you have any difficulty seeing these layers, check your slide preparation and the microscope optics for cleanliness; if you continue to have trouble, ask your instructor for help. Study carefully the formation of a pseudopod, particularly the transition from **sol** (granules in motion) to **gel** (granules fixed in position) cytoplasm. At what point in the formation of a new pseudopod does the transition take place? Make sketches if you like, and continue these observations for as long as you wish.

(*Note:* If you watch carefully, you may see an *Amoeba* ingesting one of the small, ciliated protozoans upon which it feeds. The food is engulfed or surrounded by advancing pseudopods, and a vacuole is thus formed around it. Digestive enzymes are then secreted into the vacuole. If you can find a newly formed food vacuole — the prey will still be moving inside — watch it for a while to determine how long it takes before the prey is inactivated.)

B. Film of amoeboid movement

If a film of amoeboid movement is available, make sure that you see it sometime during the period.

Before going on to the next section, use your text to help you identify the Phylum and Class of *Amoeba*, and enter it in the appropriate portion of Appendix 2.

**PART II.
WATER RELATIONS
IN NITELLA**

You will recall that a living cell is largely able to regulate its chemical make-up by actively transporting substances across the cell membrane. At the same time, there are a few substances that can freely cross cell membranes. One of these substances is water. Water relations are so important to cells and organisms that it is appropriate and essential for us to begin an experimental consideration of the cell by examining the passage of water across cell membranes.

Diffusion, you remember, is defined as the passage of molecules of any substance from a region where they are in higher concentration to a region where they are in lower concentration. Because it is so important, the diffusion of *water* across a semipermeable membrane is given a special name: **osmosis.** In this exercise, we will observe the causes and effects of osmosis in the living cell, confining our study to the large-celled fresh-water alga *Nitella.*

A. Structure of Nitella

Nitella will be found in a tray at the supply desk. Obtain one strand in a petri dish of pond water and examine with the dissecting microscope. The long pieces of "stalk" that separate tufts of lateral "branches" are called internodes, and in *Nitella* the internodes are actually single cells. A large, balloonlike vacuole in each of the cells makes it appear transparent. Bounding the *Nitella* cell, as in nearly all plant cells, there is a firm cell wall.

Place the petri dish on your compound microscope stage, being careful not to get water on the stage or objectives, and focus carefully with *low power* on one of the internodes. Observe the **chloroplasts** (structures that carry out photosynthesis) lying closely side by side, and arranged in perfect rows along the inner surface of the cell wall. The chloroplasts are actually in the cytoplasm, but the cytoplasm is pressed firmly against the cell wall by the vacuole, which occupies most of the volume of the cell. One peculiar feature is the conspicuous absence of chloroplasts in a pair of neutral lines, running in a loose spiral around the internode.

The chloroplasts are stationary in the cytoplasm, but if you focus properly just beneath this layer, you can see cytoplasmic streaming. If you look very closely, you will see that the cytoplasm is streaming in opposite directions on either side of a neutral line. At the end of the internode, in favorable preparations, the streaming particles can be observed to glide around from one side of the neutral line to the other. What functions do you think such streaming may serve? In older cells the nucleus will have divided into several portions that travel about in the streaming cytoplasm.

B. *Osmosis in* Nitella

Since *Nitella* protoplasm is hyperosmotic to[*] the surrounding environment (fresh-water pond) water tends to enter the cells by osmosis. The stiff cell wall prevents the cells from enlarging, and the protoplasm becomes rigid and tightly pressed against the cell wall. Such a cell is said to be **turgid,** or to exhibit turgidity. In contrast, if the cells are placed in hyperosmotic solutions, so that the protoplasm is hypoosmotic to the environment, water tends to leave the cells, and the protoplasm shrinks away from the cell wall. This condition is called **plasmolysis.**

1. After you have become familiar with the normal appearance and rigidity of the turgid cells (feel one with a probe to test its rigidity), set up six petri dishes, properly labeled, and containing the following solutions:

1. 0.06 M glucose	4. 0.50 M glucose
2. 0.12 M glucose	5. 0.75 M glucose
3. 0.25 M glucose	6. 1.00 M glucose

2. With a probe, transfer the *Nitella* strand through the six solutions in order of increasing concentration. Using the dissecting microscope, note the appearance of the cells in each solution, and probe them to determine the turgor level. If a dent is easily made with a probe, and if it straightens out after a few seconds, you have found the solution that approximates the internal concentration of the *Nitella* cell. This concentration is spoken of as the **isosmotic threshold** of *Nitella*. Within what range of concentration does the threshold lie?

3. Observe the plasmolysis in the 1.00 M glucose solution, and gently fold the whole strand into a ball, using a probe. Now gently pick up the folded mass with the probe and place it back into a petri dish of pond water. What happens? How fast? Why? Would you say that the *Nitella* cell is easily able to control the passage of water?

**PART III.
MODELS OF
CELL BEHAVIOR**

Biological models are simply physical or mathematical analogs of living systems, which serve as "conceptual crutches" or as aids to visualizing isolated aspects of more complicated situations. Life is a complex phenomenon, and it sometimes helps a scientist to think about explanations for the behavior of living systems if he represents the living system, or some part of it,

---

[*]You recall that "is hyperosmotic to" means "has a greater concentration of dissolved (osmotically active) particles than." Hypoosmotic and isosmotic are similarly defined.

with a non-living system that behaves in a similar way. A model is such a representation.

No biological model is meant to be a complete and detailed representation of a life process; it is intended, rather, to serve as a means of abstracting certain characteristics of the living system. In doing so, it may help a scientist to envision possible means by which the more complicated living system is operating. By manipulating the model in various ways and examining its behavior in a variety of different situations, a scientist is helped in making predictions (hypotheses) about the behavior of the living system under comparable circumstances.

You need to remember that some models are close analogs, and thus useful predictors, of the systems they exemplify, while other models are distant analogs and may, indeed, only superficially resemble the living system they are intended to represent. All, however, are fruitful food for thought; as you examine these models of cytoplasmic streaming and of semipermeable membranes, think very carefully about how far you can carry the comparisons between them and the living processes and structures they resemble.

A. Model of cytoplasmic streaming

Obtain a watch glass, partially fill it with dilute nitric acid, and place it on a sheet of white paper. Have your instructor introduce a drop of mercury into the acid; with forceps, place a large crystal of potassium dichromate about half an inch from the mercury drop. Do not disturb the preparation, but watch carefully as the crystal dissolves in the acid and diffuses outward. What happens when the potassium dichromate reaches the mercury? Observe carefully, and describe the events that occur. In what ways do they resemble the behavior of *Amoeba?* How do they differ from the behavior of *Amoeba?*

Mercury behaves in this manner toward potassium dichromate because its surface tension, which causes it to assume a "drop" shape, is disrupted by oxidation, in turn brought about by the action of potassium dichromate in nitric acid. At the random points of oxidation, the surface tension is suddenly lessened and the mercury flows out, giving rise to the streaming motion. When the surface tension is regained, there is a "backflow."

If the motion stops, you can start it again by transferring the acid, with a dropper, to a waste container (on the supply desk) and adding fresh acid and a new dichromate crystal. When you are finished, *make sure* that you discard the mercury into the waste container. If mercury is accidentally dropped on the floor or the table, make sure it is located and removed at once; if left on the floor, it gets into the dust, and may be poisonous if inhaled.

B. Model of osmosis through a semipermeable membrane

Obtain three small pieces of cellophane dialysis tubing. Tie a tight knot in one end of each tube, and partially fill each piece of tubing with about the same volume of 1.5 M sucrose solution. Knot the other end of the tubing or tie it tightly with string in such a way as to exclude air. It does not matter if a little of the sucrose solution escapes, as long as each of the tubes has roughly the same volume of fluid inside.

Now rinse the "sausages" and place one of them in each of three labeled containers, with the following solutions:

1. distilled water
2. 1.5 M sucrose
3. 3.0 M sucrose

Allow this setup to stand for ten minutes or more, while you go on to read the next section. When you return, examine and note the *appearance* and *texture* of the three "sausages."

If the sausage with 1.5 M sucrose is analogous to a *Nitella* cell, which of the three solutions most closely resembles *Nitella's* natural environment? How do you explain the texture of the "sausage" in that container? What substances in the *Nitella* cell are analogous to the sucrose inside the "sausage?"

Why is an hypoosmotic environment beneficial to most plants? What structural element of a plant cell keeps the membrane from bursting?

Which of the solutions is hypoosmotic to the 1.5 M sucrose "sausage?" What is the appearance of the "sausage" in that solution?

Explain the behavior of the "sausage" in 3.0 M sucrose. Can you set up an experiment to try regenerating this "sausage" in the same manner that you "regenerated" the *Nitella* cell?

How does the membrane used in this model differ from the membrane of a living cell?

(*Note:* Do you think an *animal* cell would behave in the same way as a plant cell toward fluids that are not isosmotic? To help you test the answer to this question, your instructor will provide lancets, slides, and cover slips, and a range of solutions varying from hypoosmotic to hyperosmotic to your blood cells. Mix a drop of each of these solutions with a drop of your blood on three separate slides; examine under low and high power. Explain the results.)

PART IV.
THE SLIME MOLD
PHYSARUM
POLYCEPHALUM

During this part of the laboratory, you will have the opportunity to work with a very unusual organism, the slime mold *Physarum polycephalum*. One important purpose of this exercise is to illustrate, by means of relatively simple observations and experiments, how questions about organisms may arise and be answered, or partially answered, generating more questions in the process. At the same time, this study will test your ability to select the best instrument, lighting arrangement, and magnification for most efficient observation. You will also find the slime mold itself a fascinating object of study.

Below you will find a series of suggested observations and experiments, from among which you may choose, according to your interest and the time available. You will not have time to do them all, and you should feel free to try out any experiments that occur to you, whether or not they are included in the list. Cooperate with others in your group to avoid unwanted repetition of experiments; keep full notes, and turn in a report (as directed by your instructor) of your experiments and observations in the next laboratory period.

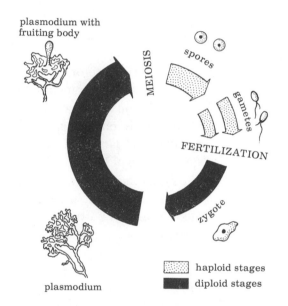

Fig. 2-2. Life cycle of *Physarum polycephalum*.

You are provided with a petri dish containing agar, a gelatinlike substance (non-nutritive for slime molds) upon which a culture of the **plasmodium** stage of the slime mold *Physarum polycephalum* is growing. Figure 2-2 shows diagrammatically the life cycle of the mold. To study the organism, it is necessary to remove the cover of the petri dish, but do not leave it off for long periods when you are not observing the organism, for the culture will dry out. If you wish to do experiments that will damage or kill the cultures (such experiments in the following list are indicated by an asterisk*), make certain that you follow your instructor's directions for checking with him before you proceed. In any case, save damaging experiments until others you wish to do have been completed. If you run into serious snags in observation or lighting, your instructor will have suggestions.

1. In nature, slime molds are found among rotting leaf mold and other decaying vegetation on the forest floor. They move slowly about, absorbing bacteria, the products of bacterial decay of vegetation, and small particles of decaying matter. How would the oat flakes on the agar contribute to nourishing the mold? If the culture were sterile, would the mold be able to feed?

2. What color is your specimen? How would you design an experiment to discover whether color is a heritable trait or whether it is caused by some other agent, such as the type of food ingested?

3. Examining your specimen only with your unaided eye, can you see any evidence that it has recently undergone any changes in shape, size, or position on the agar?

4. Examine the thin white tracks on the agar. What explanations for their presence can you suggest? How can these explanations be tested?

5. By experimenting with different methods of microscopy and lighting, you should be able to locate regions in your mold where cytoplasmic streaming is taking place. The best regions are those of relatively new growth, where there are many small, interlaced branches. The following eight questions all concern cytoplasmic streaming.

6. Devise a means of approximating your microscope's field diameter; from this information and your observations estimate the rate of streaming in millimeters per minute.

7. Is the rate of streaming consistent within a single branch? Is it the same in different branches on the same specimen?

8. If you observe closely, you will notice that the direction of streaming is constantly changing. Can you detect any rhythm or regularity in these oscillations? Does the streaming occur in both directions for exactly the same length of time?

9. Is there any consistency in rate of streaming or in rate of oscillation between your specimen and others in the class? (If you discover that the answer is "yes" after observing consistent regularities in data from other students, can you devise an explanation for the consistency?)

10. In regions where streaming is taking place, do *all* parts of the plasmodium exhibit movement? How would you describe the physical arrangement of sol and gel protoplasm in a single streaming branch?

11. Can you detect any structures that might be responsible for the movement? Where do streaming particles originate, and where do they go?

12. Can the mold stream *upward,* on a vertically oriented piece of agar?

13. Devise a means of discovering how rapidly the mold as a whole moves across the agar.

14. Do you think that the mold can sense the presence of food at a distance, or is contact with food materials made entirely by chance? Design and carry out an experiment to test your hypothesis.

15. Is *Physarum* a plant or an animal? Can you find any good reason to make the distinction?

16. Devise and carry out an experiment to discover whether the pattern of peripheral branching differs in fed and unfed specimens. Explain the results.

*17. With a dissecting needle, make a delicate puncture into a region where active streaming has been observed. If possible, make the puncture while you are examining the mold under the microscope. What is the organism's immediate response? Record your observation over a five- to ten-minute period. Is the puncture apparently repaired during this time? Does the streaming return to normal?

*18. With the edge of a clean razor blade, make a delicate, narrow cut across the edge of a branch. Examine the cut at intervals. Do the severed regions rejoin?

19. Your slime mold has been cultured in the dark, where these organisms usually grow. Cooperating with others at your table, set up a *properly controlled* experiment to determine the mold's response to window light during the interval until the next period.

*20. Heat a glass rod in an alcohol lamp and quickly bring it close to (but not touching) a streaming region of the plasmodium. What is the response?

21. Test the mold's response to cold by placing the *covered* petri dish on crushed ice for ten minutes. Describe the reaction. Can the mold recover from its response to cold? (Be careful *not* to get ice into the culture dish!)

TOPIC 3 ENZYME ACTIVITY

Biological Science, pp. 40-44, 171-174.
Elements of Biological Science, pp. 31-34, 113-115.

A cell is the site of incessant chemical activity. Energy is constantly being extracted (as ATP) from the catabolism of large organic molecules, and put to work in a remarkably wide-ranging variety of synthetic reactions and other energy-requiring activities. The mere maintenance of its own complexly ordered structure is a chemical task of staggering magnitude, but in higher organisms a cell usually has some special activity, such as hormone or special enzyme secretion, active transport, etc. If it is a green plant cell, it also carries out the complex biochemical synthesis of carbohydrates, from carbon dioxide and water. All of the cell's activities, in fact, are based upon chemical reactions, and all these reactions are made possible by enzymes.

Enzymes, as you know, are large organic molecules that belong to the class called proteins and act as catalysts in biochemical reactions. A catalyst, you recall, cannot initiate a reaction that would not have come about anyway in its absence, but it can, and does, radically affect reaction *rate*, with the result that the cell can carry out rapid and complex chemical activities at relatively low temperatures. Most enzymes are highly *specific*, i.e., they tend to accelerate only one or a group of related reactions. The result is that many different enzymes may be present in a cell and may act simultaneously without mutual interference.

In current hypotheses to explain enzyme activity, the enzyme E is pictured as combining temporarily with its substrate S (the substance upon which the enzyme acts) to form a complex ES. As a result of this temporary union, the energy required to activate the breakdown of the substrate molecule is reduced, so that it breaks apart more easily into A and B, products of the reaction. In summary:

$$E + S \rightleftarrows ES \rightleftarrows A + B + E$$

Note that the enzyme is recovered at the end of the reaction, and is therefore available to catalyze the breakdown of additional substrate molecules.

In this laboratory you will study the enzyme **catalase,** which accelerates the breakdown of hydrogen peroxide (a common end product of oxidative metabolism) into water and oxygen, according to the summary reaction:

$$2\ H_2O_2 + \text{catalase} \rightleftarrows 2\ H_2O + O_2 + \text{catalase}$$

Catalase is very widely distributed in the body, but you will isolate it from liver, where it is found in particularly high concentration, and then measure its rate of activity under different conditions, in a test tube. Rate of enzyme activity will be determined by measuring the amount of oxygen produced while the reaction proceeds in a simple closed manometer system. Read through the entire laboratory exercise and make certain that you understand it, before doing any of the experiments.

**PART I.
PREPARATION OF
THE MANOMETER
AND THE ENZYME**

In order to execute the experiments properly, you must have some idea of the limitations of the closed manometer system and an appreciation of the importance of choosing an appropriate concentration of catalase.

A. *The closed manometer system*

Study Appendix 3 carefully, making reference to Figures A3-1 and A3-2 and to the apparatus at your desk. These experiments will use the same basic setup, with the T-rigs modified slightly by the addition of two long-needled syringes (FIGURE 3-1), which permit the addition of substrate to enzyme without taking the system apart. Put the long-needled syringes into the system as illustrated, removing the escape glass plugs and moving the regular calibrated syringes into the escape tubes. Put the glass plugs away in a safe place.

Load the manometer with Brodie's manometer fluid, using the special syringe and needle supplied for this purpose. Check the system carefully for leaks as described in Appendix 3. When you are *certain* that the system is closed, go on to B.

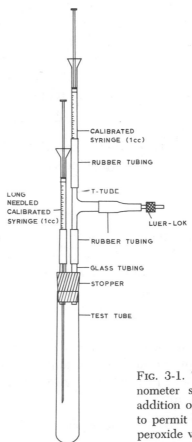

CALIBRATED
SYRINGE (1cc)

RUBBER TUBING

T-TUBE

LUER-LOK

LONG
NEEDLED
CALIBRATED
SYRINGE (1cc)

RUBBER TUBING

GLASS TUBING

STOPPER

TEST TUBE

FIG. 3-1. T-rig for the closed manometer system, modified by the addition of a long-needled syringe, to permit the addition of hydrogen peroxide without taking the system apart.

B. Extraction
of catalase

1. Put about one cubic millimeter of fresh liver into a small test tube, half full of distilled water. Rest the bottom of the tube against a table, and shred the tissue with the *unpolished* end of a stout glass rod.

2. Decant the turbid fluid into a clean 100 cc beaker, leaving the bulk of liver in the test tube.

3. Again, half fill the tube with distilled water, mix, shred again if any pink tissue remains, and decant into the beaker. Repeat until the beaker is one-quarter full.

4. Add distilled water until the beaker is half full. You have now prepared a homogenized suspension that should contain all the enzymes initially present in the liver. In this **homogenate** the concentration of the enzyme catalase should be close to the range measurable in the closed manometer system. Reserve a dropper for use *only* with this **stock catalase solution.**

5. To determine catalase concentration,
   a. put ten drops of stock solution into a clean test tube,
   b. add one drop of 3% hydrogen peroxide to the catalase, and gently agitate the tube. Watch carefully what happens, and approximate the time required for the reaction to run to completion.

If the reaction lasts less than two or three minutes, dilute the catalase with distilled water until a moderate reaction, lasting two or three minutes, is observed. (If the stock catalase is too concentrated, oxygen will be evolved to rapidly to be measured in the manometer.)

6. (Optional) Calculate the volume of oxygen evolved during the reaction.

If one drop (about 0.05 cc) of 3% hydrogen peroxide (3 grams per 100 cc solution) is completely decomposed into water and oxygen, what volume of oxygen will be evolved?

(*Hints:* The molecular weight of hydrogen peroxide is 34; one mole of any gas at 0°C (273°K) and 1 atmosphere pressure occupies 22,400 cc. This volume is increased at room temperature by the factor

$$\frac{(\text{room temperature } °C) + 273}{273}.)$$

It is helpful to know the volume of gas evolved per unit of substrate reacted, in order to determine how often the manometer fluid must be leveled.

PART II.
EXPERIMENTS
WITH CATALASE

The following experiments will be carried out in the closed manometer system. You will measure and record the rate of catalase activity under different conditions of enzyme concentration, substrate concentration, pH, and temperature.

A. Time course

To determine whether catalase activity increases, decreases, or remains constant over a period of time, measure the quantity of oxygen evolved at one-minute intervals for ten minutes after the substrate has been added to the enzyme.

1. Label the experimental manometer chamber with an "X" in red wax pencil, and add to it 12 drops of enzyme and 4 drops of distilled water. Add 16 drops of water to the control chamber. Stopper both chambers, lower

them into the water bath, and allow ten minutes for the tubes to come to the same temperature.

2. Find Table 1 on the data sheets. You will record your data from this part of the experiment in Table 1. Each time you measure the oxygen evolved by adjusting the calibrated syringes, record the time and the syringe marking. You can later fill in the "oxygen evolved" column by subtracting successive syringe markings from one another. (You are provided with similar tables for the other experiments.)

3. Adjust the plungers of the calibrated syringes (experimental to 0 cc and control to 1.0 cc — why?).

4. Fill the two long-needled syringes to the 0.5 cc mark with 3% hydrogen peroxide, and carefully tighten them into the proper tubes.

5. Level the manometer fluid in the U-tube by adjusting the appropriate calibrated syringe. Record the syringe settings next to time 0 in Table 1. You will measure the oxygen evolved, as the reaction proceeds, by releveling the manometer fluid and noting the syringe setting required to reset the fluid at each time interval. When the first syringe has been adjusted as far as it can go, use the other.

6. Add 3 drops of substrate (hydrogen peroxide) to the water in the control tube and 3 drops to the catalase mixture. This is the beginning of the experiment (Time = 0). *Make certain* that the drops of hydrogen peroxide fall straight down into the reaction mixture and do not become lodged on the side of the tube.

7. Since the manometer's volume is small, you must take syringe readings and relevel the manometer fluid at frequent intervals. In selecting times to take readings, keep in mind that (a) each running interval should be approximately one minute, if the catalase was correctly diluted, and (b) you must be able to take enough readings during the experiment to construct a significant graph. For example, two or three readings at three-minute intervals would be insufficient; five, ten, or more readings would make a good graph. Avoid resetting an extremely small fluid displacement; similarly, avoid waiting for such a long time that the displacement is very large. (*Warning:* The first few readings will have to be taken *very soon* after you mix the enzyme and the substrate; keep your stopwatch running constantly.) Continue leveling at intervals for about ten minutes, or until the reaction is completed, being sure to agitate the chambers very gently for a few seconds every minute.

8. Plot your results on Graph A. What is the shape of your curve? Explain your results.

B. Effect of catalase concentration

Repeat the procedure above, using three different concentrations of stock catalase. The experimental chamber should be thoroughly cleaned and the control chamber rinsed between runs. Perfectly clean pipettes must always be used. If in doubt about cleanliness, always clean the questionable item.

Catalase dilutions may be prepared directly in the experimental chamber, as follows:

CONCENTRATION	DROPS OF CATALASE	DROPS OF WATER
1	16	0
½	8	8
¼	4	12

For each run the control chamber should have 16 drops of water. Run the experiments just as in A above, adding 3 drops of substrate to both control and experimental chambers and recording the gas produced for about ten minutes at intervals of approximately one minute. Use Tables 2, 3, and 4 to record your data, and plot the results as separate curves on Graph A. (What concentration of stock catalase does Table 1 represent?) Be sure to label each of the curves carefully.

How does enzyme activity vary with enzyme concentration? Why?

C. Effect of temperature

Measure enzyme activity in the usual manner, at about 37°C and 10°C. Choose a concentration of catalase that seems suitable on the basis of your previous measurements. The catalase and substrate should be brought to the testing temperature *before* they are mixed. Use cold water and ice to establish and maintain a 10°C water bath; use hot and cold water from the tap to obtain a 37°C water bath. Check the temperature at the beginning and end of the experiments, and record it adjacent to the appropriate table. Record the results in Tables 5 and 6, and plot them on Graph B. Be sure to label the graph axes and each of the curves.

From these data, what can you conclude about how temperature affects enzyme activity? How would you explain the results?

D. Effect of pH

Measure enzyme activity with the reaction mixture buffered at pH 5, 7, and 9. Use the same procedure as in A, but dilute the catalase stock with buffer instead of water. Also, of course, add buffer to the control. Record the results in Tables 7, 8, and 9 and plot them on Graph C, being sure to label the curves and the axes of the graph.

What can you conclude about the effect of pH on enzyme activity? At what pH would catalase normally have to act?

PART III. ENZYME INHIBITION

Any treatment that changes the structure of an enzyme or interferes with formation of the complex between enzyme and substrate, will have an inhibiting effect upon the reaction, and may eliminate it altogether. Measure the activity of catalase after treating it with:

A. Heat

Place a small test tube containing a few cc of catalase solution into a boiling water bath for two minutes. Cool the boiled catalase, and with a *clean* dropper transfer 12 drops to the experimental chamber along with 4 drops of water. As usual, add 16 drops of water to the control tube. Level the manometer fluid, add substrate, and proceed to measure the breakdown of hydrogen peroxide in the usual manner. Record the results in Table 10, and plot them on Graph D. How does boiling affect the enzyme? Why?

B. Trypsin   To 20 drops of catalase stock in a small test tube, add 5 drops of 10% tryp-sin. (What is trypsin?) Place the test tube in a water bath at 37°C. After 30 minutes or longer, cool the trypsin-treated catalase and add 16 drops of it to the test chamber. To the control chamber, add 4 drops of trypsin and 12 drops of water. Then measure the breakdown of hydrogen peroxide in the usual manner. Record the results in Table 11, and plot them on Graph D. Explain the results of trypsin treatment.

C. Hydroxylamine   Hydroxylamine attaches to the iron atom (a part of the catalase molecule) and thereby interferes with the formation of enzyme-substrate complex. To 20 drops of catalase solution in a small test tube, add 5 drops of neutralized hydroxylamine. Agitate for one minute. Use 16 drops of this hydroxylamine-treated enzyme in the experimental chamber; add 4 drops of hydroxylamine and 12 drops of water to the control chamber. Proceed as usual; record the results in Table 12, and plot them on Graph D. Explain the results.

QUESTIONS FOR FURTHER THOUGHT

1. In this study, you extracted catalase from liver. From what other sources could you have obtained this enzyme?
2. What is the purpose of the water bath in the manometer system?
3. In the manometer experiment, you added a small amount of sub-strate to a relatively large amount of enzyme in the experimental chamber. What differences would you predict if you added a small amount of enzyme to a large volume of substrate?
4. *From your data,* what is the optimal temperature for catalase activity?
5. How could you break down hydrogen peroxide without using an enzyme?
6. What is a buffer?
7. What reaction occurs between catalase and trypsin?
8. By what mechanism does hydroxylamine interfere with catalase activity?

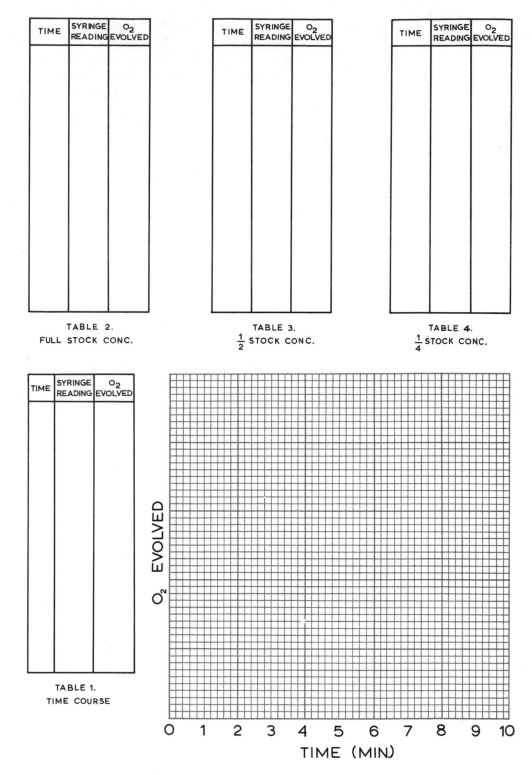

TIME	SYRINGE READING	O₂ EVOLVED

TABLE 2.
FULL STOCK CONC.

TIME	SYRINGE READING	O₂ EVOLVED

TABLE 3.
$\frac{1}{2}$ STOCK CONC.

TIME	SYRINGE READING	O₂ EVOLVED

TABLE 4.
$\frac{1}{4}$ STOCK CONC.

TIME	SYRINGE READING	O₂ EVOLVED

TABLE 1.
TIME COURSE

O₂ EVOLVED

TIME (MIN.)

GRAPH A. EFFECT OF TIME AND ENZYME CONCENTRATION

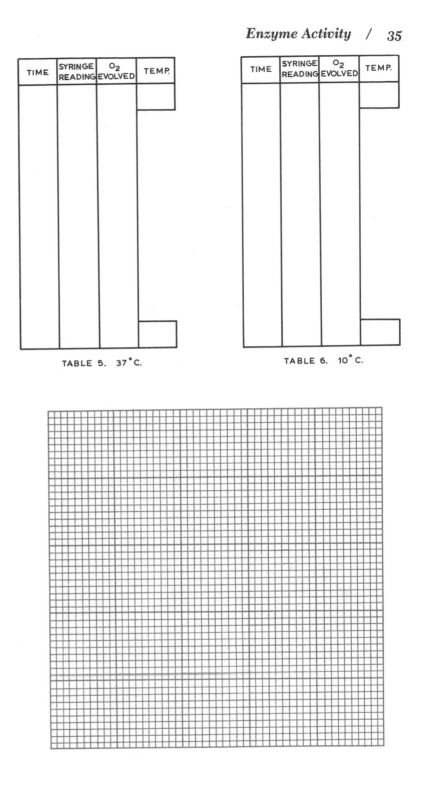

TIME	SYRINGE READING	O₂ EVOLVED	TEMP.

TABLE 5.   37°C.

TIME	SYRINGE READING	O₂ EVOLVED	TEMP.

TABLE 6.   10°C.

GRAPH B.   EFFECT OF TEMPERATURE

TIME	SYRINGE READING	O₂ EVOLVED

TABLE 7. pH 5.

TIME	SYRINGE READING	O₂ EVOLVED

TABLE 8. pH 7.

TIME	SYRINGE READING	O₂ EVOLVED

TABLE 9. pH 9.

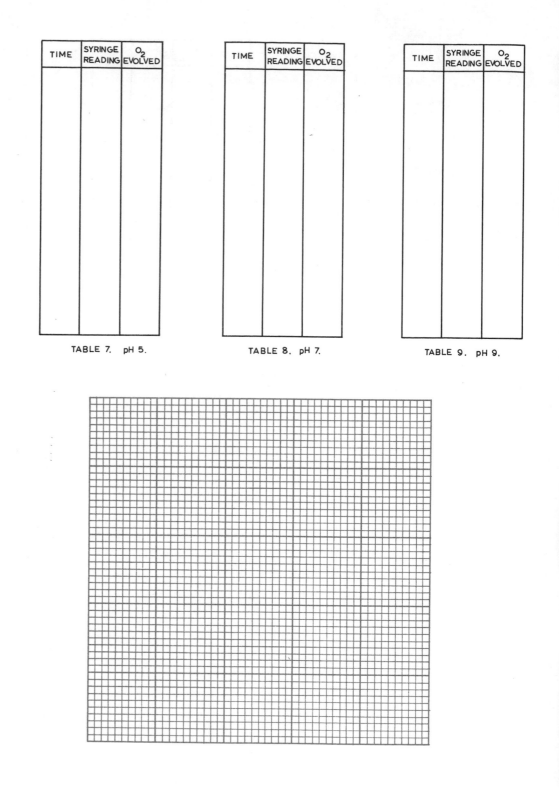

GRAPH C. EFFECT OF pH

TIME	SYRINGE READING	O₂ EVOLVED

**TABLE 10.
BOILING**

TIME	SYRINGE READING	O₂ EVOLVED

**TABLE 11.
TRYPSIN**

TIME	SYRINGE READING	O₂ EVOLVED

**TABLE 12.
HYDROXYLAMINE**

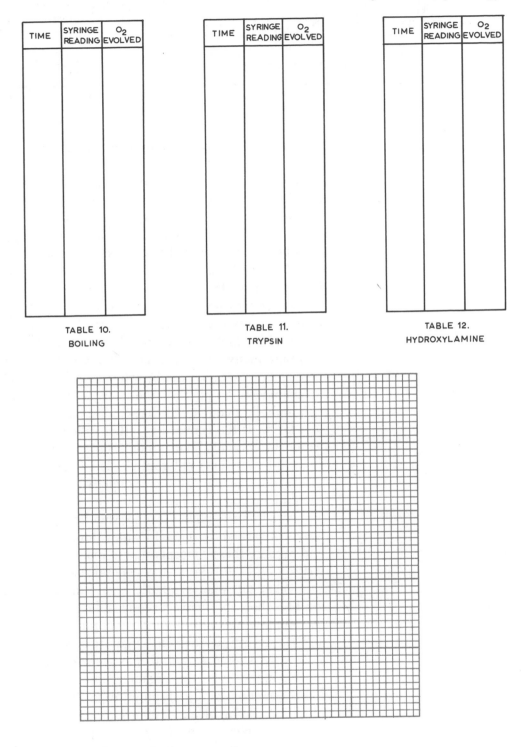

GRAPH D. EFFECT OF ENZYME INHIBITION

AUTOTROPHIC NUTRITION

Biological Science, pp. 73-74, 79-85, 99-117, 135-146.
Elements of Biological Science, pp. 53, 57-62, 72-84, 95-101.

You should be fully aware by now that organisms require a constant supply of usable energy. This is a direct consequence of the Second Law of Thermodynamics, a fundamental principle of physics which tells us that all systems have a natural tendency toward increasing disorder, a tendency that can be counteracted only by the expenditure of energy. Since the system "Living Organisms" is highly ordered, it follows that energy within this system would soon be depleted were there no external energy source. Ultimately, that external source for almost all living things is the sun. Only certain organisms, however, have the capacity to trap the energy of sunlight directly and to convert it into potential chemical energy; this energy-trapping process is called **photosynthesis.** Although terrestrial plants, the photosynthetic organisms with which you are probably most familiar, are usually green in color, there are many very important photosynthetic organisms which do not look green — for example, the brown and red algae (common seaweeds). What is common to all photosynthetic organisms, however, is the green pigment **chlorophyll,** a compound that can be activated by light. Photosynthesis, as you know, is the conversion (often called "fixation") of carbon dioxide and water into organic compounds. The energy that drives this process comes from light-activated chlorophyll. The overall conversion that takes place is shown diagrammatically in FIGURE 4-1. Careful examination of the figure should help you to understand the relationship between the energy-trapping and carbon-fixing phases of photosynthesis, as well as the fate of the various atoms of carbon dioxide and water during their conversion to carbohydrate and oxygen; these concepts should make it clear to you why water is usually written on both sides of the generalized equation for photosynthesis.

What sets photosynthetic organisms apart from the rest of the living world is their ability to synthesize all of the organic compounds they require for building and maintaining their protoplasm, from inorganic compounds and the energy of the sun. Organisms having this independent nutritional status are known as **autotrophs** (self-feeders). All other living organisms must depend upon the organic compounds synthesized by autotrophs as their source of energy (through respiration and fermentation) and of protoplasmic building blocks. Organisms having this dependent nutritional status are called **heterotrophs.** Why autotrophic organisms must be the first link in any food chain, and therefore assume a central and crucial role in the community of life, should now be clear to you. How do you think that the organic carbon synthesized by autotrophs becomes available to heterotrophs?

In this unit you will examine some of the important structural components of terrestrial plants, and make some quantitative measurements of photosynthesis. Although you will study only terrestrial plants, you should not forget the importance of marine algae as the primary fixers of carbon.

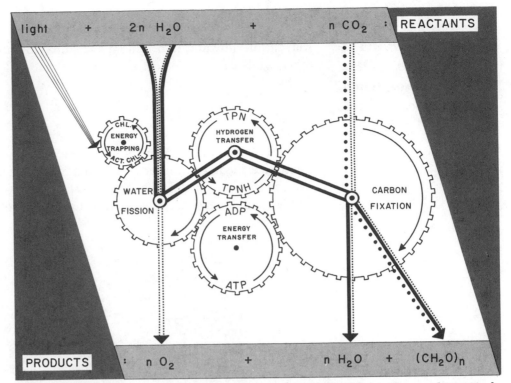

light   +   2n H₂O   +   n CO₂ : REACTANTS

PRODUCTS : n O₂ + n H₂O + (CH₂O)ₙ

Fig. 4-1. Relationship between carbon fixation and water fission in photosynthesis. Each photosynthetic unit in the chloroplast can be visualized as a tiny machine for trapping light energy and using it to synthesize carbohydrates. Light turns the *energy-trapping* gear by "exciting" chlorophyll to an activated (high-energy) state. Energy trapped as activated chlorophyll is coupled to the fission (splitting) of water, a process itself coupled both to the production of "active hydrogen" (TPNH) and to the conversion of energy into a synthetically useful form (ATP). The *hydrogen-transfer* and *energy-transfer* gears work in tandem to drive the *carbon-fixation* gear.

There is a constant flow of materials through the machinery; raw materials enter at the top and products emerge at the bottom. The number of water and carbon dioxide molecules entering the water-fission and carbon-fixation gears, respectively, depends upon which carbohydrate $(CH_2O)_n$ is being formed; the n in the formula is six for glucose $(C_6H_{12}O_6)$. Lines in the figure indicate schematically the fate of reactant atoms. Thick solid lines show the flow of hydrogen atoms; each line represents two hydrogens. Dashed lines signify oxygen flow and heavy dotted lines carbon flow.

Energy also flows constantly through the machinery; light energy is continuously converted to chemical energy, stored in the carbohydrate molecules that emerge from the "assembly line." Energy is transmitted from gear to gear by coupled chemical reactions in which energy-producing reactions drive energy-requiring ones. When one gear is smaller than another, the small gear must turn over several times for each turn of the larger gear. This reflects the greater energy input required to turn the larger gear; the energy required for the activation of one molecule of the large-gear component is equivalent to the energy contained in several molecules of the smaller-gear component. The energy-trapping, energy-transfer, and hydrogen-transfer gears represent fixed pools of molecules that are continuously recycled by being alternately energized and de-energized; the chlorophyll pool is considerably larger than the ADP and TPN pools.

The water-fission and carbon-fixation gears, however, represent the continuous flow of molecules through a sequence of reactions; each product molecule that leaves the chloroplast is replaced by reactant molecules, and the sequence is repeated. Water fission takes place only in the light and involves the intermediate formation of "active water." This active form drives the energy-transfer and hydrogen-transfer gears; oxygen gas is evolved in the process. The splitting of two water molecules leads to the evolution of one molecule of O_2 concomitant with the production of two ATPs and two TPHNs. If the cell contains adequate raw materials and chemical energy, carbon fixation can proceed in darkness. In this process, light energy (now ATP) and the hydrogens of water (now TPHN) are used in bonding n CO_2 molecules to one another in a chain, forming a molecule of carbohydrate. The extra oxygen atom contained in each CO_2 molecule is disposed of by the formation of water. (*Courtesy Steven C. Carlin.*)

PART I.
THE PLANT FORM
A. Absorption of
raw materials; roots

Almost all the water and minerals required by a plant are absorbed by the roots, while carbon dioxide, the key raw material in carbohydrate synthesis, is absorbed principally by the leaves.

ROOT HAIRS

1. Obtain a radish seedling from the supply desk, or examine those provided in a petri dish at your table. These seedlings have been grown in such a way that the roots are free from soil particles. If you examine it closely, you will see that the part of the root closest to the tip is covered with numerous slender and delicate threadlike structures, the **root hairs.** These hairs, which

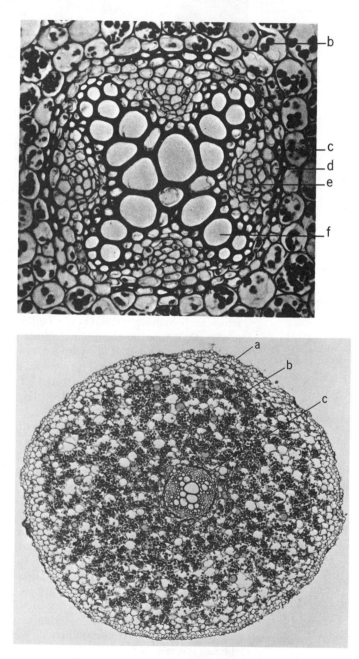

FIG. 4-2. Cross section of buttercup (*Ranunculus*) root. (*Top: Courtesy General Biological Supply House, Inc., Chicago. Bottom: Courtesy Thomas Eisner, Cornell University.*)

are extensions of epidermal cells, greatly increase the absorptive area of a root. Water and dissolved inorganic salts of such elements as potassium, nitrogen, phosphorus, sulfur, calcium, magnesium, and iron all enter through the root hairs. Dissolved gases, such as the oxygen required for the respiration of root cells, also enter through the root hairs. Make a wet mount for observation with the compound microscope. How thick is the cell wall? What are the methods by which water and dissolved minerals are absorbed by the roots?

CELLULAR STRUCTURE OF THE ROOT

2. Examine demonstrations of several different types of root systems. Each demonstration will be accompanied by an explanatory card. Then examine a slide of a cross section of buttercup (*Ranunculas*) root. Locate on the slide and label in FIGURE 4-2 the following regions:

a. The **epidermis** forms the continuous protective outer layer of the plant body; the cells are generally not very large, and their walls, particularly the outer ones, are rather thick.

b. The **cortex** is inside the epidermis but outside the vascular region; it forms most of the bulk of the young root. The cells of the cortex are **parenchyma,** a tissue type composed of relatively unspecialized cells that are usually spherical, ovoid, or cylindrical in shape. They frequently have thin cell walls and there are often numerous intercellular spaces between the cells. Note the many stored starch grains in the parenchyma cells.

c. The **endodermis** is a single layer of cells just inside the cortex. Endodermal cells have thickened cell walls and form the outermost portion of the **stele,** or vascular cylinder. What is the function of the endodermis?

d. The **pericycle** is a layer of thin-walled cells just inside the endodermis. These cells retain the ability to divide, and give rise to branch roots.

e. The **phloem** tissue lies inside the pericycle as small, compact cells between the arms of the much larger cells in the middle. These cells function in the conduction of food materials in the plant.

f. The large, thick-walled cells in the middle, whose arms prevent the phloem from making a continuous ring, are the **xylem vessels.** They function in upward conduction of water and minerals, and for support. (We shall examine both phloem and xylem tissues in more detail in a later laboratory.)

B. Movement of materials

Water and minerals absorbed by the root hairs pass into or between adjoining cortex cells, through the cells of the endodermis and pericycle, and into the xylem of the root.

At the head of your table, you will find a beaker containing celery stalks that have been immersed in a solution of 0.5% aqueous basic fuchsin* for

* A powerful red dye. *Handle with care!*

an hour. Place one of the stalks on a folded paper towel and make *thin* cross sections through the stem, several inches from the bottom. Mount the slices on clean slides in a small amount of water, but without cover slip. Examine under your microscope.

Which tissues are dyed? How far up the stem did the dye move? (If the tissues are not transparent, make several more cross sections or a well-placed longitudinal section to determine the answer to this question.)

C. Manufacture of food; leaves

While the stem and root function in internal transport and in support, the leaf is the organ in which the critical process of photosynthesis takes place. It must therefore be adapted to expose large numbers of chloroplasts to sunlight. It must also provide a moist cell membrane surface for exchange of gases with the environment, yet prevent excessive water loss. Finally, it must contain extensions of the vascular system of the stem to move the products of photosynthesis from their site of production to non-photosynthesizing parts of the plant, and to move water and minerals into the leaf. As you examine a representative leaf in cross section, ask yourself which of its structural features are of particular importance in making it an efficient site for photosynthesis.

THE LEAF

At the supply desk, obtain a fresh leaf from the plant *Ligustrum* (privet hedge) or some other small, stiff-leaved plant. Section it with a fresh, sharp razor blade, as shown in FIGURE 4-3. Make a wet mount of several such sections, add a cover slip, and examine under low power of the compound microscope. Find a region on one of the slices that seems suitable for microscopic study. (*Note:* These observations may be carried out on a prepared slide if your instructor so directs.)

The leaf is composed of a complicated arrangement of cells. You will probably first notice sections of vascular bundles, recognizable on the whole leaf as **veins.** Cross sections of the stem cut the vascular regions perpendicularly, while in the leaf the veins run in a complex pattern, and the plane of the section often cuts them obliquely, giving a "smeared" appearance. Look for an exact cross section of a vascular bundle — the main "mid rib" of the leaf is usually the easiest place to find one. A vascular bundle is composed

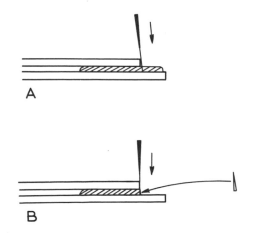

FIG. 4-3. Method for making hand sections. The material to be sectioned is placed between two glass slides. As the top slide is moved back to expose the material, sections are made with a razor blade. (A) Make first slice on a slight angle. (B) Make another slice perpendicular to the slide. A wedge is thus prepared; some regions will be too thick, others too thin, and still others will be the proper thickness for microscopic observation.

of two types of vascular tissue: xylem and phloem. These are of course continuous with the same tissues in the stem and the root. Between the veins lie extensive regions of photosynthetic tissue. Observe (from the top of the leaf down):

> a. The **cuticle,** a waxy layer on the upper and lower leaf surfaces, but in *Ligustrum* thicker on the upper surface. What is its function? Why is it thicker on the upper surface?

> b. The **upper epidermis.** How many cells thick is the upper epidermis? Does this layer contain chloroplasts?

> c. The **mesophyll,** that region between the upper and lower epidermal layers, excluding the vascular tissues. Note the chloroplasts; where within individual cells are they located? The mesophyll is the primary photosynthetic tissue of the plant. It has two distinct regions: **palisade mesophyll,** cylindrical cells nearest the upper epidermis, and **spongy mesophyll,** a region of irregularly shaped cells beneath the palisade layer, containing many intercellular spaces.

> d. The **lower epidermis.** Do these cells contain chloroplasts? Locate pairs of very small cells called **guard cells,** bordering **stomata** (pores) leading to intercellular spaces in the mesophyll. Are there chloroplasts in the guard cells? How do the guard cells function?

THE CHLOROPLASTS Make a wet mount of a leaf of the water plant *Elodea.* If possible, use a very small young leaf from the tip of the plant. Under low power, the cell walls and chloroplasts of the leaf cells can be seen easily. Under high power, notice that the leaf consists of two cell layers, and that the cells on one side are larger than those on the other. If the leaf is not now mounted so that the larger cells are uppermost, remove the cover slip and turn the leaf over.

Watch a chloroplast as it is moved about in the cytoplasm of one of the larger cells. You should be able to determine its shape as it moves about and assumes several different positions. If the cell is in good condition, the cytoplasm should stream continuously. In addition to the chloroplasts, you will see many smaller particles moving through the cytoplasmic strands.

Record your observations in the form of drawings. A large drawing of a whole cell, with arrows to indicate the direction of flow of cytoplasm, should be accompanied by a smaller drawing showing the shape of an individual chloroplast. (*Note:* Consult Appendix 1 for drawing directions.)

How does the structure of an *Elodea* leaflet compare with the structure of the *Ligustrum* leaf? Are there stomata in *Elodea*? How would you explain the differences in structure?

Chloroplasts contain several different green or blue-green **chlorophyll** pigments, and also yellow or orange pigments called **carotenoids.** FIGURE 4-4 reveals that the chloroplast has a complex internal membranous organization; each group of membranes, which looks like a stack of different coins, is called a **granum.** The lower inset shows a granum and the upper inset diagrams the structure thought to characterize the layers within the granum. You will note that chlorophyll is bound between the layers.

What photosynthetic reactions occur in the chloroplast? Can any of

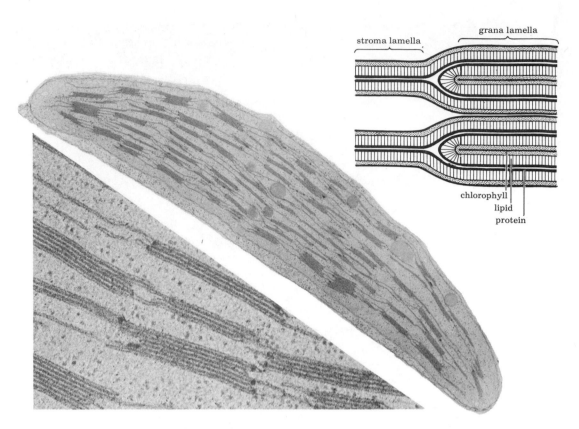

FIG. 4-4. Structure of the chloroplast. The large central electron micrograph shows the structure of a whole chloroplast from a cell of tobacco leaf. Lower and upper insets, respectively, show grana within the chloroplast and a diagram of a small section of the granum membranes. (*Micrographs, courtesy Herbert W. Israel, Cornell University. Diagram of lamellar structure, modified from A. J. Hodge*, et. al., J. Biophys. Biochem. Cytol., *vol. 1, 1955.*)

them occur in chloroplasts outside a living plant? Can any of them occur in chloroplasts that are not intact? Would you consider an isolated chloroplast a living unit?

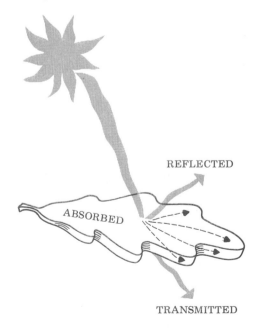

FIG. 4-5. Light striking an object, such as a leaf, may be reflected, absorbed, or transmitted.

PART II.
PLANT PIGMENTS

When a substance absorbs visible light of certain wavelengths (colors), while reflecting and/or transmitting light of other wavelengths (FIGURE 4-5), we see the substance as colored, the actual color depending upon the light reaching our eyes; such substances are called **pigments.** The absorption of light energy by a molecule leads momentarily to the formation of an "excited" or high-energy form of the molecule, but in most cases this energy is dissipated without doing any useful work. Plants, however, "trap" the energy absorbed by the chlorophylls and couple this trapping process to the synthesis of carbohydrates. Other plant pigments, such as the carotenoids (e.g., [orange] carotenes and [yellow] xanthophylls) absorb wavelengths of light different from those absorbed by the chlorophylls; this energy is then transmitted to the chlorophylls.

A. Extraction of pigments with organic solvents

The chlorophylls and carotenoids have been extracted from leaves by your instructor, using hot acetone, an organic solvent in which these pigments are soluble. The extract was then filtered to remove cellular debris.

1. Obtain about 1 cc of the acetone extract in a test tube (about 1 cm level) and add to it about twice its volume of petroleum ether and a few drops of water. Observe that the ether and acetone are mutually insoluble (immiscible) and that the water falls through the upper ether phase and dissolves in the lower acetone phase. Cork the tube or cover it with a plastic film (such as Parafilm). Keeping your index finger over the top of the tube, invert it several times. This allows droplets of one phase to "fall through" the other, so that there is a large surface area of contact between the immiscible solvents. The pigments, or in general any dissolved compounds, will be "partitioned" (separated) between the phases according to their relative solubility in the two phases. In your experiment all of the pigments will dissolve in the petroleum ether layer, since the presence of water in the acetone phase makes the pigments practically insoluble in this phase. What does this tell you about the nature of the pigment molecules? Although the carotenoids are also extracted into the ether phase, the ether will appear green because the darker color and greater quantity of the chlorophylls will mask the carotenoids. You will separate these two classes of pigments in the next operation.

2. Decant the petroleum ether layer into a second test tube and add 1 cc (20 drops) of 30% potassium hydroxide (KOH) in methyl alcohol. Chlorophyll is soluble in the methanolic KOH, whereas the carotenoids are not. Note the color of each layer after this separation has been accomplished. Record your observations as a sketch of the test tube, showing the different colored layers.

B. Separation of pigments by paper chromatography

Having just separated the chlorophylls and carotenoids by a simple extraction procedure, you will now separate these pigments by paper chromatography, a technique more widely used than solvent extractions for separating mixtures of compounds; it is especially useful when the compounds to be separated are very closely related chemically, and hence difficult to separate, or when only very small amounts of sample are available. By the methanolic KOH extraction, you were only able to separate the pigments of the leaf into two classes, but by paper chromatography you will be able to distinguish various pigments within each class.

In chromatographic separations, the compounds to be separated are streaked along a line near one end of a piece of filter paper. After being allowed to dry, the same end of the paper is placed in a dish or tray of the chosen solvent — petroleum ether in this case — in a closed chamber where the atmosphere has been saturated with the same solvent. In the procedure you will be using, the filter paper will be rolled into a cylinder, so that it will stand in a dish of solvent. Another common procedure, however, is to hang the paper from a tray of the solvent that is mounted in the upper part of the chamber. In both cases, the solvent will move along the paper by capillary action, ascending in the former case and descending in the latter.

Paper chromatography takes advantage of rather subtle differences in the partitioning, or relative solubility, of various substances between a stationary water phase and a mobile organic phase. The greater the solubility of a given pigment in the mobile ether phase relative to its solubility in the stationary water phase, the greater its tendency to remain in the organic phase and the further its ascent. Subtle differences between the partitioning behavior of various pigments — differences that are often not great enough to permit separation of the pigments by a simple extraction — are greatly accentuated by the relative movement of the two phases. Thus, the different pigments in the original mixture will travel at different rates, and the original band will separate into a number of different bands. Each band will move a certain fraction of the total distance traversed by the solvent front; this fraction is a characteristic of the particular pigment making up the band, and will always be the same whenever this pigment is chromatographed in the same solvent system. If a different mobile phase or a different type of stationary phase were to be used, the distance traveled by each compound would differ from that in the original system. By judicious choice of mobile solvent and supportive medium it is possible to separate the components of almost any mixture of compounds.

To separate substances other than pigments, the chromatogram must be "developed" in a reagent that differentially colors the separated substances, or it must be observed under ultraviolet light, where many compounds may be seen either as dark absorption bands or as bright fluorescent bands. One of the most important uses of chromatography is to identify unknown substances. This is done by comparing the mobility of the unknown with reference standards in a number of different solvent systems; if the unknown migrates at the same rate as a given standard in all the solvent systems, then the unknown and the standards are probably identical compounds. You can now appreciate why chromatography has proved itself an exceptionally useful technique for separating and identifying compounds present in very small quantities.

1. At the supply desk, you will find a container of a concentrated extract of leaf pigments in petroleum ether, prepared by your instructor. Obtain a small amount of this extract in a small beaker or dish, and take it to your desk. With a capillary pipette, apply a *thin* line of pigment (see FIGURE 4-6) 2 cm from the long end of a sheet of filter paper. Allow the solvent to evaporate from the paper, and apply the pigments again, to the same

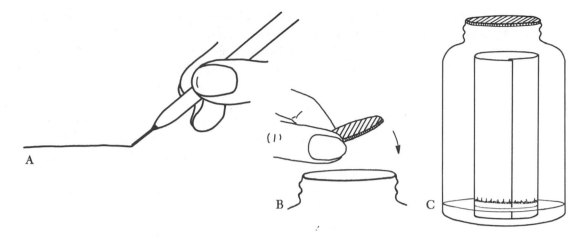

FIG. 4-6. Paper chromatography. (A) The manner in which pigment is applied to the chromatography paper. (B) Placing lid on the chamber. (C) The chamber with a chromatogram running within it.

line. Repeat this operation as often as necessary (at least five times) to produce a dark green line, allowing the strip to dry between applications.

2. In a dry chromatography chamber (quart Mason jar) put petroleum ether to a level of 1 cm.

3. Roll the filter paper sheet into a cylinder, with the long sides forming the top and bottom circumferences and the short sides stapled together (FIGURE 4-6). Insofar as possible, avoid handling the filter paper with anything but very clean fingers; oils and other substances with which the solvents may come into contact will interfere with their movement across the paper.

4. Place the cylinder into the chamber, pigment line down, in such a way that the paper does not touch the sides of the jar. Place the lid on the chamber, and wait twenty minutes or more for the pigments to separate. How many pigments were there in the extract? On the basis of color, try to identify them. Remove the chromatogram, allow it to dry standing up, take out the staples, and place in your laboratory guide. Each student should make a chromatogram for his own notebook.

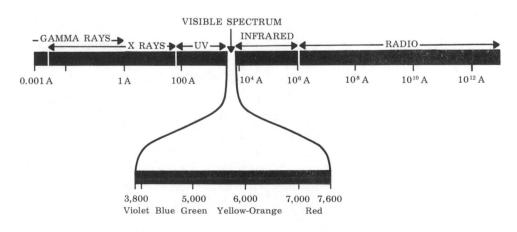

FIG. 4-7. Portion of the electromagnetic spectrum. Visible light constitutes only a small part of the total spectrum. Within the visible spectrum, light of different wavelengths stimulates in us different color sensations. Wavelengths are given in angstroms.

C. *Absorption of light*
by chlorophyll

Visible light, which constitutes a small part of the spectrum of radiant energy (Figure 4-7), can be separated into a variety of colors if passed through a prism. These colors correspond to the different wavelengths of light present in the visible range of the spectrum, and are interpreted as the colors violet, blue, green, yellow, orange, and red, with the wavelength increasing from violet to red. As mentioned before, a pigment is colored because of its ability to absorb certain wavelengths, while reflecting others. In this section, you will study the absorption of light by plant pigments.

A spectroscope (an aimed prism) has been set up at each table with a dissecting lamp in such a way that you may analyze the spectrum of rays transmitted by substances. Between the spectroscope and the lamp, place the following, one at a time:

1. pieces of colored glass (green, red, blue, yellow, etc.)
2. a container of acetone extract of chlorophyll
3. a container of acetone
4. a leaf

In each instance, note the effect on the spectrum and record your observations.

Which colors are *absorbed* by chlorophyll? Which are *transmitted*? What wavelengths of light correspond to the colors absorbed by chlorophyll? What wavelengths are known to be maximally effective in photosynthesis? What wavelengths of light are minimally effective? Why was colored glass used? Why was a container of pure acetone used? How does light absorption by the chlorophyll solution differ from that of the leaf? Why should chlorophyll have a different absorption spectrum when in solution? (*Hint:* When chlorophyll is excited in solution there is no coupling mechanism for using this energy to drive other processes, as there is in the leaf. How do you think the absorbed energy is dissipated in solution? Careful close-range examination of the chlorophyll solution with a light beam shining on it at an angle will help you to find a partial answer to this question; what color do you see?)

PART III.
MEASUREMENT OF
PHOTOSYNTHESIS

While one thinks of a plant as a photosynthetic device, it is important to remember that it respires continuously, as any organism must, in order to obtain energy for its activities. A part of the oxygen released during photosynthesis is required for normal respiration, and when the plant is not carrying out photosynthesis it must obtain respiratory oxygen from the environment. The consumption of oxygen in the dark may be used as a measure of respiration. In the light, with both photosynthesis and respiration proceeding simultaneously, the oxygen evolved is the difference between these opposing reactions. Under optimal conditions, the rate of oxygen production by photosynthesis is ten to twenty times greater than the rate of oxygen consumption by respiration. Hence, under such conditions, the oxygen *measured* during photosynthesis represents 90% to 95% of the total oxygen produced.

To measure oxygen production, we will employ the closed manometer system. If you did not use the manometer in Topic 3 (Enzyme Activity), please read Appendix 3 carefully before you do this experiment.

1. Put three or four sprigs of *Elodea* into the experimental tube, with the cut ends upward and near the top. Fill the tube to within 5 mm from the top with 2.5% sodium bicarbonate solution, thoroughly aerated before use. Fill the control tube with the same amount of sodium bicarbonate solution. Sodium bicarbonate serves as a carbon-dioxide source, according to the following reaction:

$$NaHCO_3 \xrightleftharpoons{H_2O} NaOH \quad + \quad CO_2$$

It is necessary to aerate the bicarbonate solution, so that it will be completely saturated with air. (Your instructor has done this for you prior to the laboratory period, by bubbling air through the bicarbonate solution.) Why is it necessary to saturate the bicarbonate solution with air?

2. Set up the manometer and use a 150-watt lamp as a light source.

3. Fill a battery jar or other flat bottle with cold water, and place between the manometer and the light source, to act as a heat filter. This will prevent large temperature fluctuations in the manometer.

4. Allow the whole system fifteen minutes to equilibrate, so that all parts will come to room temperature. All parts of the system should be open to the atmosphere.

5. Now close the system (as described in Appendix 3) and read the changes in volume every two minutes, recording your data in the left-hand "light" table. After the rate of oxygen evolution is *constant* for several consecutive readings, note the time on the table and continue reading for twenty more minutes. As in the other manometer experiments, take readings off the calibrated syringes.

6. Shut off the light and continue to measure for twenty minutes, placing your data in the center table marked "dark."

7. Turn on the light again and continue reading for about another twenty minutes, recording data in the right-hand "light" table.

Plot the data in three separate, properly labeled curves on the graph. Be prepared to interpret and explain your data.

Why can you not measure the *total* oxygen produced in photosynthesis?

What factors influenced the rate of photosynthesis in your experiment?

Design experiments to demonstrate how rate of photosynthesis is influenced by each of the following: light intensity, light quality, temperature, and concentration of carbon dioxide.

What are the possible sources of error in using the manometer?

If there is time, your instructor may be able to provide materials so that you can carry out some of the experiments you designed above.

QUESTIONS FOR
FURTHER
THOUGHT

1. What functions does the root cortex serve?
2. What is the function of the endodermis?
3. In part IB, by what method did the fuchsin travel up the stem?
4. What physical characteristics make the leaf an effective photosynthetic organ?
5. What special adaptations does the leaf have for gas exchange?
6. Suggest an hypothesis to account for the fact that leaves of different plants have different shapes.
7. What is the role of the carotenoids in photosynthesis?
8. What colors of the visible spectrum are absorbed by the leaf?

TIME	SYRINGE READING	O$_2$ PROD.

LIGHT

TIME	SYRINGE READING	O$_2$ PROD.

DARK

TIME	SYRINGE READING	O$_2$ PROD.

LIGHT

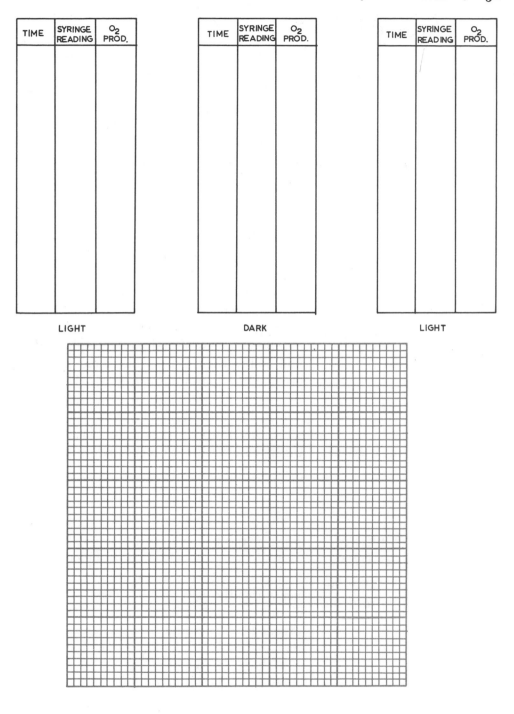

O$_2$ PRODUCTION IN PHOTOSYNTHESIS

HETEROTROPHIC NUTRITION

Biological Science, pp. 154-171.
Elements of Biological Science, pp. 104-113.

Heterotrophic organisms, as you recall, are those that are incapable of synthesizing their own organic compounds from simple inorganic nutrients, and that must therefore rely largely upon prefabricated organic molecules. Since the majority of such molecules are too large and insoluble to pass through semipermeable membranes, they must be broken down into their smaller, absorbable subunits.

While we are all familiar with **digestion** as the breakdown of nutrients within our own digestive tracts, it is well to remember that the term describes *any* breakdown of a complex compound, such as protein, fat, starch, or compound sugar, into its constituent building blocks (amino acids, fatty acids, glycerol, and simple sugars), and that such processes regularly occur *within* most cells of both heterotrophs and autotrophs in the course of normal metabolism. Thus, while almost all organisms carry out digestion of organic compounds intracellularly, heterotrophs must also carry out digestion before they are able to absorb organic nutrients from the environment.

In the laboratory today, you will examine some of the varied means by which heterotrophic organisms secure nutrient materials, and some of their digestive adaptations for the processing of organic nutrients prior to absorption.

PART I. NUTRIENT PROCUREMENT
A. Engulfment of food by Protozoa

Most of the Protozoa, and a variety of cells in multicellular organisms, for example, white blood cells, feed by engulfing smaller organisms directly into temporary vacuoles. These food vacuoles circulate in a characteristic manner within the cells, and enzymes are secreted into them for digestion. Some Protozoa, such as *Amoeba*, can form vacuoles anywhere on the cell surface; others, like *Paramecium*, have regions of the cell specialized for vacuole formation.

Make a wet mount of *Paramecium* culture into which you have stirred a drop of methyl cellulose and a tiny bit of Congo-red yeast mixture (just enough to color the mixture pale pink; toothpicks are available for mixing). Examine under low and high powers, with reference to Figure 5-1. Locate the **cilia**, fine cytoplasmic threads projecting from the surface of the animal. Very long cilia sweep minute particles (stained yeast, in this case) along an **oral groove** and into a **gullet**, at the end of which food vacuoles are formed. Locate a more or less stationary animal and watch the collection of yeasts at the end of the gullet, formation of a food vacuole, and the manner in which the vacuole circulates inside the cell. If you are a patient observer, and the animal you choose is quiet enough, you may be able to map the route taken by the vacuole. Very observant students will find that the color of the vac-

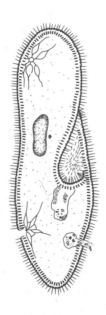

Fig. 5-1. *Paramecium.*

uole changes on its route through the cytoplasm. Your instructor will have more information on this reaction between the dye on the yeasts and the materials within the food vacuole secreted by the *Paramecium*. (What sorts of substances would these be?)

Vacuoles of this kind are formed anew each time food is ingested; these animals lack permanent structures to digest food. Thus, food vacuoles are the functional counterparts of the digestive systems of multicellular animals.

Is this intracellular or extracellular digestion?

Do you think *Paramecium* and *Amoeba*, living in the same habitat, ever compete directly for food?

B. Predation by multicellular organisms

In the broadest sense, a predator is an organism that is free-living and feeds on other living organisms. This definition, of course, embraces organisms like *Paramecium*, which one might not ordinarily think of as a predator, and it also includes animals that may be predators when prey is available but that resort to some other mode of feeding when prey is absent.

Some predators are **carnivores,** feeding on animal prey; others, that feed on plant material, are **herbivores.** Organisms that feed on plant and animal materials are called **omnivores.** How would you classify *Amoeba*? *Paramecium*? *Homo sapiens*?

Carnivores often have interesting adaptations for capturing prey, as we shall see in this and several subsequent laboratories. Prey animals, on the other hand, may be adapted to avoid capture: Birds, for example, have very sharp eyesight, and some of the insects upon which they feed are so colored as to blend into their surroundings and thus avoid capture. To catch an animal, a carnivore usually must move rapidly; an herbivore, on the other hand, does not need to be speedy unless the speed is an adaptation to avoid predators.

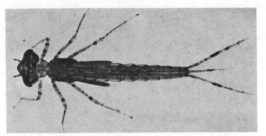

FIG. 5-2. Damsel fly larva. (*Courtesy Carolina Biological Supply Company.*)

DAMSEL FLY LARVA:
AN ANIMAL CARNIVORE

1. Obtain a damsel fly larva (the immature stage of a winged insect; FIGURE 5-2) in a culture dish of pond water, and some *Tubifex* worms in a paper cup. With forceps, offer one of the worms to the larva, or drop it just in front of the larva. *Describe* the initial response, the capturing of prey, and, insofar as possible, the process of ingestion. What senses do you think are normally involved in the location of prey by the damsel fly larva? How could you set up an experiment to test your hypothesis?

BLADDERWORT:
A PLANT CARNIVORE

2. A very few plants supplement their inorganic nutrients with animal prey, and some of these plants are dependent upon predation as a source of nitrogen (in amino acids). We shall examine the common bladderwort *Utricularia* (FIGURE 5-3), a pond plant that is curiously adapted to capture small aquatic animals.

FIG. 5-3. *Utricularia* with bladders. (*Courtesy Carolina Biological Supply Company.*)

From the supply desk, obtain some *Utricularia* in a culture dish of pond water. Examine it with a dissecting microscope. If you like, individual bladders can be examined with a compound microscope, if mounted in a depression slide or on a slide with the cover slip raised with vaseline. Note the numerous small bladders on the plant. Introduce some mosquito larvae or several *Tubifex* worms into the dish with the bladderwort, and watch what happens under the dissecting microscope. (*Note:* If the animals do not come near the bladders by chance, hold one of the worms near a bladder with forceps.) What happens when one of the animals contacts one of the bladders?

See if you can stimulate an unfed bladder with pressure from one of your clean dissecting instruments.

If a film of predator-prey interactions is available, make sure that you see it.

Look up the Phylum or Division and the Class of each of the animals and plants that you have so far studied in this period, and enter each of them in the appropriate space in Appendix 2.

PART II. VERTEBRATE DIGESTION

You will now examine digestive adaptations in a number of animals, concentrating study on the digestive tract in a representative of the subphylum Vertebrata, the group to which all backboned animals, including men, belong. Of the several classes of vertebrates, the one in which we are most naturally interested is that to which we ourselves belong: Mammalia, or mammals. The mammal that you will study as representative of its class is the pig, in this case, a **fetal** pig (i.e., one taken from its mother's uterus before birth). Classify the pig in Appendix 2 before you proceed.

Inasmuch as this is your first exercise with the fetal pig, we will not be concerned with the digestive tract alone, but will also look briefly at some external characteristics of the animal. Likewise, as the dissection proceeds, your attention will sometimes be called to internal structures not directly related to the alimentary canal, but more easily examined now than at a later stage in your work.

A. Introduction to dissection
TOOLS AND GENERAL PROCEDURES

1. Examine the contents of your dissecting kit. A **scalpel,** if one is in your kit, or a single-edged razor blade, is used to make initial cuts through epidermal layers. On rare occasions, when a broad cutting edge is required, it is used to incise or remove some organ. The scalpel or razor blade is *never* used in routine dissection, and if you have formed the habit of using it, you should break that habit at once. **Scissors** are the primary cutting instruments, but they may also be used for spreading tissues, by inserting the closed points and then opening them to push the tissues to either side. The **forceps** is used as a probe, as a manipulator, and as a means of lifting structures to be cut so that underlying areas are not damaged. The **probe** is used for lifting, pushing aside, feeling (where a structure may not be clearly visible), and in tracing the course of ducts. It is also effective in tearing loose bits of connective tissue. The **teasing needle** is useful for punctures or the few other occasions when a sharp-pointed instrument is necessary. Be sure to keep your instruments clean and dry, and they will remain serviceable for a long time.

In carrying out your dissections, you should be very careful to follow the directions in this guide, and those given verbally by your instructor. Dissection is to be undertaken with great care, and is a procedure to help you locate, expose, and learn the anatomy of an animal. It is *not* a random cutting-up of a carcass. Always remember that you should do as little cutting as possible, because the more you cut, the more you will alter the structural relationships among the parts. It is your objective always to learn the *original* organizational patterns, not some new arrangements created as artifacts of your dissection. You will have to cut out connective tissues, in order to see the relationships between organs, but you should always know *what* you are cutting and *why* you are cutting. If you keep this rule in mind, you should not have any difficulty preparing a good dissection, at the same time avoiding damage to your specimen.

You are encouraged to compare notes with others as you work, but the actual dissection on the specimen should be your own. Since you will need to have the specimen in adequate condition for future study, work with care. When you have finished today, place your animal back in its plastic bag, tie the bag tightly closed, and attach a tag on which you have written your name and laboratory section. Follow your instructor's directions for cleaning up and disposing of waste tissue.

ANATOMICAL TERMS

2. In order to utilize fully the directions given in this and succeeding laboratories, you will need to learn a few adjectives used to denote orientation and direction in anatomical work. Be sure you understand the following, and refer back to this list whenever necessary:

right and left	—	the *pig's* right and left
anterior	—	toward the head
posterior	—	toward the tail
caudal	—	toward the tail
dorsal	—	toward the back; up
ventral	—	toward the belly; down
lateral	—	toward the side (right or left)
medial	—	toward the middle or center
proximal	—	near a specified point of reference
distal	—	away from a specified point of reference
pectoral	—	referring to the shoulder region
pelvic	—	referring to the hip region

B. External features of the fetal pig
BODY REGIONS

1. The body of a mammal is arbitrarily divided into a series of readily identifiable regions: head, neck, trunk, and tail. The trunk consists of an anterior thorax (protected by the ribs) and a posterior abdomen.

THE HEAD

2. Locate the eyes, nostrils, mouth, and external ears or pinnae (singular, pinna).

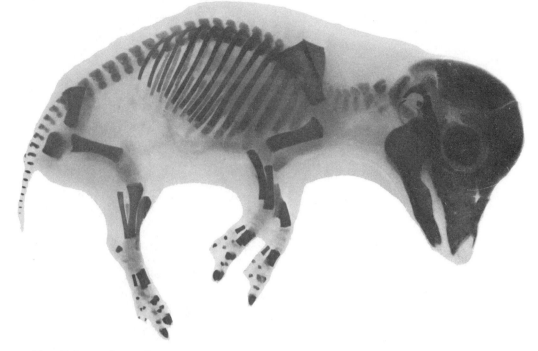

Fɪɢ. 5-4. Fetal pig, cleared to show developing bones. (*Courtesy Carolina Biological Supply Company.*)

THE TRUNK

3. a. Examine the four legs. Feel their various joints and parts, making reference to Fɪɢᴜʀᴇ 5-4 where necessary, and determine which parts of the forelimb correspond to your upper arm, elbow, lower arm, wrist, and hand. Do the same for the hind limb, corresponding it to your leg. How many toes are there on each foot? How many toes touch the ground when the pig walks? Note that the pig is **ungulate** (hoofed) and that it walks on its toenails.

b. Note the **umbilical cord,** arising from the ventral portion of the abdomen. Blood vessels, visible in its cut end, carried nutrients, oxygen, and waste materials between the circulatory system of the embryo and that of the embryonic portion of the placenta, where they were exchanged with the maternal part of the placenta, the uterus, by diffusion.

c. Locate the small **nipples.** Determine their location and the number of pairs. Nipples are the external openings from the mammary glands, a distinctive feature of mammals.

d. **Hair** in some places on the body is another characteristic of mammals that helps to distinguish them from the other classes of vertebrates. What body coverings do other classes of vertebrates, such as fishes, amphibians, reptiles, and birds possess?

e. Just under the tail is the posterior opening of the digestive tract, the **anus.** In females, an opening that serves both genital and urinary systems, the **urogenital opening,** is located just ventral to the anus, and a small, fleshy **papilla** can be seen projecting from it. In females, the urinary and genital systems separate later in development, and the single opening becomes two. In males, the single urogenital opening is located just posterior to the umbilical cord; the duct leading to it runs forward from between the legs in a long muscular tube, the **penis,** which can be felt under the skin. What sex is your specimen?

C. The digestive tract

Animal digestive tracts vary widely, and the differences are often closely correlated with the type of food consumed. You will examine the fetal pig's digestive tract in detail, but you should note other animals on demonstration.

THE ORAL CAVITY

1. a. Open the **jaws** as widely as you can without cutting, and identify the **oral cavity,** with the **tongue** on its ventral surface and the **hard palate** forming its dorsal surface. The hard palate separates the oral cavity from the **nasal cavities.**

b. Examine the jaws on a demonstration skeleton of a mammal, and note how they fit one another. Are both jaws movable?
   We know from study of fossils that the first vertebrates lacked jaws. The early vertebrates were sluggish, mud-grubbing types that lived along the bottoms of ponds and streams. Early descendants of these jawless vertebrates developed a predatory mode of life, and in adaptation to it, evolved jaws for seizing and holding prey. From this stock eventually evolved fishes, amphibians, reptiles, birds, and mammals, all of which have retained the jaws in a more or less modified form. On demonstration is a **lamprey,** one of the few extant members of the ancient class of jawless vertebrates. Though jawless, the lamprey itself is modified in that it is no longer a mud-grubber as an adult, but has developed a parasitic mode of life.

c. With heavy scissors, starting at one corner of the mouth, cut through the tissue *and* bone posteriorly to the base of the tongue, along a line from the corner of the mouth to the bottom of the ear. (*Note:* For this and all subsequent cuts, follow the dotted lines in FIGURE 5-5; do not be afraid to cut through the pig's jawbone — the cut is necessary to expose deeper parts of the oral cavity.) Repeat this operation on the other side, and spread the jaws so that the oral cavity is exposed to view. Examine the tongue carefully, and look at the bones on a skeleton in the region where the tongue would be attached. How is it attached? What is the role of the tongue in feeding? Can you think of any animals in which it is adapted for other functions?
   Locate the papillae scattered over the surface of the tongue; these contain taste buds.

d. Examine the hard palate in more detail, and locate its posterior border. Posterior to the hard palate is the **soft palate,** which contains no bone. In humans, a small posterior extension of the soft palate, the **uvula,** hangs down into the throat. Locate the uvula on your neighbor.

e. Locate the upper and lower **gums** and the few teeth that have probably erupted. Those that have erupted will probably be the third pair of **incisors** and the **canines.** Cut into the gums of your pig and

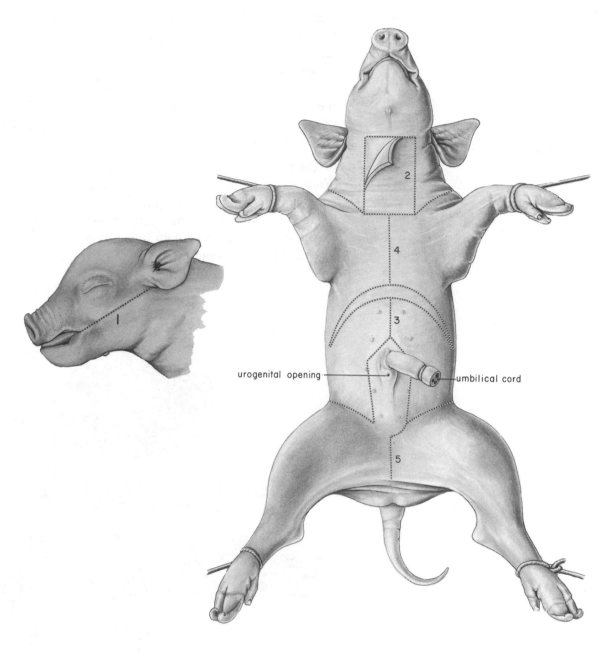

Fig. 5-5. Male fetal pig. Dotted lines show all subsequent cuts to be made into major regions of the body.

determine how many embryonic teeth are present; remove them and observe their characteristics. For what particular functions are the teeth of each type specially adapted?

One major function of teeth is mechanical breakup of large food masses, which provides a greater surface area upon which the enzymes involved in chemical digestive processes can act. Animals have developed many other means of mechanical breakup: examine the demonstration specimens of bird, cockroach, earthworm, and cray-fish, showing the organs these animals have developed for physical breakup of food.

See also demonstrations of teeth modified in ways different from those of the pig. In each instance, try to correlate the differences with the mode of feeding and the type of food utilized. Mechanical break-up is, of course, only one of the many functions of teeth. What are some other functions?

f. The most posterior part of the oral cavity is known as the **pharynx** (FIGURES 5-6 and 5-7). This cavity, a common passageway for the food and respiratory canals, is delimited ventrally by the caudal base of the tongue and dorsally by the caudal border of the soft palate. (Be sure that you have cut far enough into the jaw, or you will not be able to spread the jaws properly to see the pharynx.) Note the opening into

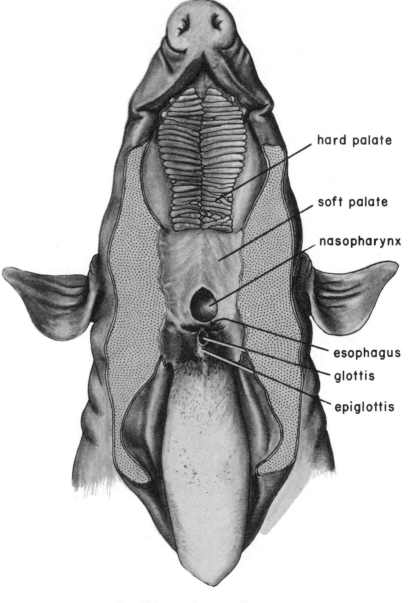

hard palate

soft palate

nasopharynx

esophagus

glottis

epiglottis

FIG. 5-6. Fetal pig, oral view.

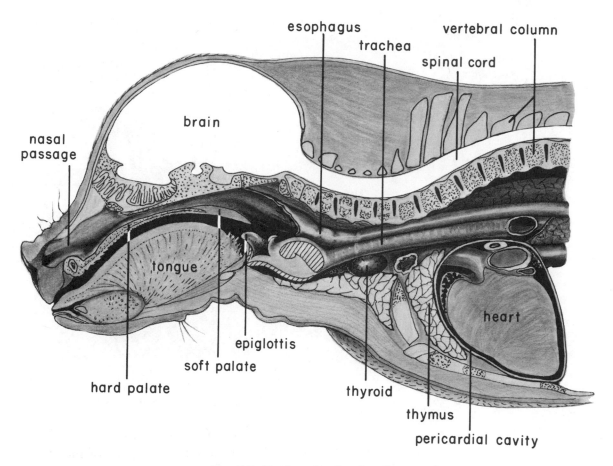

Fɪɢ. 5-7. Fetal pig head and neck, sagittal section.

the pharynx of the nasal passages; food and respiratory canals cross in the pharynx.

g. Identify the **epiglottis,** a median flap at the base of the tongue. The epiglottis partially covers an opening, the **glottis,** which is the entrance into a passageway leading to the lungs. Posterior and dorsal to the glottis, find the opening into the **esophagus,** the tube of the digestive tract that leads through the neck and thorax to join the stomach. Pass the end of your probe into the glottis and into the esophagus, and be certain that you can differentiate them.

THE NECK REGION
(FIGURE 5–8)

2. Turn the pig ventral side up and remove skin from an inch-square area from the middle of the lower jaw posteriorly to the chest (FIGURE 5-5). When the skin is off, you will have exposed muscles and glands. The muscles in this region occur mostly in the form of thin ribbons, with the fibers lying lengthwise. A pair of large **thymus glands** will be exposed, lying among the muscles. You should be able to distinguish muscle tissue from gland tissue in any future dissection.

The thymus glands are not part of the digestive system, but function in the young mammal to produce the line of cells that enable the animal to manufacture antibodies as a defense against disease.

Using a probe or forceps, separate the superficial muscles and glands

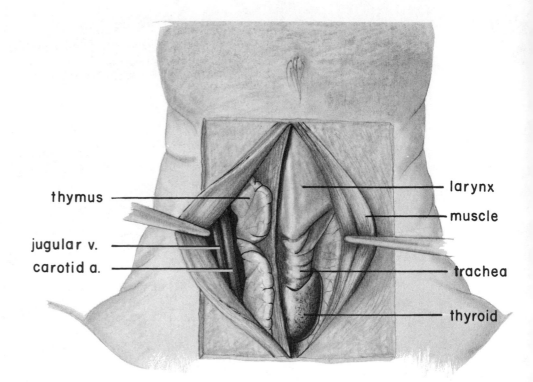

thymus

jugular v.

carotid a.

larynx

muscle

trachea

thyroid

FIG. 5-8. Fetal pig neck region, ventral dissection.

from one another without cutting them. Probe down into the deeper layers of the neck. Medially, beneath several strips of muscle, will be found the hard-walled **larynx** and **trachea,** parts of the respiratory passage to be examined later in more detail. Dorsal to the trachea, probe for the esophagus; fingers are useful instruments for this part of the dissection. One easy way to be certain that you have located the esophagus is to open the mouth, probe well down into the esophagus, and then feel in the neck for the tube containing the hard probe. Be *certain* that you understand the relationships of the food and air passages in the head and neck.

THE THORACIC REGION

3. The esophagus continues posteriorly through the cavity of the thorax without changing significantly. We shall therefore not open the thorax at this time, but shall examine its contents later in connection with the circulatory and respiratory systems.

THE ABDOMINAL REGION

4. a. The abdominal region is that area, posterior to the ribs, in which the ventral body wall has no bony support. The body wall in this region encloses a large **peritoneal cavity.** Most of the digestive, excretory, and reproductive organs are found within the peritoneal cavity, together with some organs assigned to other systems. Each of the three systems mentioned above has one or more ducts which pass to the exterior through a ring of bones called the **pelvic girdle.** Examine this region on a skeleton.

b. Take two pieces of soft cord and tie them to the ankles of each leg in such a way that the animal lies in the dissecting pan ventral side up, with the strings under the pan (FIGURE 5-5). This will hold the specimen in position; at the end of the period, slip the strings off the ends of the pan, leaving them attached to the animal so that you can position it quickly at the beginning of your next dissection.

c. In opening the peritoneal cavity, use the umbilical cord as a landmark, and follow the dotted lines in FIGURE 5-5. Be careful to leave a median strip of tissue posterior to the umbilical cord, and to avoid cutting structures that lie directly in the midline. Cut through the body wall carefully, starting with a scalpel or razor blade, and continuing with scissors. Pin back the lateral flaps at the side of the abdomen.

d. Examine one of the lateral flaps; it is composed of three separable layers of tissue, which you may already have noticed as you cut through the skin. The outermost layer is the skin; a middle layer consists of abdominal muscle; and the thin, transparent innermost layer is an epithelial tissue called **parietal peritoneum.** This tissue lines the inside of the wall of the entire peritoneal cavity; a similar tissue, **visceral peritoneum,** invests all the visceral organs in the cavity. Parietal and visceral peritoneum are connected by thin sheets of tissue called **mesentery;** several mesenteries suspend and support the visceral or-

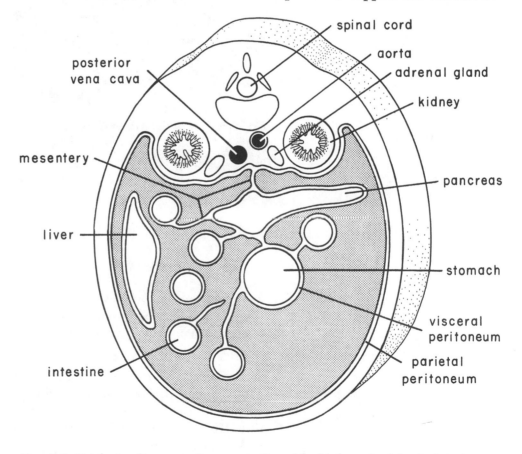

FIG. 5-9. Fetal pig, diagrammatic cross section of body through abdominal region, to illustrate the relationship between parietal peritoneum, visceral peritoneum, and mesenteries.

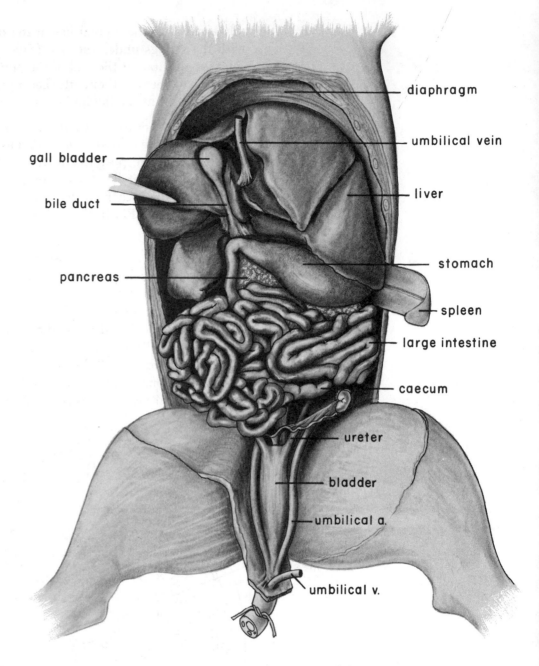

gall bladder

bile duct

pancreas

diaphragm

umbilical vein

liver

stomach

spleen

large intestine

caecum

ureter

bladder

umbilical a.

umbilical v.

FIG. 5-10. Fetal pig abdomen, ventral dissection.

gans, and serve as bridges (between the body wall and the organs) for the passage of blood vessels, nerves, and various other ducts. The relationship between parietal and visceral peritoneum and the mesenteries is illustrated in FIGURE 5-9.

e. Examine the umbilical cord, and make a fresh cut across it if the end is damaged. The **umbilical vein** runs anteriorly from the umbilical cord to the **liver,** if you did not cut it when you opened the body wall. (If you *did* cut it, locate the cut ends on your specimen and on FIGURE 5-10, before you proceed.)

f. The liver is the largest organ in the abdomen. Note that its anterior surface is smoothly convex and fits snugly into the concavity of the

diaphragm, a muscular partition that separates the thoracic and abdominal cavities. Push the liver aside with your fingers, and examine the diaphragm. (*Note:* If the peritoneal cavity of your specimen is partially filled with dark, brownish material, take your animal to the sink and rinse it thoroughly before you continue to dissect. This material is clotted blood, which filled the peritoneal cavity when one or more of the blood vessels associated with abdominal organs burst during injection of the circulatory system. It does not harm the specimen or make it less useful for dissection, but it should be thoroughly cleaned out.)

g. Push the liver aside and identify the **stomach,** a large sac dorsal to the liver on the left side. Locate the point at the anterior end of the stomach (near the midline of the body) where the esophagus penetrates the diaphragm and then almost immediately joins the stomach. At its posterior end, the stomach makes a curve to the right and narrows to join the anterior end of the **small intestine.** The constriction at the junction between the stomach and the small intestine is called the **pylorus.** Attached to the stomach by a mesentery is a long, flat reddish organ, the **spleen,** which is not a part of the digestive system. What is its function?

What digestive functions take place in the stomach?

Does any absorption take place in the stomach?

How is the stomach related to the phenomenon of discontinuous feeding?

How would your own feeding habits have to be modified if your stomach were removed?

h. Examine more closely the anterior end of the small intestine, called the **duodenum.** The liver and the duodenum are connected in the pyloric region by the **bile duct,** which runs in the mesentery stretching between the two organs. Under a dissecting microscope, the bile duct may be traced into the liver. On its way to the liver, it gives off a branch to the **gall bladder,** a small greenish sac embedded in the liver on the underside of one of the right lobes. How does the liver aid in digestion, and what is the role of the gall bladder? What are some of the liver's non-digestive functions?

i. Lift the stomach and locate the **pancreas,** a light-colored diffuse gland lying in the mesentery between the stomach and the small intestine. The pancreas has a duct which empties into the small intestine near the pylorus, and which may be seen with very careful dissection. What enzymes are secreted by the pancreas, and what is their action? What other function has the pancreas?

j. With a scalpel or razor blade, make an incision into the posterior end of the stomach, and carry the cut through the pylorus a short distance into the duodenum. Find the **pyloric sphincter** muscle. What is its function?

Although the embryonic animal is not feeding, small bits of epithelial tissue on its surface become dislodged and wash into the mouth. These bits of tissue, along with small amounts of bile from the liver, pass through the digestive tract during embryonic development. You may notice a considerable amount of this greenish material in the stomach and duodenum.

k. From the pyloric region, the small intestine runs posteriorly for a short distance and is then thrown into an irregular mass of bends and coils held together by a common mesentery.

What enzymes are secreted by the small intestine, and what is their action?

Where does most absorption take place?

What is the chief function of the large intestine?

In what animals might this function be relatively unimportant?

The distal end of the small intestine joins the **large intestine** posteriorly, in the left side of the abdominal cavity (right side in man). A blind sac, the **caecum,** will be found projecting from the large intestine near the point of juncture. In man there is a smaller end to the caecum; what is it called?

l. Follow the main portion of the large intestine, known also as the **colon,** as it runs from the point of juncture with the small intestine into a tight coil, then out of the coil anteriorly, then posteriorly again along the midline of the dorsal wall of the abdominal cavity. In the pelvic region, the digestive canal is called the **rectum;** it will be seen more clearly later. Posteriorly, the digestive tract opens to the outside through the anus.

PART III. MODIFICATIONS OF DIGESTIVE TRACTS

Animal digestive tracts frequently include modifications that serve to increase the internal surface area. Such modifications are common in most vertebrates, but are generally more common in herbivores than in carnivores. Why?

Another frequent modification involves structures for food storage at the anterior end of the digestive tract, in addition to the stomach.

A. Villi

An exceedingly common method of increasing the effective digestive and absorptive area in the intestine is the presence of fingerlike extensions of the inner layer of the small intestine. These projections are called **villi** (singular, villus). Open the small intestine of your fetal pig, and examine the villi under a dissecting microscope. Also study the demonstration slide, showing a section through a mammalian intestine. In some vertebrates, such as the reptiles, no villi are present, but the same effect is achieved by the presence of innumerable folds of the intestinal lining.

Invertebrates have similar adaptations; for example, earthworms have a dorsal invagination of the intestinal wall, the typhlosole, and sea anemones have a digestive cavity divided by a number of partitions that serve to increase surface area. These and other animals may be on demonstration.

B. Intestinal length

Great increase in absorptive area is often attained by increase in intestinal length. Thus the intestine of many animals is very long and much coiled; this condition, as previously noted, is more typical of herbivores than of carnivores.

On demonstration are several animals showing different intestinal lengths. As you examine them, try to correlate what you see with what you know, or can find out, about the digestive and dietary habits of the animal. For example, you will see the intestine of an adult frog and that of a tadpole, which is the young of the same species of frog. What do the differences in intestinal length suggest regarding the changes in feeding habits of this frog as it develops? What *other* changes in the frog's anatomy suggest dietary changes?

C. Caeca
You have already seen the caecum of the fetal pig, which is a diverticulum of the proximal end of the large intestine. This structure is usual in mammals and birds, and serves little function in most carnivores. In some herbivores, however, such as rodents, horses and most other ungulates, the caecum is a large, thin-walled sac containing vast quantities of microorganisms which help to break down material not completely digested in the small intestine, including some cellulose.

Such extensions of the digestive tract may occur at any point, not just the junction mentioned above. For example, examine the **pyloric caeca** in fish, and **gastric caeca** in cockroaches on demonstration.

D. Spiral valve
Many primitive fish and sharks have a curious **spiral valve** which increases surface area in the intestine. Examine the specimen on demonstration. How does the valve function?

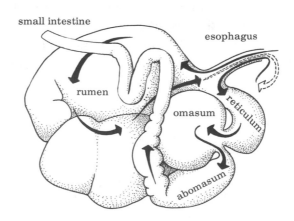

FIG. 5-11. The four "stomachs" of a ruminant. Food is swallowed initially into the rumen and reticulum (the drawing here shows it going only into the rumen), where it is fermented by microorganisms. It is then regurgitated as the cud for more chewing (white arrow) and then reswallowed, this time bypassing the rumen and moving quickly along a fold in the wall of the reticulum into the omasum and then into the abomasum, which is the true stomach.

E. Crops   The **crops** of birds and some other animals, including numerous invertebrates, are expanded portions of the anterior end of the digestive tract, usually the esophagus; they function in food storage. Examine the demonstration specimens.

F. Additional   Cows and other ruminant animals are commonly said to possess four stom-
"stomachs" of   achs. Actually, only one of the four chambers is a true stomach; the others
ruminants   are usually expanded portions of the esophagus. Microorganisms in some of
the accessory chambers aid in digestion of substances like cellulose, which
would otherwise be indigestible (FIGURE 5-11).

QUESTIONS FOR   1. What are the digestive functions of the liver?
FURTHER   2. The liver, as you know, has many general metabolic functions in
THOUGHT   addition to those concerned directly with digestion. Can you suggest adaptive advantages to having so many important functions localized in a single
organ? Can you think of any disadvantages in such an arrangement?
3. Discuss and compare digestive adaptations in herbivores and carnivores.
4. List all the organisms observed in this laboratory, and make an hypothesis about the natural diet of each. Be able to back up your assertions with
evidence from observations.
5. Many kinds of animals have gizzards. What do their diets have in
common?
6. Trace a molecule of absorbable nutrient from the point of ingestion
in a pig to the point of storage in the liver, naming as many structures as you
have learned up to this point.
7. Trace a particle of nonabsorbable material through the digestive system of a pig from ingestion to defecation, naming all structures involved in
the passage.
8. Explain the function of the gall bladder.
9. How is an animal's size related to its metabolic rate?

TOPIC **6** # GAS EXCHANGE IN ANIMALS

Biological Science, pp. 183-192.
Elements of Biological Science, pp. 119-124.

We have seen how high-energy organic compounds are made by autotrophic organisms and have examined some of the methods by which such materials are procured by heterotrophs. The largest amount of energy (as ATP) can be gleaned from these compounds if they undergo complete cellular respiration; for this to occur, oxygen is usually necessary. Cells must therefore have a continuous supply of oxygen, and a means of releasing carbon dioxide. Since exchange of gaseous raw materials and waste products of metabolism is a basic problem facing all organisms, we are not surprised to find that special organs have evolved, adapted especially for carrying on exchange of such materials with the environment.

It should be remembered that, ultimately, the critical process in gas exchange is diffusion of the gases into and out of individual cells. Such diffusion is possible, of course, only across a moist membrane. It follows that an organism, in order to have sufficient gas exchange with its environment, must have a surface of moist cell membranes exposed to the environment, and exposed in such a way that the chances of drying out are minimized. Another requirement is that the respiratory surface be large enough, relative to the size and activity of the organism, to allow for an exchange of sufficient magnitude to satisfy the organism's needs. In view of the first requirement, this may sometimes be difficult to accomplish. Furthermore, such a large, thin, moist surface is often very fragile, and easily suffers mechanical damage. There has generally been selective pressure for the evolution of protective devices, particularly when the respiratory surface is an evaginated one.

You will see this week a variety of ways in which animals have evolutionarily met the requirement for an adequate respiratory system. The diversity of types becomes less bewildering if one bears in mind, while studying each of them, that it merely represents one way of solving the basic problems that every such system must fulfill:

1. the need for a respiratory surface of adequate dimensions
2. the need for keeping the surface moist
3. the need for protecting the surface from mechanical injury
4. in the case of many multicellular organisms, the need for a method of transporting gases between the sites of exchange with the environment and the more internal body cells

Respiratory surfaces (excluding very small organisms where the whole body surface is a respiratory surface) may be grouped into two general categories: **invaginations** (inpocketings) of the body surface, and **evaginations** (outpocketings). Which of the two would you expect to find more commonly among terrestrial organisms? Why?

In this unit, we shall examine a few of the systems specialized for gas exchange, and then make quantitative measurements of the exchange in one of these systems, using the closed manometer.

PART I. GAS EXCHANGE SYSTEMS OF INVERTEBRATES A. Phylum Annelida
EARTHWORM

1. The earthworm burrows in moist soil. Gas exchange depends upon a moist environment and upon a well-developed capillary network, close to the skin. Gases diffuse across the thin, moist skin to the bloodstream that carries oxygen throughout the body. Carbon dioxide is released to the environment in the same manner. Though the skin is moistened by mucous glands of the epidermis and by fluid from excretory pores located near the grooves between segments of the worm, the animal depends upon the moist terrain where it burrows, and is unable to live elsewhere. Examine the earthworms on demonstration.

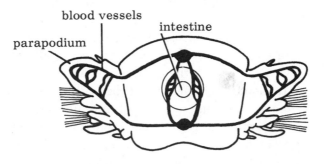

FIG. 6-1. Cross section through one segment of *Nereis*. Notice that the flaplike parapodia on each segment are richly supplied with blood vessels.

MARINE CLAM WORM

2. This marine relative of the earthworm has evolved a means of increasing the surface area available for gas exchange. On each segment there is a pair of flattened lateral lobes (**parapodia;** FIGURE 6-1). The parapodia are highly vascularized. Gases are thus enabled to diffuse across the thin skin to and from the circulatory system. *Nereis* spends most of its time in a burrow at the bottom of the sea, in sand or mud. The parapodia flap gently in a uniform fashion, creating a current through the burrow, and thus changing the water constantly.

Examine the demonstration specimens of *Nereis*. Why would this animal require a greater surface area to accomplish the same amount of gas exchange as the earthworm?

B. Phylum Arthropoda (joint-footed animals)
CRAYFISH

1. Obtain a living crayfish in a small bowl of pond water. To handle a crayfish, immobilize it by pressing the index finger down on the carapace (FIGURE 6-2) and grasping the body just behind the claws, with the adjacent pair of fingers. An animal held in this position is harmless.

Quickly orient yourself to the animal by locating the **head** with the **eyes,** two sets of sensory **antennae,** and several small jointed mouthparts; the **thorax,** bearing eight pairs of appendages (three maxillipeds and five walking legs, the first of which is much larger than the rest); and the **abdomen,** bearing five pairs of jointed **swimmerets** and a terminal flipper.

The entire outer body surface is covered by a hard, chitinous **exoskeleton,** which protects the animal from many of its enemies and from mechan-

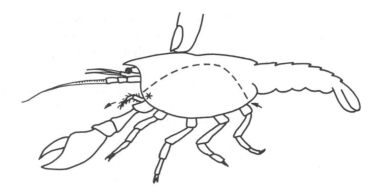

Fɪɢ. 6-2. Crayfish. Pressure from index finger immobilizes the animal; dotted line shows cut to be made in carapace; asterisk indicates position of gill bailer; arrows indicate direction of respiratory current.

ical injury. One large section of exoskeleton is an expansion of the plates of the back, and extends low on the sides of the thorax. The back plates and their lateral extensions together form the **carapace,** which covers a **gill chamber** on each side.

Pick up the specimen. With fine scissors, make a cut in the carapace on one side as shown by the dotted line in Fɪɢᴜʀᴇ 6-2. (This procedure is comparable to cutting fingernails.) Observe the feathery **gills,** attached at the base of each appendage in the thoracic region. Each gill branches into numerous filaments. Remove a small portion of a gill and examine under your compound microscope. Observe how the surface area is increased. How does gas exchange occur in the crayfish?

The carapace covers the gill chamber, protecting the fragile gills and isolating them from the environment. Thus, the gill chamber is open only at the anterior and posterior ends. Water is driven through it by the action of a part of the second maxilla, one of the mouthparts. We therefore call this appendage the **gill bailer.** It may be seen through the semitransparent carapace in the region just behind the obvious mouthparts, labeled in Fɪɢᴜʀᴇ 6-2. Its action may be seen in more detail if part of the carapace covering it is removed (this is most easily done on the side where a cut has already been made).

The current created by the gill bailer may be easily observed, as follows:

a. Obtain a few dropperfuls of India ink in a beaker.

b. Draw up half a dropperful in a clean dropper, wipe the excess from the outside of the dropper, and *without touching the animal,* release some of the ink underneath the posterior ventral portion of the carapace, as indicated by the rear arrow in Fɪɢᴜʀᴇ 6-2. (*Note:* If the animal moves away, gently hold it still with pressure on the carapace; however, the demonstration works best if the animal is unrestrained.)

c. Look for ink currents emerging anteriorly, in the gill bailer region.

TERRESTRIAL ARTHROPODS (INSECTS)

2. Most aquatic arthropods, like the crayfish, carry out gas exchange by means of gills. Many arthropods, however, have become terrestrial in the course of their evolution, and for them, gills are quite obviously unsatisfac-

tory. Different land-adapted arthropods have solved the exchange problem posed by their environment in several different ways, but one pattern, unique to the Phylum Arthropoda, has appeared several times. It is employed successfully by insects, and we shall examine it today.

The hazard of desiccation of the respiratory surface rules out evaginated structures for air-breathing animals. Of those with invaginations, animals with lungs depend upon a well-developed circulatory system to transport oxygen from the lungs to the body cells and to convey carbon dioxide in the opposite direction. This is not the case with such terrestrial arthropods as insects, centipeds, millipeds, etc. The invaginated gas-exchange system in these animals, known as a **tracheal system,** is based on an entirely different strategy. Here, the air enters through small openings in the body wall known as **spiracles,** and is carried from the spiracles by a system of branching tubules called **tracheae.** The tracheae usually branch repeatedly until the finest tubules reach individual cells or groups of cells. At the very end of the tracheae, then, are the moist cell membranes across which respiratory gas exchange occurs. In the tracheal system, the enormous surface area required is usually achieved by repeated subdivision of the air tubes, not by convolution of surface and branching of tubes within a restricted structure such as a lung; the circulatory system, therefore, plays little or no role in tracheal gas exchange.

Obtain a specimen of a large insect, and a small, clean waxed dissecting pan. Partially fill the pan with insect Ringer's solution (a fluid isosmotic to insect tissue) and carry out the following dissection under the fluid, with the aid of your dissecting microscope. If there is not enough fluid to cover the insect after it is in the pan, add more Ringer's.

a. Anesthetize the insect with carbon dioxide or otherwise, as specified by your instructor.

b. Quickly remove the wings and legs.

c. With fine scissors, make a longitudinal cut (see FIGURE 6-3A) through the dorsal surface, close to each lateral margin, from the posterior end of the abdomen to the anterior end of the prothorax. Bend back or remove this section in one piece. The heart will adhere to the back, and will probably continue to pulsate. Observe it adhering to the middorsal body wall, and examine the set of muscles that

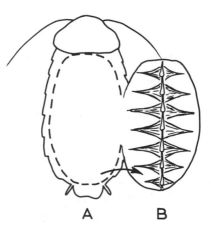

FIG. 6-3. Dorsal view of large insect with dorsal section hinged open. (A) Outlines where cut is to be made. (B) Dorsal surface overturned, showing heart.

A   B

extend from it to the body wall (FIGURE 6-3B). In the following laboratory, we will discuss the operation of the circulatory system.

d. As you study the viscera in detail, it will be necessary to break or tear some of the white **fat bodies** away from the tracheal tubes. In this fresh specimen, the tracheae are filled with air and are silvery in appearance. It may also be necessary to remove certain muscles that are in the way. Locate and identify as many of the visceral organs as possible.

e. Remove a section of trachea at the point where it leaves the body; carefully examine the spiracle. Also examine a wet mount of a piece of large trachea from just inside the body wall; note the prominent spiral thickenings. What is their function?

f. Under the dissecting microscope, follow a branching trachea as far as possible into the body. Carefully remove a section at this point, make a wet mount, and examine. How does this region differ in appearance from the large trachea?

***PART II. GAS
EXCHANGE
SYSTEMS IN
VERTEBRATES
A. Fish***

Observe a living fish in an aquarium. Notice the shape of the body and the manner in which the animal moves through the water. Does the fish have a neck? If so, is it of any use? Notice the position of the eyes. Can both eyes see the same object simultaneously? Can the eyes move? Are there eyelids?

1. Notice the gills on either side. Each set of gills is protected by a flap, the **operculum.** Watch an operculum carefully: how does it function? Water is drawn in through the mouth, into the pharynx, while the operculum is closed (FIGURE 6-4). Then valves at the front of the mouth close as the oral cavity contracts; water is forced across the gills and out behind the open operculum.

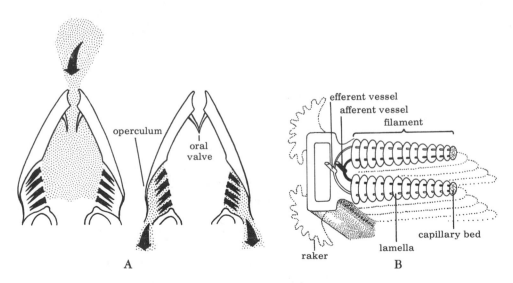

FIG. 6-4. Gills of fish. (A) Longitudinal section through the oral cavity and gill region. (B) Structure of a gill. Each gill is composed of many filaments, each subdivided into numerous lamellae.

2. Obtain a preserved fish in a dissecting tray. Examine the **nostrils**, located just behind the blunt anterior end of the snout, and observe the two small apertures on either side. The interior aperture is guarded by a flaplike valve. Do the nostrils open into the mouth? What, in your opinion, is their function? Observe the **oral valves** at the front of the mouth.

3. The gills may be seen by lifting and removing the operculum. Notice the branchial filaments of the gills, gill arches, and gill clefts. FIGURE 6-4B diagrams the structure of a gill. Carefully remove a gill arch and observe the bony gill rakers and the soft branchial filaments. The gill rakers prevent food from passing out of the gill slits. Examine a gill more closely under the dissecting microscope.

4. To observe the interior of the mouth, make a cut from the corner of the mouth through the middle of the operculum, and open the mouth by spreading the jaws. Note the course followed by the respiratory water current and the route taken by food. How can the mouth cavity be distinguished from the pharynx? Is there a tongue?

FIGURE 6-4A is labeled with arrows showing the method by which water is passed over the gills.

B. Amphibians

Salamanders and frogs carry out gas exchange in a variety of ways: gills, lungs, skin, or mouth lining. The larvae generally occur in aquatic habitats and possess gills; adult stages may retain gills or may lose them. Of those that lose the gills, some then depend exclusively on diffusion across the thin moist skin, while others supplement this surface exchange with lungs, which enable them to exploit a semiterrestrial environment.

OBSERVATION OF LIVING FROGS

1. Observe the mechanism of respiration in a frog (in a terrarium) that has been left undisturbed for some time. Slight oscillations of the floor of the mouth cavity increase or decrease pressure in the mouth cavity; air is drawn in and pushed out through the nasal passages. The mouth breathing movements are interrupted at regular intervals by strong movements of the floor of the mouth, which are accompanied by movements of the nostrils and the sides of the body. The latter movements are associated with the forcible expansion of the lungs. Be ready to describe the movements you observe.

2. Place a frog in a jar filled to the top with ice water; close the jar tightly. After five minutes, can any movement or breathing be observed? What type of gas exchange does this demonstrate? Remove the frog, and observe recovery. Explain the results.

3. The instructor will demonstrate the croaking of a frog, by lifting up the back of the forelegs with the thumb and index finger. How is the sound produced? In a male frog the skin at the top of the forelimbs puffs in and out during croaking. These inflated regions mark the position of the vocal sacs. Of what advantage to the frog is croaking?

DISSECTION OF A FROG
RESPIRATORY SYSTEM

4. Before you begin, make sure that you have available clean dissecting instruments, pins, cotton, and a small beaker of frog Ringer's solution (isosmotic to frog tissues).

(*Note:* In order to handle the frog for these observations and the subsequent parts of this exercise, its nervous system should be inactivated either by anesthesia or by destroying the central nervous connections. Because anesthesia tends to wear off, and to affect parts of the body with which we would like to work, the latter procedure, called **pithing,** is usually preferred. This brief operation effects a complete loss of all the animal's sensory capacities, as well as responses that would interfere with your work. Hence, the animal presumably feels no discomfort except for a pin prick at the beginning of the operation, while for a considerable length of time its physiological reactions are unimpaired. Your instructor will supply you with a pithed animal or will show you how to pith if you wish to try. The method is described in Appendix 5.)

a. Expose the respiratory system in the following way. Place the pithed frog ventral side up in a dissecting pan, with pins through the palms of the outstretched forelegs. Note that the frog's skin is not attached to the muscles along most of the ventral side, and may thus be lifted up. Carefully hold up the skin with forceps and with scissors cut from the pelvis to the throat region, along the midline (see FIGURE 6-5). At each end, cut to the right and to the left, making a pair of flaps.

Observe the many blood vessels in the skin. Why is a heavily vascularized skin of advantage to the frog?

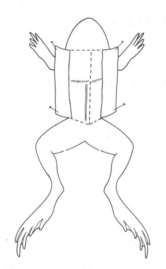

FIG. 6-5. Method of opening frog. The skin flaps have been pinned back, exposing the muscle layers and the ventral abdominal vein.

b. Again with scissors, make a very careful cut (FIGURE 6-5) through the muscles forming the abdominal wall. *Avoid* the ventral abdominal vein, in the midline, although it will be necessary to cut some minor vessels. Extend the cut anteriorly, keeping the scissors-point *up*. When you near the pectoral girdle, it is especially important to keep the scissors-point up to avoid injury to the heart, which lies directly beneath. Cut through the bony sternum and pectoral girdle, and up toward the throat.

You may need to respread the forelimbs of the animal to expose the internal organs. Follow again the course taken by cuts in the skin, and turn back the muscular flaps.

c. Open the mouth wide, making small cuts at its corners. Examine the back of the mouth and the pharynx, which, as in mammals, serves as a common passageway for air and food. Notice the openings in the roof of the mouth. Where do they lead? Examine the floor of the pharynx and the opening of the trachea, the glottis. Place the small end of a pipette, with the bulb removed, into the glottis. Blow into it, and inflate the lungs. This is an example of **positive pressure breathing.** What color are the lungs? Why?

d. The inner surface of the lung, often difficult to see because the lung contracts when cut, bears a complex network of ridges which project into the interior and look like an irregular honeycomb. All of these ridges are abundantly supplied with blood vessels, fed by the pulmonary artery. The main substance of the lung is connective tissue containing elastic fibers, traversed by a network of capillaries. Its cavity is lined by a layer of epithelium, and its outer surface is covered by peritoneum. Examine the lung under a dissecting microscope. You may find trematode parasites (what phylum?) attached to the inner epithelial lining.

e. When inhaling, the frog depresses its throat; because the mouth is shut, air is drawn in through the nostrils. The floor of the mouth is then raised, and the nostrils are simultaneously closed. Since the esophagus is contracted except during swallowing, only one route remains for air to follow — into the glottis. Exhalation is due to a contraction of the elastic fibers and muscles of the lungs, accompanied by opening of the nostrils.

f. At the anterior end of each lung is a small, thin-walled chamber called the **laryngo-tracheal chamber.** The croak originates in this area. The walls of this chamber and the edges of the glottis are supported by cartilages; the largest of these bound the glottis to the right and left. The mucous membrane on the inner, adjacent faces of the cartilages is raised into a pair of horizontal folds, the **vocal cords.** By means of muscles, these folds can be brought into either a parallel or a divergent position. When they are parallel, the air passing to and from the lungs sets their edges into vibration and gives rise to the characteristic croak. Its pitch can be slightly altered by stretching or relaxing the cords.

g. Remove the lungs, including the laryngo-tracheal chamber and glottis, and place them in a small petri dish of Ringer's solution. With a razor blade, cut the respiratory system longitudinally in half. This incision should pass exactly through the slit of the glottis, so that each part contains half of the laryngo-tracheal chamber with one lung attached. Examine one of the halves, and locate the vocal cords. Identify the opening into the pharynx and the opening into the lung; label the parts on FIGURE 6-6, and add to it any details revealed by your dissection.

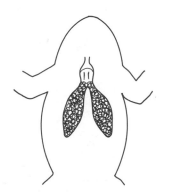

Fig. 6-6. Frog respiratory system.

C. Mammals   With the exception of certain forms that have secondarily returned to the water, mammals have become entirely emancipated from the aquatic environment, and they all carry out gas exchange by means of lungs, restricted invaginations. All mammals depend upon a blood circulatory system to transport gases between the moist respiratory surface of the lungs and the individual body cells.

1. On your fetal pig, once again locate the nostrils and the inner end of the nasal passages, just behind the soft palate. To examine the nasal passages more carefully, use a scalpel or razor blade to cut across the snout about one inch back from the nostrils. Removal of the tip of the snout will expose the nasal passages, separated from each other by a bony and cartilaginous partition called the nasal septum. The floor of the passages is formed by the hard palate anteriorly and the soft palate posteriorly. Along the lateral walls of the passages are large folds, which cause eddies in inspired air and thereby serve to increase the contact of the air with the extensive epithelial surface within the nasal passages. Why is such contact important?

2. Review your understanding of the pharynx as a common passageway for air and food. Also review the location of glottis, epiglottis, trachea, and esophagus.

3. Probe into the dissection already made in the neck, and carefully clean all muscle tissue from the surface of the hard-walled **larynx.** (The walls are hard because they contain cartilages.) The larynx is the chamber whose anterior opening, the glottis, was seen on the ventral side of the pharynx.

Posteriorly, the larynx connects with the **trachea,** identifiable as a tube whose walls are supported by numerous rings of cartilage. Clean and examine it carefully. Do the cartilage rings extend all the way around the trachea? What is their function? Toward the posterior part of the trachea, on its ventral surface, find the **thyroid gland.** It appears much darker in color than other organs in this region. The thyroid is a part of the endocrine system, not the respiratory system, but it is most easily located at this time. What is its function?

4. Referring to FIGURE 5-5, p. 59, cut through the wall of the thorax slightly to the left of the midventral line, extending the cut from the pectoral girdle posteriorly through the last rib. Locate the posterior boundary of the thoracic cavity by reaching through the incision with a finger and feeling for the **diaphragm.** Then make two additional cuts (FIGURE 5-5), one to the left and one to the right, through the chest wall just anterior to the dia-

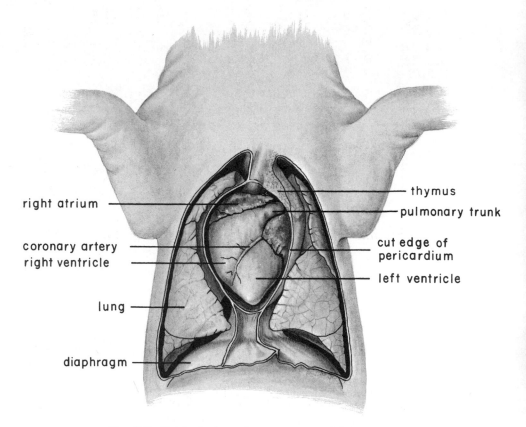

right atrium

coronary artery
right ventricle

lung

diaphragm

thymus

pulmonary trunk

cut edge of
pericardium

left ventricle

FIG. 6-7. Fetal pig thoracic cavity, ventral dissection.

phragm. Extend these cuts well toward the back. The chest wall may now be folded back like two doors, and the thoracic cavity exposed (FIGURE 6-7). In folding back the right flap, you will need to tear the thin membranes that divide the thoracic cavity into three compartments: the left **pleural cavity,** containing the left lung, the right pleural cavity, and the **pericardial cavity,** containing the heart.

5. Examine the lungs; locate the four lobes of the right lung and the three lobes of the left lung. You will find that the lungs in the fetal pig are compact and firm, rather than spongy. This is, of course, because breathing had not yet begun in these specimens, and because the lungs are hardened by the preserving fluid. Your instructor may be able to provide a demonstration of fresh lung tissue.

6. You will be unable to examine the trachea and the two **bronchi** into which it branches, because these tubes are dorsal to the heart and its attached vessels, which we will examine in detail later. At that time, the elements of the circulatory system will be removed, exposing the bronchi to view.

The bronchi subdivide repeatedly, forming bronchioles. The ultimate subdivisions terminate in blind sacs called **alveoli** (singular, alveolus). The alveoli are the real functional units of the lung; it is across their moist walls that gas exchange actually takes place. FIGURE 6-8 shows the relationship between the air-transporting tubes and the alveoli.

7. Examine the diaphragm and identify the central tendon, which is a membranous area in its middle. Muscles, arranged radially around the rim of the diaphragm, attach it to the chest wall. The chest wall itself has an

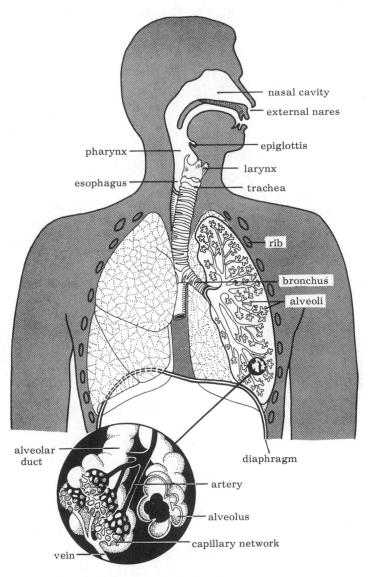

FIG. 6-8. Respiratory system of man.

outer layer of skin, a middle layer of ribs and rib muscles, and an inner, thin membranous layer, the parietal peritoneum. The surface of the lungs is covered by visceral peritoneum. The heart is similarly invested in membranes.

Be sure you thoroughly understand the roles of the rib cage and the diaphragm in breathing. At the demonstration desk, a model has been set up to help you grasp the concept of **negative pressure breathing,** as performed by mammals. You should be prepared to contrast this with positive pressure breathing, which has already been seen in the frog.

Look up the Phylum and the Class of each of the animals that you have studied in this period, and enter each of them in the appropriate space in Appendix 2.

**PART III.
MEASUREMENT OF
RESPIRATION**

In order to understand the importance of oxygen in cellular respiration, one needs to measure the amount of oxygen consumed by an organism in a given amount of time. To measure respiration, we shall use the closed manometer system, the same system used in the units on enzyme activity and photo-synthesis.

The oxidation of glucose may be summarized by the equation:

$$C_6H_{12}O_6 + 6O_2 \longrightarrow 6CO_2 + 6H_2O + energy$$

As the equation shows, the quantity of oxygen consumed is equal to the quantity of carbon dioxide released. Since carbon dioxide, like oxygen, is a gaseous component of air, the volume of air in the experimental tube of the manometer will remain constant unless the carbon dioxide produced by respiring cells is absorbed or in some manner retained in solution. A granular soda lime (solid sodium hydroxide and calcium hydroxide) is therefore added to the system to absorb carbon dioxide.

The measurements will be made on two of the large insects of the type you previously dissected. We use this insect for two reasons: its size and its relatively high oxygen consumption for an animal of its weight.

1. Review your understanding of the closed manometer system. If necessary, refer to Appendix 3.

2. The respirometer (experimental tube) should be assembled with a perforated container of soda lime at the bottom and two large insects above the soda lime (FIGURE 6-9). Be sure that the test tube is dry *before* introducing the animals, and make certain that the animals cannot come into contact with the soda lime.

3. Assemble the control tube with the same quantity of soda lime; the perforated container is not necessary. Fill with paper the same space as that taken up by the insects in the experimental tube.

4. With the escape tubes open, connect the experimental and control chambers to the manometer block. Wait five minutes for the tubes to come to the same temperature in the water bath.

5. Set the experimental syringe at 1 cc and the control syringe at 0 cc (Why?). Put the glass stoppers simultaneously into the escape tubes. Set the manometer fluid to the starting level by means of the appropriate calibrated syringe.

6. Read the changes at two-minute intervals for thirty minutes. As in the other manometer experiments, return the fluid to the starting level on the block each time you take a reading from the syringe. Record the syringe calibration changes in the table provided. As the experiment proceeds, note

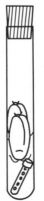

Fig. 6-9. Large insects and container of soda lime in experimental tube.

the activity of the animals (from "none" to "extreme"), and record in the table each time you take an "oxygen consumed" measurement.

7. Using your data, plot the oxygen consumed against time. Be sure to label the axes properly. What was the total oxygen consumption? What is the average oxygen consumption over a two-minute period? With an estimate of the weight of the animals, given to you by the instructor, calculate the volume of oxygen consumed per gram of tissue per hour.

QUESTIONS FOR FURTHER THOUGHT

1. How does an insect meet each of the requirements for an adequate gas-exchange system?

2. Does the crayfish have an invaginated or an evaginated gas-exchange surface?

3. In your observations of the crayfish (IB), what did you decide were the functions of the appendages? What do these highly specialized adaptations tell you about the animal's habitat, diet, manner of gas exchange, etc.?

4. What is a Ringer's solution? Why is it used?

5. What are the functions of the rib cage and diaphragm in mammalian breathing?

6. Compare positive and negative pressure breathing.

7. From your data on insect respiration, what can you conclude about the relationship between the animal's activity and the amount of oxygen consumed?

8. What factors regulate the rate of gas exchange in a mammal?

9. What role does atmospheric nitrogen play in gas exchange?

10. Compare the energy expended by an aquatic animal and by an air-breathing animal to obtain the same volume of oxygen.

TIME	SYRINGE READING	O₂ CONS.	ACTIVITY none → ext.

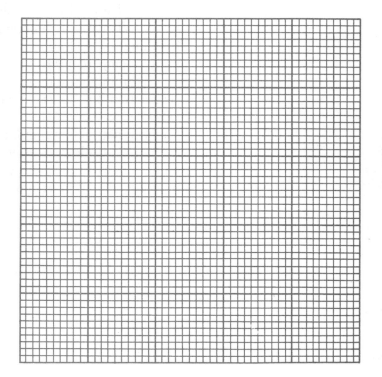

O₂ CONSUMED DURING RESPIRATION

GAS EXCHANGE AND TRANSPORT IN PLANTS

Biological Science, pp. 176-183, 193-212. *Elements of Biological Science,* pp. 116-119, 124-137.

The cell's requirement for a supply of oxygen and nutrients, and for a means of disposing of wastes, makes necessary some sort of mechanism whereby these materials may be 1) exchanged with the environment and 2) transported from the site of exchange to other parts of the organism. You have already examined photosynthesis, a variety of mechanisms of nutrient procurement in heterotrophs, and gas exchange in animals. Future laboratories will deal with transport and excretory systems in animals. In this laboratory, you will study the mechanisms of gas exchange in green plants, and the structural adaptations of land plants for internal transport.

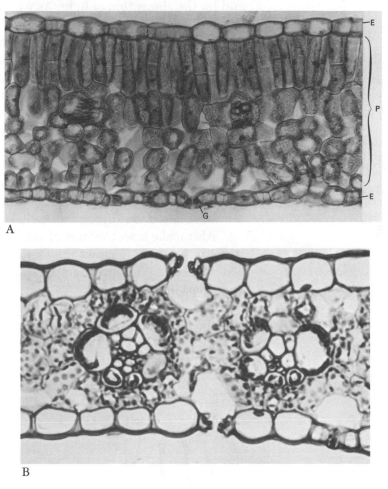

A

B

Fig. 7-1. Leaf cross sections. (A) Photograph of privet leaf (E-epidermis, G-guard cell, P-parenchyma). (B) Photograph of corn leaf (notice stomata on top and bottom of leaf). (*A: Courtesy Thomas Eisner, Cornell University. B: Courtesy Carolina Biological Supply Company.*)

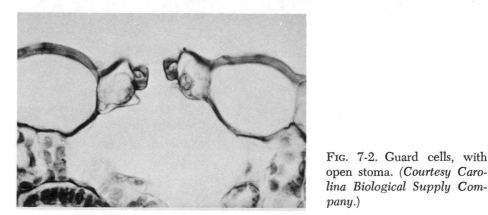

FIG. 7-2. Guard cells, with open stoma. (*Courtesy Carolina Biological Supply Company.*)

PART I.
GAS EXCHANGE

While individual isolated cells of a plant would be unable to control water loss, the leaf as a whole is adapted to maintain diffusional loss of water at an appropriate level, while providing as much area as possible for gas exchange and for the absorption of light. As we shall see, the leaf has turned diffusional loss of water, **transpiration,** to advantage by utilizing it as a means of upward transport of water. On the other hand, should the supply of water from below run lower than that necessary to replenish what is lost by transpiration, special groups of cells, the **guard cells,** bounding the **stomata** in the epidermis, can regulate the degree to which the interior of the leaf is open to exchange of materials with the atmosphere.

A. Stomatal structure

Review the relationship of a stoma to the leaf interior, and the structure of the guard cells, by examining the photomicrographs (FIGURE 7-1, A and B) of cross sections through privet (*Ligustrum*) and corn (*Zea*) leaves. FIGURE 7-2 shows a section through a single stoma of a corn leaf.

Also make a wet mount of epidermis from leaves supplied by your instructor. FIGURE 7-3 shows the method for removing the epidermis from the leaf. Label the photomicrographs; prepare and label a drawing of guard cells and stoma in *surface* view.

Notice in the photomicrographs that each stoma leads into a maze of chambers in the mesophyll. It is across the moist walls of these mesophyll cells that exchanges of water, carbon dioxide, and oxygen can take place. Why do the mesophyll cells not dry out?

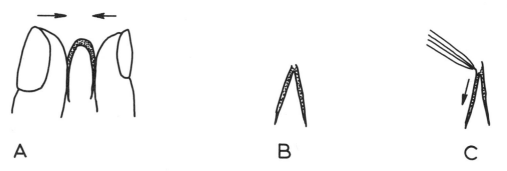

FIG. 7-3. Method for removing epidermis from leaf. (A) Pinching the leaf. (B) Leaf after epidermis cracks. (C) Removing epidermis with forceps.

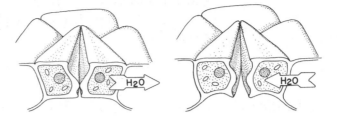

Fig. 7-4. Guard cells. Left: A closed stoma. Right: An open stoma.

B. Mechanism of stomatal action

The size of the stoma is regulated by changes in shape of the guard cells. When the guard cells are turgid, the stoma is open (Figure 7-4); when the guard cells are flaccid, the stoma is closed. Guard cell shape is controlled by water content; this, in turn, is influenced by the concentration of osmotically active particles in the guard cells. The most important solute is sugar, the concentration of which is changed by photosynthesis and by the conversion of starch into sugar or sugar into starch (sugar is osmotically active, but starch is not). An increase in sugar content would cause water to be drawn into the guard cells, increasing their turgor and opening the stoma. Recent theories of stomatal activity suggest that such alterations in the concentration of osmotically active solutes are probably supplemented by some energy-requiring mechanism.

Conditions of high humidity or dryness may be simulated by mounting small pieces of leaf epidermis in tap or distilled water and in 5% salt (NaCl) solution. Examine guard cells and stomata under high power under these two conditions, and record your observations in the form of sketches. Explain the results.

How do these conditions differ from those to which the plant would normally be exposed?

C. Adaptations to special habitats

The privet leaf, which you have examined twice before, is called a **mesophyte,** which is a shorthand way of saying that it is adapted to a moderate sort of habitat, neither very wet nor very dry. Extreme conditions of environment have sometimes engendered adaptations for gas exchange quite different from those of the privet leaf.

1. Examine a cross section of oleander (*Nerium*) leaf. Oleander is a **xerophyte,** a plant adapted for life in a very dry habitat. Note the prominent cuticle, thick epidermis, and stomata located in pits lined with epidermal hairs. Explain the functions of these adaptations, and compare this leaf with that of the privet.

2. Study a cross section of the pondweed (*Potamogeton*) leaf. *Potamogeton* is a **hydrophyte,** a plant adapted for life in a very wet habitat. Note the stomata in the upper epidermis only, very thin cuticle, chloroplasts in the epidermis, and very large air chambers in the spongy mesophyll. Which of these adaptations would you consider to be evidence that this is a floating plant?

3. Your instructor will supply a leaf cross section from a plant of un-

known or unspecified habitat. Study its structure carefully, determine the type of habitat to which it is adapted, and be able to give reasons for your choice.

D. Lenticels   Gas exchange in stems is accomplished partly through exchanges with the vascular tissues and partly by means of **lenticels,** loose groups of cells in the otherwise impervious epidermis or corky layer of a woody stem. Examine lenticels with a dissecting microscope. Figure 7-5A shows them on a maple (*Acer*) twig and Figure 7-5B in cross section of an elderberry (*Sambucus*) stem.

A   B

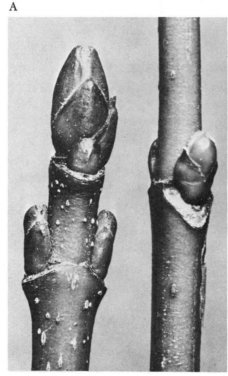

Fig. 7-5. Lenticels. (A) On maple twig. (B) In cross section of elderberry. *(Courtesy Carolina Biological Supply Company.)*

PART II.
VASCULAR TISSUE
OF THE STEM

Many of the smaller organisms and some bigger ones with large, generally distributed surfaces for exchange have no special transport systems, because diffusion and osmosis alone are sufficient for movement of substances across relatively short distances. In the case of the larger animals and most large terrestrial plants, however, an efficient transport system, involving the mass flow of materials, is a physiological necessity. Local molecular traffic may still involve diffusion, osmosis, and active transport, but major distances are usually bridged by movements of fluids through a more or less well-defined system of internal channels. In this section, we will consider the structural basis of transport in vascular plants.

The movement of fluids in plants occurs, for the most part, in two distinct types of **vascular tissue.** One of these, the **xylem,** usually carries water and dissolved inorganic materials from the roots upward, where they are utilized in photosynthesis and other metabolic activities. The other vascular tissue is **phloem,** and it functions to transport nutrients, particularly the organic products of photosynthesis, from one part of the plant to another. In addition to these predominantly vertical transporting systems, groups of undifferentiated (parenchyma) cells called **rays** serve in the larger plants for lateral transport across both xylem and phloem.

Xylem and phloem are both complex tissues; both consist of several different types of cells, some of which are alive and some of which are dead when functionally mature.

A. Corn stem   Before you study vascular systems in detail, it will be helpful to examine a relatively simple plant, in order to gain some idea of how the vascular tissues are related to the plant as a whole, and to see what other sorts of cells we shall encounter in examining the vascular components of other plants.

Study a cross section of corn (*Zea*) stem under the dissecting microscope, and low and high powers of the compound microscope. Locate and label in the accompanying photomicrographs (FIGURE 7-6):

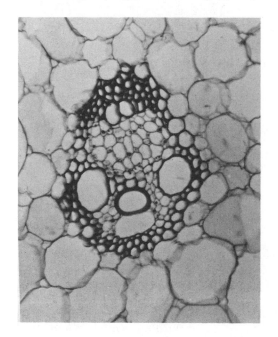

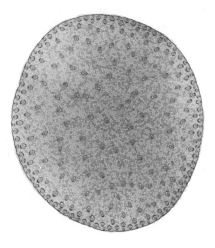

FIG. 7-6. Monocot (corn) stem cross section. (*Courtesy Thomas Eisner, Cornell University.*)

1. **Vascular bundles,** scattered throughout the stem but concentrated near its edge.

2. Thick-walled **sclerenchyma** cells, surrounding each vascular bundle.

3. Large, open, thick-walled **xylem vessels,** usually two or three in number, toward the center of the stem.

4. **Phloem tissue,** located toward the periphery of the stem, containing thin-walled cells of two sizes:

   a. **sieve tube cells,** which function in transport of organic substances, and

   b. **companion cells,** smaller cells adjacent to the sieve tube cells.

Scattering of the vascular bundles throughout the stem is characteristic of the **monocotyledons** (a subclass of the flowering plants, Angiosperms; see Appendix 2 or your text), of which corn is a representative. Note how the bundles are oriented, with phloem always to the outside, and xylem always to the inside. The tissues always differentiate in this manner, but the reason why is unclear. Can *you* think of any explanation for this arrangement?

The sclerenchyma cells, thick-walled and usually stained a deep red, are supportive elements. Can you find sclerenchyma in the stem outside the vascular bundles? Where in the vascular bundles is this cell type most heavily localized?

Compare the thickness of wall and size of lumen (interior space) of sclerenchyma cells and xylem vessels.

The bulk of the corn stem, and not a few cells in the vascular bundles themselves, consists of parenchyma, thin-walled undifferentiated cells without secondary thickenings.

Before considering the other types of arrangement of vascular tissues, we should examine briefly the structure of the transporting elements in xylem and phloem.

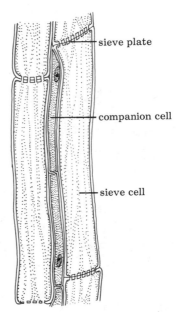

sieve plate

companion cell

sieve cell

FIG. 7-7. Longitudinal section of sieve tubes and companion cells.

B. The structure of phloem and xylem

PHLOEM

1. The transporting cell type in phloem tissue is the sieve tube cell, an elongate living cell without a nucleus and with many perforations in its end wall. In close association with the sieve tubes are smaller, elongate companion cells, which do have nuclei and probably exercise some metabolic control over the functioning of the sieve tube cells.

Examine a longitudinal section through a pumpkin (*Cucurbita*) stem. Locate the sieve tube cells and the companion cells, and identify as many of the other cell types as you can. Draw and label the sieve tube cells and the companion cells; how do they compare to the diagrammatic drawing in FIGURE 7-7?

XYLEM

2. The transporting cells in the xylem are of two main types, **tracheids** and **vessels,** but within each of these groups there is wide variation. Both types are dead at maturity, and have in common thickened secondary walls and a more or less cylindrical shape (FIGURE 7-8). Tracheids are spindle-shaped cells, closed at the ends, and have a smaller lumen than vessels, which are often stouter and tend to be heavily perforated or even open entirely at

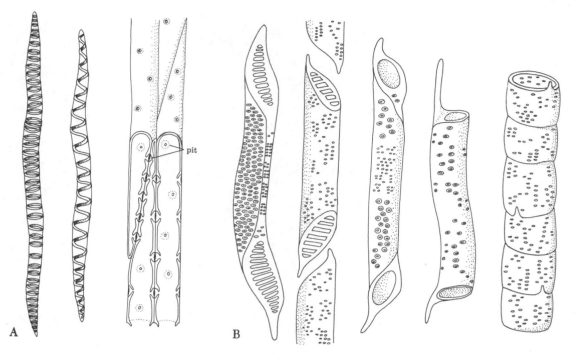

Fig. 7-8. Xylem. (A) Tracheids of three different types. (B) Vessels of five different types, with the more primitive on the left and the more advanced on the right. In the more advanced cells, all traces of the end walls have disappeared. (*A: Modified from V. A. Greulach and J. E. Adams*, Plants: An Introduction to Modern Botany, *Wiley, 1962. B: Modified in part from A. J. Eames and L. H. MacDaniels*, An Introduction to Plant Anatomy, *McGraw-Hill Book Company, 1947.*)

the ends. Tracheids often have **pits** in the walls, and abut against one another more or less diagonally; vessels may have perforated walls, but they abut directly end-to-end, forming columns. (You should be able to see the latter condition in the pumpkin stem section; it also shows up in FIGURE 7-8.)

Examine slides of macerated wood (wood treated so as to dissociate the cells); compare a conifer like pine (*Pinus*) with a hardwood like basswood or ash (*Tilia* or *Fraxinus*). Which cells are common in the conifer? In the hardwood? Locate as many different types of vessels and tracheids as time allows. If there is enough time, your instructor may give you specimens and directions for macerating your own wood specimens. In some of your slides, you may find tissues or cells not identifiable as tracheids or vessels. What are they?

C. Vascular tissues in other stems You have already examined corn, a monocot, which has vascular bundles scattered throughout the stem. Some dicots (the other group of flowering plants; see Appendix 2) also have vascular bundles, but these are almost always organized into a *ring* at the periphery of the stem (compare FIGURES 7-9 and 7-10, of the buttercup *Ranunculus* and the alfalfa *Medicago*). The significance of this arrangement will be apparent shortly. Most monocots and many dicots, like the two illustrated here, are known as **herbaceous** plants, since their stems remain soft and succulent, in contrast to **woody** plants. (Do you notice any features of the buttercup stem that suggest it is

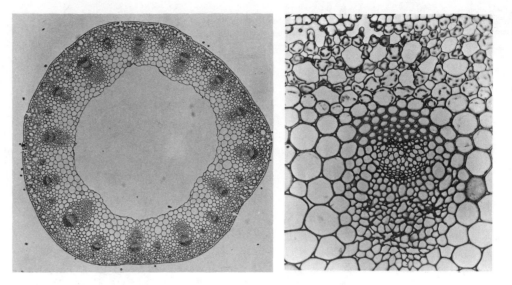

FIG. 7-9. Cross section of stem of buttercup, a herbaceous dicot. Left: Low power. Right: High power. (*Courtesy George H. Conant, Triarch Incorporated, Ripon, Wisconsin.*)

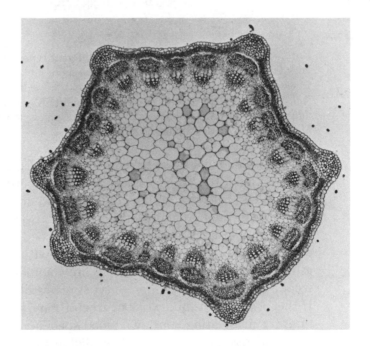

FIG. 7-10. Cross section of stem of alfalfa, a herbaceous dicot. (*Courtesy Thomas Eisner, Cornell University.*)

adapted to carry out photosynthesis?)

All the tissues in the corn, buttercup, and alfalfa, and those in a great many other herbaceous plants, are known as **primary tissues.** By definition, primary tissues are those that differentiate directly from the **apical meristems,** the shoot tip and root tip of the plant. Primary tissue growth is sufficient for these plants, since their relatively small stature does not require any substantial increase in girth.

Plants that *do* increase substantially in girth, and often live through many growing seasons, increase their diameter by reproduction of cells from a second type of meristem, the **vascular cambium.** This tissue can be seen in primitive form in the alfalfa stem, but usually the xylem and phloem are joined into a more complete ring, and the cambium, another complete ring,

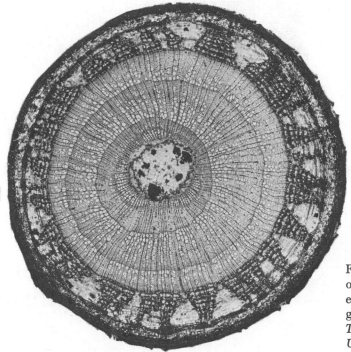

Fig. 7-11. Cross section of basswood stem at the end of the third year of growth. (*Courtesy Thomas Eisner, Cornell University.*)

lies between them. As the cambium divides, new cells are added to the xylem, on the inside, and to the phloem, on the outside. This **secondary tissue** is added to the stem or root in each growing season, and the organ's diameter and strength are thereby augmented. Would the arrangement of vascular tissues in corn be adapted to the production of secondary tissues?

Examine a cross section of basswood (*Tilia*), a woody dicot, with the dissecting microscope and under low power of the compound microscope. Label FIGURE 7-11 as you identify the parts of this stem's anatomy.

1. From the center of the stem outward, the bulk of the stem consists of xylem tissue, arranged in **annual rings.** Each season's growth appears in a ring because of the difference in size of tracheids and vessels laid down at the end of one growing season and the beginning of the next. In a temperate climate, the beginning of the growing season is usually wet, and the xylem elements are large, known as **spring wood.** The end of the growing season is frequently marked by drought; the xylem elements formed in this part of the growing season are much smaller, and are called **summer wood** (compare FIGURES 7-11 and 7-12).

How old is your stem? Can you think of conditions under which annual rings would *not* be formed?

2. The innermost xylem ring is primary tissue, formed by the apical meristem; internal to the xylem, at the center of the stem, there is usually a small amount of parenchyma tissue called **pith.**

3. Each of the xylem rings outside the primary xylem was deposited by the vascular cambium. See if you can distinguish tracheids and vessels in the xylem.

4. Careful observation will reveal that the xylem is traversed by radially oriented groups of parenchyma cells; these are called **rays,** and they function in moving materials laterally through the stem.

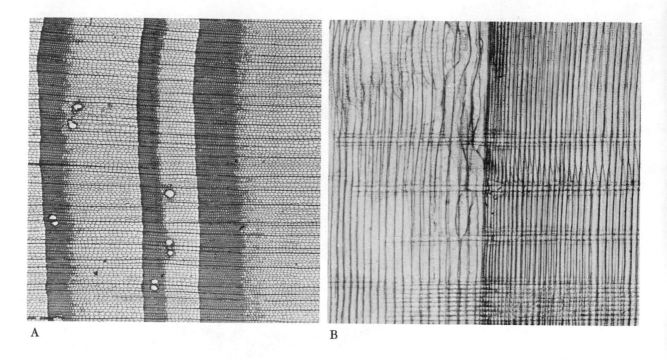

A   B

Fig. 7-12. Spring and summer woods. (A) Cross section of larch. (B) Radial section of pine. Notice the fluctuation in tracheid size between spring and summer wood. (*Courtesy U. S. Forest Products Laboratory.*)

5. The vascular cambium, a ring of tissue one cell thick, is best located under high power just outside the youngest ring of xylem.

6. Tissues just outside the cambium are functional phloem; except for the thick-walled sclerenchyma cells, cell types usually cannot be distinguished very easily. The older phloem, further out in the stem, is dead and does not function in transport. How *does* it function?

As one might expect, addition of cells on either side of the cambium causes the xylem and the phloem to increase in diameter. As cells are added to the xylem, on the inside of the cambium, the cambium expands to accommodate the increase in stem diameter. The older phloem, however, is fixed in size, and cannot expand. As it is pushed outward by the advancing cambium, it finally breaks. Fill-in tissue is supplied by expansion of phloem parenchyma and expansion of the rays in the phloem, both of which are living cells capable of division. The breaking and fill-in is, however, evident in the appearance of the phloem: the triangular groups of cells in the phloem are partly old broken phloem, youngest to oldest, from the interior, and partly phloem ray and parenchyma expansion. Figure 7-13 shows a small portion of this region under high power.

7. As a stem matures, a number of changes also occur in its outer surface. The original **epidermis** is lost through weathering, and in the underlying **cortex** a new meristematic region, called the **cork cambium**, forms. This meristem produces cork cells, which become impregnated with a waxy substance and then die, thus forming an outer rough, water-repellent layer of the stem. As the cork breaks and sloughs off because of pressure from the vascular cambium and from weathering, the cork cambium is continually renewed a little deeper in the stem. Eventually, of course, the original cortex

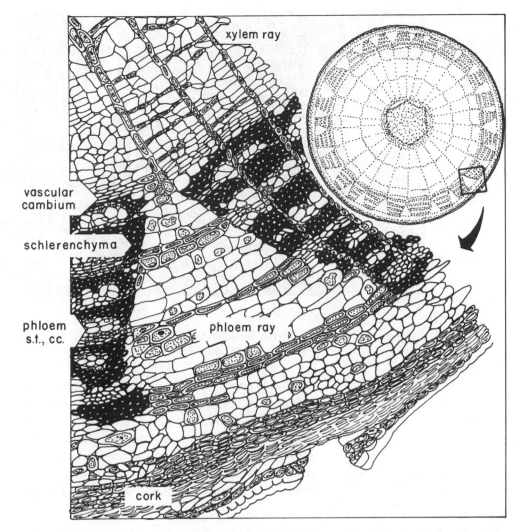

FIG. 7-13. Cross section through basswood *(Tilia)* phloem. The high-power drawing illustrates the progressive change in phloem from the vascular cambium to the outside of the stem.

weathers away completely; after this has occurred, new cork cambium forms from phloem parenchyma.

8. Examine radial and tangential sections of *Tilia* stem (if necessary refer to FIGURE 1-5B, p. 6). Locate and identify as many different vascular elements as you can. Label the photographs of these sections in FIGURE 7-14.

Remember that although we have used the stem as a basis for discussion, roots and leaves contain continuations of the transport system we have been studying, and that transport is no less important in those organs than it is in the stem.

PART III. EXPERIMENTAL STUDY OF TRANSPORT

Of the two transport systems in plants, xylem is far less refractory to experimental manipulation, and is much better understood, both anatomically and functionally. Currently available evidence suggests that water and mineral transport results almost entirely from physical forces which natural selection has shaped to the plant's advantage. One of these physical forces is the re-

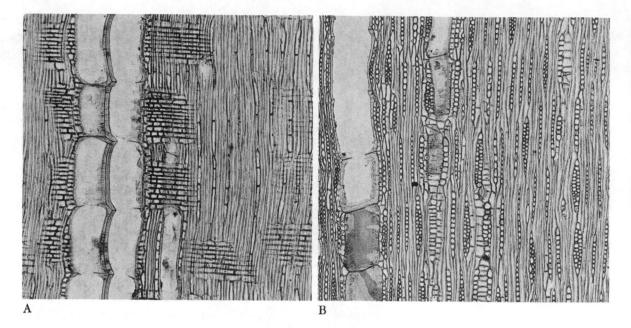

Fig. 7-14. Sections through persimmon wood. (A) Radial. (B) Tangential. (*Courtesy U. S. Forest Products Laboratory.*)

markable cohesiveness of water molecules, which serves to keep water in the xylem in a continuous column, where other fluids would break under the negative pressure. The other force is simply cell-to-cell differences in osmotic pressure, which may tend to move water into or out of the xylem. One small source of osmotic pressure gradient is root pressure, the result of osmotic differences between the root cell fluid and the soil; the major source, however, is transpiration, which you will study experimentally during this period.

A. Measurement of transpiration rate

1. Several months prior to the laboratory period in which this experiment is to be performed, you or your instructor should have prepared the experimental plant in the following manner. Take a cutting from a *Coleus* plant and put the stem through one hole of a large two-hole stopper. Put the cut end of the stem into moist sand or potting soil in a flowerpot, and keep it moist and in sunlight until the cutting has regenerated a substantial root and shoot system.

2. Remove the plant gently from its pot, immerse the root end in a large container of water, and swirl to wash the sand or soil from the roots. Fit the plant into a pint fruit jar which you have previously filled about two-thirds with tap water; press the stopper in to make a very tight fit.

3. Set up the rest of the apparatus as illustrated in Figure 7-15. Make certain that all connections are very snug. Allow the entire apparatus to come to room temperature.

4. When you are ready to measure the transpiration rate, use a small syringe to insert a drop of manometer fluid or colored water at the open tip of the pipette. The rate of water loss can be calculated in milliliters per hour by measuring and recording changes in the position of the fluid drop over known periods of time. Observe the approximate rate of movement for a little while, and then decide on convenient intervals at which to make posi-

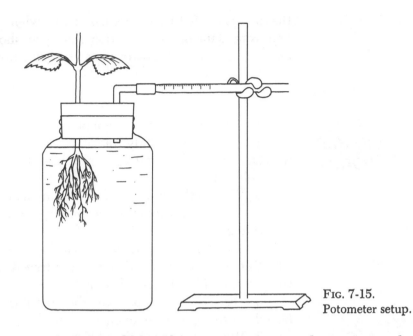

FIG. 7-15.
Potometer setup.

tion and time readings. The precise volume and time interval you use are unimportant so long as you get sufficient data with which to construct a graph, and so long as the measurements you do take are accurate and consistent with one another. After you have obtained several consecutive readings that agree well with one another, average them and record that figure as the normal transpiration rate.

5. With a strong light bulb, increase the light intensity near the plant, and continue to take readings for ten or fifteen minutes. Then remove the light and take "normal" readings again.

6. Record these data in the table as you go along; when you finish, plot them together on the graph, indicating the axes clearly and marking the curve to show where the light intensity was changed.

B. Additional experiments

Your instructor may specify, or you may wish to design, other experiments to measure the transpiration rate under different conditions. Some suggestions follow:

1. Compare transpiration rates under different conditions of illumination intensity or quality or "wind" (from a fan).

2. Compare the transpiration rates of plants that have had the stomata blocked (by coating one or both leaf surfaces with vaseline) with plants that have not had the stomata so blocked.

3. Remove part of the root system from your original plant, and compare the plant's transpiration rate under these conditions with what you recorded when the root system was intact.

C. Rate of fluid movement in the xylem

If there is time, you may wish to try to estimate the rate of fluid movement in the xylem. To do this, make a cross section of the stem of your experimental plant as described in FIGURE 4-3, p. 42. Measure the diameter of your low-power microscope field, and from this information make an estimate of the area occupied by the lumen of the xylem vessels. With this knowledge, and your data on transpiration rate, you should be able to calculate the approximate rate of flow of water through the xylem.

1. What direct role does photosynthesis play in guard cell operation?

2. Conifers have only tracheids, no vessels. Can you suggest some way in which tracheids might serve as a useful adaptation for a plant living in a very cold habitat?

3. What is the function of sclerenchyma in the vascular bundles of corn?

4. A corn stem shows no secondary growth. How does it manage to increase in girth?

5. A nail is driven into a ten-year-old tree at three feet from the ground. The tree is fifteen feet tall. When the tree is twenty feet tall, how far from the ground will the nail be? Why?

6. Discuss and compare the theories of translocation in the phloem, and be able to give evidence for each.

7. What happens to the phloem of a woody stem as the stem ages?

8. Discuss the leaf structures that help to adapt a xerophyte like oleander (Nerium) to its environment. Compare these adaptations to those of hydrophytic and mesophytic plants.

9. Describe the manner in which the age of a woody twig can be determined from external and internal observations.

10. Discuss the processes involved in the uptake of water and minerals by a plant root.

11. Compare and contrast secondary xylem and secondary phloem with regard to meristematic origin, location in the plant body, cellular characteristics, probable method of movement of materials, and general function in the life of the plant.

PIPETTE READING	H$_2$O TRANSPIRED
AVERAGE RATE:	

NORMAL INTENSITY

PIPETTE READING	H$_2$O TRANSPIRED
AVERAGE RATE:	

HIGH INTENSITY

PIPETTE READING	H$_2$O TRANSPIRED
AVERAGE RATE:	

NORMAL INTENSITY

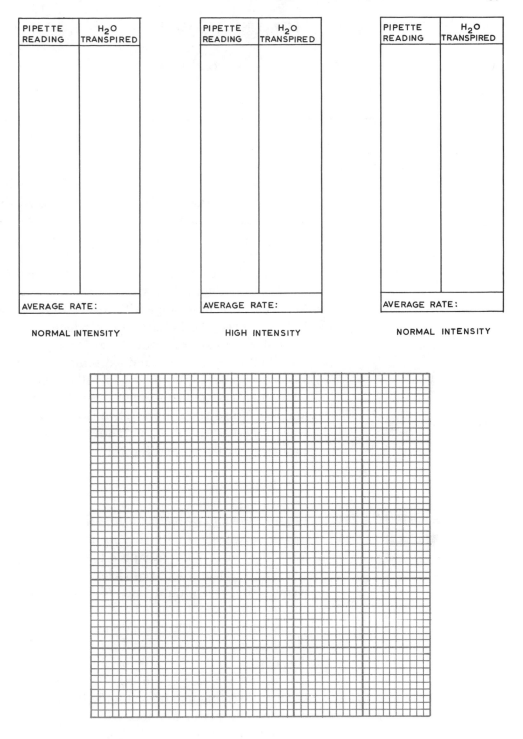

TOPIC **8** TRANSPORT IN ANIMALS

Biological Science, pp. 212-236.
Elements of Biological Science, pp. 137-148.

Association and cooperation of diversely specialized cells is what makes possible complex organisms like elephants and earthworms and vascular plants. The integrity of any organism depends upon mechanisms for functionally knitting together its widely separated parts, thereby promoting smooth and efficient operation of the whole system. We may properly regard any fluid transport mechanism, regardless of its structural basis, as an integrating mechanism; it not only carries raw materials and by-products of metabolism between sites of exchange with the environment and other parts of the body, but also serves as the medium in which a variety of chemical messages — hormones — are rapidly circulated throughout the organism.

You have already examined some adaptations for fluid transport in the vascular plants. In this unit, you will consider the anatomy of the heart and circulatory system in a mammal and, by way of comparison, in several other animals. The structure and chemistry of blood will also be a topic of this week's laboratory. In the next unit, you will have an opportunity to study capillary circulation and the control of heartbeat in a frog.

PART I. THE MAMMALIAN HEART AND GREAT VESSELS A. Fetal pig

In your previous dissection of the thorax, you exposed the heart, lying within the pericardial cavity. If you have not already done so, carefully clean away the pericardium to expose the heart fully to view.

The right and left **ventricles** compose the major portion of the heart visible from the ventral side. Anterior to the ventricles, you can see the darker-colored **atria** (singular, *atrium*). The atria are more readily visible if the heart is pushed to one side and examined dorsally.

The fetal pig heart may be sectioned frontally, separating ventral from dorsal side, and examined for the internal anatomy of the chambers and valves, but you should do this (following the instructions in B below and using a dissecting microscope) only if a beef heart is not available.

Remove any membranes still clinging to the heart or to the vessels immediately associated with it. Locate the following vessels:

ARTERIES

1. Pulmonary trunk, leaving the ventral side of the heart from the top of the right ventricle, then passing forward diagonally before branching into right and left **pulmonary arteries**, which carry blood to their respective lungs. Trace one of the pulmonary arteries as far as you can into the lung.

2. **Aorta**, arising from the anterior end of the left ventricle, just dorsal to the origin of the pulmonary trunk, then passing a short distance forward and bending to the *animal's* left as the **aortic arch.**

3. **Ductus arteriosus,** a short, stout vessel leading directly from the pulmonary trunk to the aorta.

The ductus arteriosus carries blood only in the fetus, and serves to shunt blood away from the pulmonary circulation, which is not fully functional in the fetal animal. It usually closes, through muscular contraction of its walls, at birth; sometimes, however, it fails to close. What symptoms would you expect to observe in a young mammal whose ductus arteriosus has failed to close?

VEINS

4. **Anterior vena cava,** bringing blood from the upper parts of the body and entering the right atrium at its anterior end.

5. **Posterior vena cava,** bringing blood from the lower parts of the body and entering the right atrium at its posterior end.

6. **Pulmonary veins,** returning blood from the lungs to the left atrium.

Note the difference in appearance and in texture between veins and arteries, and compare their structure with that of a capillary in FIGURE 8-1.

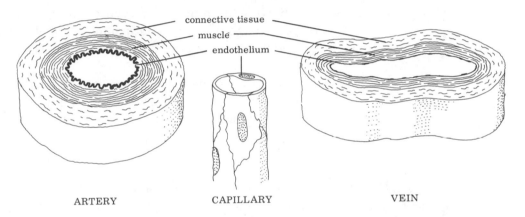

connective tissue
muscle
endothelium

ARTERY   CAPILLARY   VEIN

FIG. 8-1. A comparison of the walls of artery, vein, and capillary. Arteries and veins have the same three layers in their walls, but the walls of veins are much less rigid and they readily change shape when muscles press against them. Capillaries have walls composed only of a thin endothelium.

B. Beef heart

On a beef, calf, or sheep heart (preferably fresh), locate all the chambers mentioned in A above. Make well-placed incisions with a scalpel or razor blade, and cut the chambers open far enough to see the internal structure clearly. Locate as many of the major vessels as are still present on the heart.

Rinse the heart free of clotted blood, if any is still present. Carefully examine the interior of all four chambers, locating the following valves and other features:

1. Right atrioventricular (tricuspid) valve
2. Left atrioventricular (bicuspid, mitral) valve
3. Pulmonary semilunar valve (in the base of the pulmonary trunk)
4. Aortic semilunar valve (in the base of the aorta)

How do the tricuspid and bicuspid valves differ? Notice the muscular bands attaching the valve flaps (cusps) to the ventricle walls. What is the function of these bands?

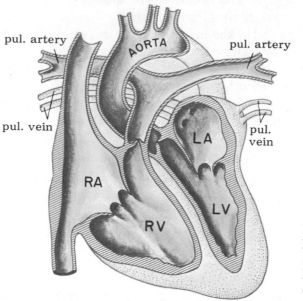

FIG. 8-2. Blood flow in the human heart. (*Modified from B. G. King and M. J. Showers,* Human Anatomy and Physiology, *Saunders, 1963.*)

Review the function of each of the valves, and the passage taken by blood through the heart of a mammal. Label FIGURE 8-2 with arrows showing the direction of blood flow. How do the hearts of other vertebrates differ from that of mammals?

Sometime during the period, if materials are available, your instructor will use the fresh beef heart to demonstrate the action of the heart valves, coronary circulation, and a simulated coronary thrombosis.

PART II. SYSTEMIC CIRCULATION IN THE FETAL PIG

The systemic circulation serves all of the body, except for the lungs; it consists of arteries, branching from the aorta all along its length, veins, returning blood from all over the body to the right atrium, and capillaries between the arteries and veins. Although you will examine a few of the major arteries and veins to get an idea of the pattern of systemic circulation in a mammal, do not forget that the exchanges between the blood and the tissues are the principal function of the circulatory system, and that such exchanges occur across the walls of the capillaries. The major vessels and the heart function only to deliver blood to the capillary beds.

Trace each of the vessels mentioned as far as you can.

A. Arteries (FIGURE 8–3)

1. The first branches off the aorta are a pair of **coronary arteries,** which leave the aorta as it crosses dorsal to the pulmonary trunk; one of them is visible on the ventral surface of the heart. These arteries supply capillary beds in the heart wall, since its needs are not met by blood passing through its lumen. The next branch from the aorta divides promptly into right and left **common carotid arteries,** serving the head and neck, and into the artery serving the right forelimb. The artery branching to the left forelimb arises directly from the aortic arch.

2. Locate the descending aorta in the middorsal line. It gives off several major branches:

a. a short vessel supplying the stomach, duodenum, liver, and spleen (trace these tributaries into the respective organs as far as you can)

b. a long, unpaired trunk, arising a short distance posterior to the previous artery, which branches to the pancreas and small intestine

c. a pair of **renal arteries,** one supplying each kidney

Near the posterior end of the aorta, paired arteries branch to the hind legs, and paired **umbilical arteries,** the largest branches from the aorta in the fetus, carry blood from the fetus to capillaries in the embryonic portion of the placenta.

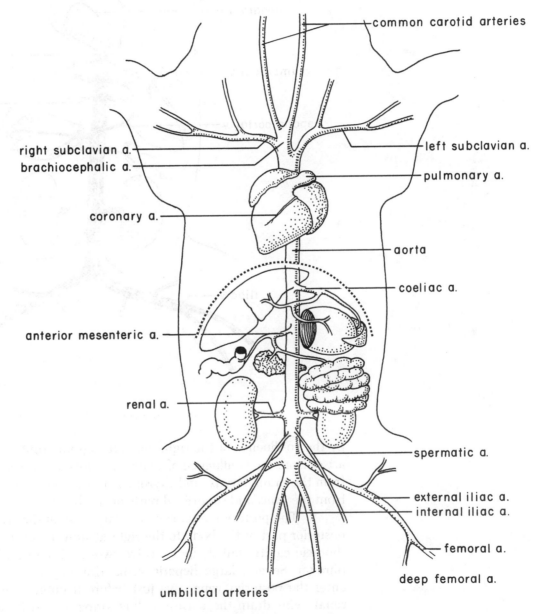

FIG. 8-3. Fetal pig, arterial system.

B. Veins
(**FIGURE 8–4**)

1. The anterior vena cava returns blood to the heart from the head, neck, thoracic limbs, and a substantial part of the thoracic wall. It is formed

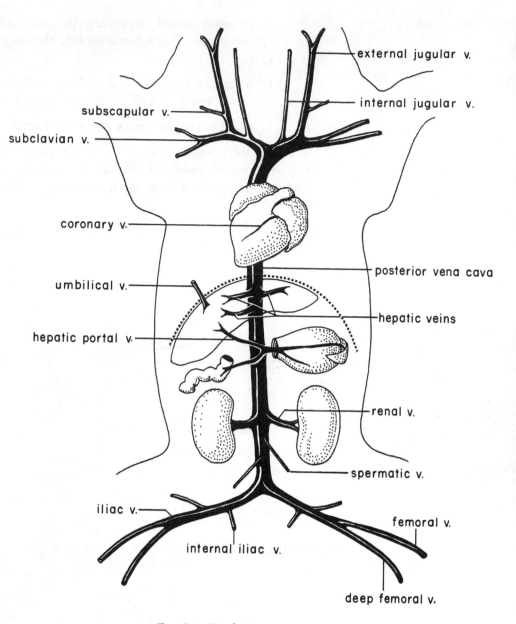

subscapular v.

subclavian v.

coronary v.

umbilical v.

hepatic portal v.

external jugular v.

internal jugular v.

posterior vena cava

hepatic veins

renal v.

spermatic v.

iliac v.

internal iliac v.

femoral v.

deep femoral v.

FIG. 8-4. Fetal pig, venous system.

by the confluence of the right and left **jugular trunks,** each of which in turn arises from the confluence of two main roots, an internal one draining blood from the brain, larynx, and thyroid, and a larger external one draining blood from the arms and superficial parts of the head.

2. The posterior vena cava is a large vessel that returns blood from the posterior part of the body to the right atrium. It receives no branches in the thoracic cavity, but it does receive two or three small veins from the diaphragm. Several large **hepatic veins,** draining capillary beds in the liver, enter the posterior vena cava just before it enters the liver. Right and left **renal veins** drain the corresponding kidneys, and lie adjacent to the renal arteries. A pair of veins draining the hind legs unite in the posterior part of the peritoneal cavity to give rise to the posterior vena cava; these veins form from a number of tributaries that came from the hind limbs and other posterior parts of the body.

3. The **umbilical vein,** which you probably cut when entering the peritoneal cavity, passes through the umbilical cord, and runs into the liver. Some of the blood in it enters the liver capillaries, but a majority passes directly through the liver into the posterior vena cava. The relationship between the hepatic veins, umbilical vein, and posterior vena cava may be seen by making a careful frontal section of the liver, separating ventral from dorsal side, with a razor blade or scalpel, but this should be done *only after completing the next section.*

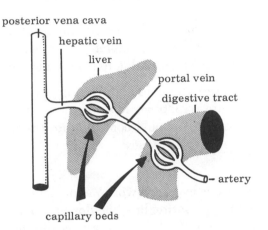

FIG. 8-5. The hepatic portal circulation. Blood from the intestinal capillaries is carried by the portal vein to a second bed of capillaries in the liver.

C. Hepatic portal system

A **portal system** (FIGURE 8-5) is defined as a group of veins that leads from one capillary bed to another. Veins leading from the spleen and intestinal capillaries to those in the liver constitute the hepatic portal system. The significance of this circulatory arrangement is apparent if one considers the important role of the liver in processing and storing materials absorbed from the intestine.

Vessels from the spleen and stomach converge to form a vein that unites with the vein draining the intestine, forming the hepatic portal vein. The hepatic portal vein runs dorsal to the pylorus and enters the liver. These vessels are most easily found by breaking the mesenteries that hold the stomach and pancreas, lifting the stomach anteriorly, and separating it from the pancreas.

PART III. CIRCULATION IN INVERTEBRATES

Many invertebrate animals whose gas exchange, transport, and excretory demands are low because of small body size and/or low metabolic rate, do not have special circulatory systems. Those animals that *do* have a special system have one of two types: a **closed** system, in which the blood never leaves distinct vessels, or an **open** system, in which the blood is to various degrees free to circulate in the tissues outside of vessels. The type of system an animal can be expected to have depends upon a number of factors, including its evolutionary history, its size, and its circulatory requirements. The open system is characteristic of all arthropods and most molluscs, while most other invertebrates with circulatory systems, and all vertebrates, exhibit the closed system.

A. Open circulatory system: Arthropoda

1. Under the dissecting microscope, examine a living, transparent crustacean arthropod such as the beach hopper (*Gammarus*) or the sand shrimp (*Crangon*). If you have a small animal, draw it into a dropper and close the dropper with a bit of vaseline or your finger while making observations; if you have a larger one, place it in a covered petri dish.

The heart, located in the center of the animal's dorsal side, can be seen beating. Locate vessels if you can. If an anesthetic (such as MS-222) is available, use it to quiet your specimen and then examine it more carefully in a petri dish under the low power of the compound microscope, being careful to avoid getting water on the stage or objectives. What course does the blood follow?

2. Your instructor may provide demonstrations of the heart in a living insect or in the crustacean *Daphnia*. The heart and arterial system of a crayfish or lobster, injected with a dye, may also be demonstrated.

Vessels are sometimes abundant in open systems, as in the crayfish and lobster, and are sometimes less common, as in insects and *Daphnia*. What factors do you think would contribute to the need for many distinct vessels to deliver blood to the tissues?

B. Closed circulatory system: Annelida

1. Obtain a live earthworm, and identify the dorsal side by the presence of a dark, pulsating line, the **dorsal blood vessel.** Insensitize the animal by placing it in 35% alcohol for a few minutes; it will writhe vigorously at first, but after a few minutes the writhing ceases and the worm responds only slightly to gentle probing. Take the worm out *as soon as* it no longer

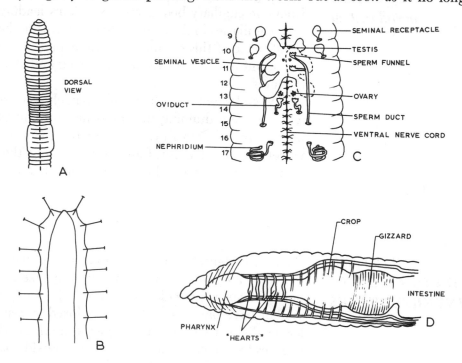

Fig. 8-6. Earthworm. (A) Outline of incision. (B) Method of pinning the worm to the pan. (C) Reproductive system and parts of the nervous and excretory systems, all in dorsal view with segments numbered. (D) Anterior parts of the circulatory and digestive systems in lateral view.

responds to the probing (several minutes); if left in the alcohol for a longer period, it may expire.

2. Pin the worm *dorsal* side up through the third or fourth segment and again through the body about three inches behind the anterior end. Fasten the specimen to a dissecting pan in such a way that the worm will be observable when the pan is later placed under a dissecting microscope. (If you pin it in the middle of the pan, the specimen often will not be visible with the dissecting microscope.)

3. With fine scissors, make an incision in the dorsal midline of the worm, beginning just in front of the **clitellum** (FIGURE 8-6) and proceeding anteriorly. Cutting tiny blood vessels in the body wall is unavoidable, but keep the scissors-point *up*, to avoid damage to fragile underlying structures.

4. When the worm is opened dorsally up to the anterior pin, spread the body wall, breaking the septa between segments, and pin the body wall to the dissecting pan as illustrated in FIGURE 8-6. Cover the dissection with Ringer's solution.

5. Locate the dorsal blood vessel, partly embedded in the top of the intestine, which runs most of the length of the worm. The dorsal blood vessel is the main pumping device, and is supplied with valves, the action of which you can see by examining the vessel with a dissecting microscope. Which direction does blood move in the dorsal vessel?

6. In segments seven through eleven (count from the anterior end), most of the blood in the dorsal vessel passes ventrally through five pairs of **aortic arches** (or "hearts"), which supplement the action of the dorsal vessel in moving blood through the system.

7. Remember that blood travels through the worm without ever leaving the vessel system. In what regions of the worm's body would you expect to find the most well-developed capillary circulation? Examine tributaries of the dorsal and ventral blood vessels to see if your hypothesis is correct.

(Take this opportunity to review, with the aid of FIGURE 8-6, as many other structures in the earthworm as time permits. When you have completed your dissection, unpin the worm, clean, rinse, and return the pins to the container at your desk, and rinse and dry the dissecting pan. Before going on to other parts of the laboratory, look up the Phylum and Class of each of the animals studied in this laboratory so far, and enter their names in the appropriate spaces in Appendix 2, except in cases where the animal has already been classified.)

PART IV. BLOOD

The first of the following experiments and observations should be carried out by individual students; the others may be done by individuals or as demonstrations, as time permits and the instructor directs.

A. Microscopic examination of blood

Clean a microscope slide with 70% alcohol. With a *sterile* lancet (open one yourself — *don't* use one already opened or employed by someone else!) obtain a drop of your own blood (alternatively, beef blood may be used if available) and let it fall on one end of the slide. Make the puncture in the

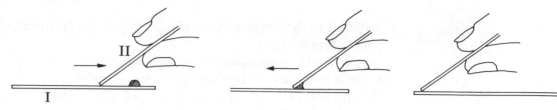

FIG. 8-7. Method for making a blood smear.

end of your index finger, and squeeze the finger toward the end if enough blood to make a large drop does not promptly appear. Discard the lancet — into the wastebasket, or as directed by your instructor.

Take a second slide, and place it at an angle on the first slide, as shown in FIGURE 8-7. Move slide II toward the blood drop until the slide touches the drop; blood will flow along the edge of slide II. Then move slide II back along slide I, drawing the blood out into a thin film or *smear* across slide I. Allow slide I to dry; place it in a petri dish, and add enough Wright's stain to cover the blood smear. *Count* the number of drops of stain you add. After one minute, add an equal number of drops of distilled water. Leave the stain and water on for three more minutes, and then rinse it off by dipping the slide in a beaker of clean tap water. Allow the stained slide to dry thoroughly, smear side up, and then examine under the microscope.

Most of the cells you see will be **erythrocytes**, red blood cells. The **leucocytes**, white cells, are rarer, scattered among the erythrocytes. The platelets are destroyed in making the slide, and will not be visible. Try to identify as many different kinds of leucocytes as you can, and draw them. Your instructor will be able to provide references that will enable you to identify these cells by name, if you are interested.

B. The composition and clotting of blood

You will be provided with beef blood to which sodium oxalate has been added. Sodium oxalate binds calcium ions in the blood, and thus prevents clotting, since clotting requires calcium ions. The oxalate does not otherwise interfere with these experiments.

1. Place some oxalated blood in a test tube and allow it to stand for a few minutes undisturbed. Alternatively, it may be put into a centrifuge tube and centrifuged. Notice that a red mass sinks to the bottom of the tube; what is in this mass? The clear fluid that remains is blood **plasma**; what does it contain?

2. Remix the separated blood, or obtain some fresh oxalated blood, add calcium chloride solution, and watch the reaction. Addition of an excess of calcium ions (more than can be bound by the available sodium oxalate) will bring about prompt clotting. After clotting has occurred, centrifuge again or allow to stand. How does the remaining **serum** differ from plasma?

C. The effect of oxygen and carbon dioxide

Bubble air into 15 ml of oxalated blood in a test tube; observe the color that develops. Then bubble carbon dioxide into the blood; note the color change, and again bubble air into the tube. How do you explain the changes in color? What reactions were taking place? How are oxygen and carbon dioxide normally carried in the blood, and under what conditions are they released?

D. The buffering
action of plasma

Maintenance of proper pH is of utmost importance to an organism, since even small shifts in the hydrogen ion concentration can cause death. A stable pH is maintained by **buffers**, substances that tend to prevent major shifts in pH by binding or releasing hydrogen ions as the solution deviates from the "standard" for that buffering solution. Review the activity of buffers in your text, and carry out the following experiment, which demonstrates the action of buffers in the blood plasma.

1. In one beaker, place 3 ml of plasma obtained by centrifuging oxalated beef blood; in a second beaker, place 3 ml of distilled water. Make sure the beakers are properly labeled.

2. Use Hydrion test paper to determine the approximate pH in each beaker, using a different piece of paper for each test. If there is a difference in pH between the two beakers, adjust the pH of the distilled water by adding 0.1 N hydrochloric acid (to reduce pH) or 0.1 N sodium hydroxide (to raise pH), drop by drop, testing after each addition, until the water is the same pH as the plasma.

3. When the plasma and the water have the same pH, add 0.1 N hydrochloric acid to the plasma drop by drop, swirling and testing after each addition, until the pH drops *two* units. Record the number of drops required for this change. Repeat this procedure with the water, again counting the number of drops required for a two-unit alteration in pH. Which of the solutions requires more acid to alter its pH? Why?

E. Hemolysis

Maintenance of the proper osmotic pressure of the blood is of great importance. What organs help control the osmotic pressure of the blood? The following experiment demonstrates the effects of a change in osmotic pressure on the erythrocytes.

Place 1 ml of oxalated blood into each of three test tubes, and number the tubes. Into tube 1, place 10 ml of isosmotic (0.9%) saline solution. Into tube 2, place 10 ml of distilled water, and into tube 3, place 10 ml of 5% saline solution. Mix the contents of each tube by shaking gently.

Do you notice any difference in the appearance of the tubes? Allow them to stand for five minutes, stir again, and make a wet mount of each. Examine under the microscope; record your observations in the form of sketches. Explain the results.

QUESTIONS FOR
FURTHER
THOUGHT

1. Trace a drop of blood in the fetal pig from the right atrium to the umbilical vein, by the shortest possible route.

2. Of what advantage is it to the insect to pump blood forward in the aorta?

3. What is the significance of the hepatic portal system?

4. The blood of a human being serves as a medium of internal transport for many substances, yet it is critically important to the organism that fluctuations in the content of this fluid be held to a minimum. Consider such materials as water, glucose, amino acids, oxygen, carbon dioxide, inorganic salts, and urea, and indicate

a. where each enters the blood
b. where each leaves the blood
c. measures the body can utilize to raise the concentration of each
d. measures the body can utilize to lower the concentration of each
(Don't forget the roles of hormones in these regulatory functions.)

TOPIC 9 HEARTBEAT AND BLOOD FLOW

Biological Science, pp. 213-224.
Elements of Biological Science, pp. 138-145.

In this period you will have an opportunity to observe a living circulatory system and to demonstrate basic aspects of the control and activity of the heart. The main objectives of this exercise are:

1. to observe blood flow in the capillaries
2. to study the heart in action and map out its activities
3. to demonstrate the effect of temperature on the rate of heartbeat of an amphibian
4. to study the nervous control of the heart, and to test the effect of some chemical transmitter substances and other drugs on heartbeat

PART I. OBSERVATION OF BLOOD FLOW

In the circulatory system of the preserved fetal pig, you observed arteries as they left the heart, and veins as they returned. In the living system, you will see an actual circulation of blood, a circulation that was only inferred from the structure of the vessels in the pig.

The major function of the circulatory system is served by capillaries, which connect the arteries and veins. In the frog, there are three areas transparent enough to enable you to see blood flowing in the capillaries: the web of the foot, the mesentery, and the tongue. Of these, the web is most convenient and can be examined under the microscope without injury to the animal.

A. Preparation of the frog

1. The frog may be anesthetized or wrapped in a towel, but it is more easily handled if the nervous system is inactivated by pithing. If you would like to pith a frog yourself, inform your instructor, and consult Appendix 5.
2. Place the frog belly-down on a balsa frog board (FIGURE 9-1) so that

FIG. 9-1. Frog attached to balsa board. The web is stretched and pinned over the hole; a rubber band holds the board to the microscope stage.

the tip of the foot just reaches over the circular hole. Stretch a part of the web over the hole, pinning so that a portion of the web is fixed taut. Now place the board on the stage of a compound microscope, fastening it to the stage with a stage clip, rubber band, or masking tape. Study the web portion with transmitted light; keep the web moist with frog Ringer's solution.

B. Observations within the capillary net

1. The **chromatophores** are black pigment cells in the skin. Some of them are irregularly branched, others are compact. The expansion and contraction of the pigment is under hormonal control; the hormone primarily responsible is called **intermedin,** and is secreted by the pituitary gland. The color of the frog is light when the pigment is contracted and dark when it is expanded.

Can you see both expanded and contracted chromatophores in the same field?
Why is regulation of color important in frogs?
Can you think of other animals that utilize color in this way?
Why is regulation of color *not* important in man?

2. There is a close network of blood vessels in the web, lying deeper than the pigmented layer.

a. **Arteries** run mainly toward the free edge of the web, constantly diminishing in size as they break up into branches. The blood flows from larger to smaller branches, but remember that appearances are reversed under the microscope.

b. **Capillaries** are the very numerous tiny vessels, frequently branching.

c. **Veins** are formed by the ultimate union of capillaries, and increase in size as they unite with one another. Blood in veins flows from smaller to larger vessels, but remember about optical inversions.

d. **Lymph vessels** are found in nearly all parts of the body as delicate, thin-walled branching tubes. Lymph vessels may sometimes be seen in favorable preparations; they originate in lymph capillaries, interwoven with, but without direct connections to, blood capillaries. These vessels pick up fluids exuded from the blood in its passage through the tissues, and eventually return the fluid to the circulatory system.

The blood-flow current is marked by erythrocytes, red blood cells. The flow is most rapid in the arteries, and slowest and most uniform in the veins. Trace the movements of the red blood cells through the capillary net.

PART II. HEARTBEAT

When one considers the large amount of blood that is pumped through the extensive system of vessels, it is easy to understand why a strong pumping device is necessary. You will now oberve this pump in action, map the path taken by blood passing through the heart, and examine some factors that influence the rate of heartbeat.

***A. The frog heart
and its action***

1. Remove the frog from the balsa board and place it ventral side up on a moist paper towel in a dissecting pan. Make a slit through the *skin only*, from the pelvis to the throat, along the midline. At each end, cut to the right and left, making two flaps. The abdominal vein and muscles will now be exposed. Following the cuts made in the skin, but *avoiding* the ventral abdominal vein, cut the muscle of the abdominal wall. Anteriorly, keeping the scissors-point *up*, cut through the pectoral girdle and up toward the throat. Spread the forelimbs and pin them, thus exposing the heart.

Now observe the heart. Use cotton moistened in frog Ringer's solution to sponge away blood and to keep all exposed organs moist at all times.

2. With fine scissors and forceps, very carefully split open the pericardium, the sac covering the heart, to expose the heart (FIGURE 9-2).

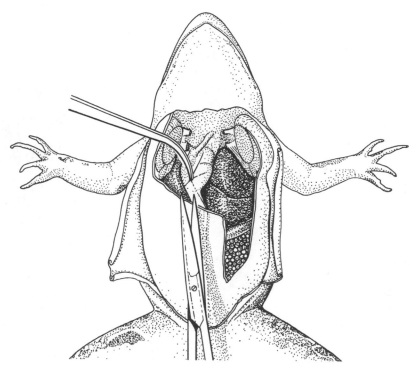

FIG. 9-2. Dissection just prior to opening pericardial sac.

Note the following:

a. ventricle (a single, posterior, cone-shaped, thick-walled region)

b. arterial trunk (arising from the right side of the base of the ventricle and dividing anteriorly into two aortic arches)

c. atrium (a thin-walled, rounded chamber anterior to the ventricle and interiorly subdivided into two chambers, right and left)

d. sinus venosus (an elongated, thin-walled sac, lying beneath the atria; it receives blood from the major veins and transmits it to the atria)

In bird and mammal hearts, the sinus venosus is not present as a separate chamber.

After examining the chambers of your frog's heart, label the dorsal and

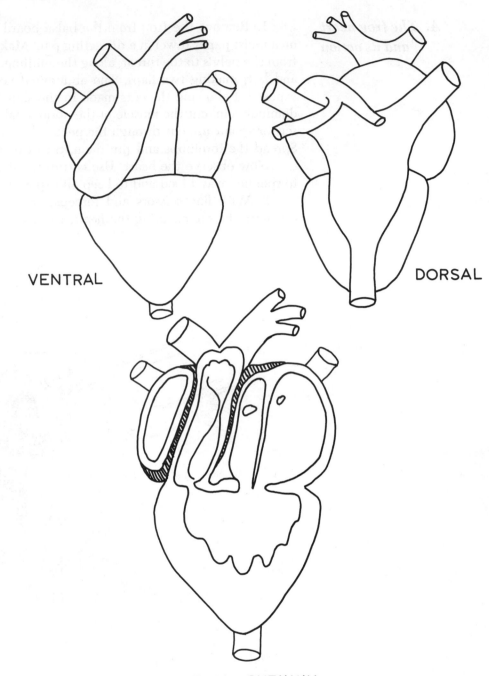

VENTRAL

DORSAL

VENTRAL CUTAWAY

Fig. 9-3. Frog heart, three views.

ventral views of Figure 9-3, indicating sinus venosus, left atrium, right atrium, ventricle, arterial trunk, aortic arches.

3. Watch the heartbeat carefully; it is a regular series of alternating contractions. The two atria contract together; the ventricle follows, and the arterial trunk contracts last. Gently raise the ventricle to see the sinus venosus; it will contract immediately before the atria.

The sequence of events in one heartbeat is as follows. The sinus venosus receives blood from the **venae cavae**, the major veins of the systemic circulation, and passes it on to the right atrium. The left atrium receives blood from the lungs (high in oxygen content) via the **pulmonary veins.** Both atria empty into the single ventricle, from which blood is forced by contraction of the ventricle wall. Because the blood only remains in the heart a short time, the two kinds of blood do not mix to any substantial extent. Venous blood on the right side of the ventricle is nearest the outlet and enters the arterial trunk first. There is partial separation of the streams within the arterial trunk, venous blood tending to go to the pulmonary arteries and oxygenated (pulmonary) blood tending to enter the systemic circulation. There is some mixing of the circulations, and there is thus some difference in efficiency between this type of circulation and that of a mammal, in which there is complete separation of the pulmonary and systemic circuits. Can you suggest a reason why the incomplete separation in amphibians might be functionally advantageous to their way of life?

Label, with arrows, the cutaway ventral view in FIGURE 9-3, indicating the direction of blood flow through the heart.

When you have seen all the structures of the heart and have carefully observed its beat, take this opportunity to review the other internal organs of the frog.

B. The effect of temperature on rate of heartbeat

In vertebrates, a distinction is commonly drawn between cold-blooded animals (fishes, amphibians, and reptiles) and warm-blooded animals (birds and mammals). Because such terms tell us nothing about how the temperature of the blood is regulated, we prefer to use the terms **poikilothermic,** applying to animals whose temperature is regulated by the environment, and **homeothermic,** applying to animals with intrinsic temperature regulation.

1. Count the number of heartbeats of your frog per minute at room temperature. Take the frog's temperature by placing the bulb of a laboratory thermometer carefully under the lobes of the liver.

2. Gently remove the frog on its towel, and cover the bottom of the dissecting pan with crushed ice. Make a frog-sized shallow depression in the ice, and place the frog, still on its towel, into the depression, ventral side up. Quickly insert a thermometer under the liver, bracing it with some object so that you do not have to hold it throughout the experiment, and so that the weight of the thermometer does not interfere with normal heartbeat. Record the heartbeat rate and the temperature at very frequent intervals until the temperature has dropped to 1°C or 2°C. It is very important that the recording be started promptly while the frog is on the ice, and that it be done at fairly rapid intervals, since the frog is poikilothermic and it quickly responds to the cooling treatment.

3. When the frog's temperature reaches 1°C or 2°C (or becomes stationary at 3°C or 4°C for several minutes), remove the thermometer and empty the pan of ice. Reinsert the thermometer as before, and brace it. Record the heartbeat rate and the temperature as the frog's body temperature returns to room temperature.

In your report, plot the frog's heartbeat rate against body temperature, and discuss the results of the experiment with regard to what you already know about the frog's changeable metabolic rate. How might these circulatory responses to cold temperatures be of advantage to the animal?

C. Nervous control of heartbeat

Although the heart is capable of beating in the absence of nerves, the rate of heartbeat can be controlled by two sets of nerves that originate in two different parts of the central nervous system. One set is from the sympathetic nervous system, and the other from the parasympathetic. In time of danger or fear, the sympathetic system prepares the body for "fight or flight" by stimulating accelerator nerves that increase the heartbeat rate. The causative agent is a chemical substance (noradrenalin) released by the accelerator nerves.

The increase in heartbeat rate and in blood pressure would be harmful if prolonged. When the danger is over, the parasympathetic system slows down the heart. Vagus nerves of the parasympathetic system originate in the medulla of the brain and terminate in the sinus node, located on the sinus venosus.

1. The instructor will demonstrate the effects of vagus stimulation. One of the vagus nerves is ligatured (tied with thread) at a point near the brain, and is cut away from the medulla. The nerve may then be lifted out of the body and stimulated electrically, without stimulating any other organ. Ask the instructor what voltage he is using, and how many impulses per second. What are your observations? Stimulation of the vagus causes nerve endings at the sinus node to release the chemical transmitter acetylcholine.

THE EFFECT OF TRANSMITTER SUBSTANCES

2. To study the chemical mediation of the heartbeat rate, it will be necessary to remove the heart. With very fine thread, tie off the veins and arteries, and remove the heart by cutting as widely as possible around the sinus venosus, veins, and arteries.

a. Place the excised heart in a depression slide of Ringer's solution. After recording the normal rate of beat, add three or four drops of 5×10^{-4} M solution of acetylcholine. Record the rate of beat until a perceptible change occurs; if necessary add more acetylcholine solution.

b. Rinse the heart thoroughly by dipping it in fresh Ringer's solution; allow the beat to return to normal. This should occur within a few minutes.

c. Transfer the heart to a second depression slide, to which you have added three or four drops of 5×10^{-5} M adrenalin solution. The effects of adrenalin are very much like those of noradrenalin. Record the heartbeat rate, and carefully watch the nature of the beating. Does adrenalin affect the heart in any way other than rate of beat?

d. Return the heart to the acetylcholine solution, and record the beat rate again.

e. Rinse the heart with Ringer's solution, and wait five minutes for it to return to normal. The effect of adrenalin wears off more slowly than that of acetylcholine.

f. Now place the heart in a third slide with drops of 5×10^{-5} M solution of atropine in Ringer's solution. After thirty to sixty seconds, record the rate of beat. Add two to four drops of acetylcholine, and record again. Describe the results.

(*Note:* If at any time during the experiments the heart fails, try to restore it with gentle massage. If this does not work, attempt restoration by "shock treatment" — 6V at 10 impulses per second from your instructor's stimulator should be enough.)

Acetylcholine exerts its effect principally in the sinus venosus area. Adrenalin affects other cardiac cells as well as those in the sinus venosus area. As you have seen, adrenalin also affects the amplitude (intensity) of contraction. How might you record the amplitude of heartbeat? Atropine appears to inhibit the action of acetylcholine in the manner of an enzyme inhibitor, by combining with the acetylcholine receptor site on the postsynaptic membrane in the sinus venosus.

QUESTIONS FOR FURTHER THOUGHT

1. What is the function and operation of the lymphatic system?

2. Explain the differences between blood flow in the veins and the arteries.

3. Explain the chemical changes occurring at the sinus venosus after stimulation of the frog's vagus nerve.

4. Why does the heart continue to beat after it is removed?

5. On the basis of your experience with the frog heart, suggest some probable difficulties which might be encountered in human heart transplants.

TOPIC **10** THE REGULATION OF BODY FLUIDS

Biological Science, pp. 244-255, 303-307.
Elements of Biological Science, pp. 154-161, 187-191.

Regulation of the osmotic concentration of body fluids is of little importance to the lower marine invertebrates, whose tissues and body fluids are isosmotic to sea water, and to a majority of plants, which can tolerate wide fluctuations in the composition and concentration of body fluids. For most other organisms, the ability to maintain a chemically stable and constant internal environment, usually quite different from the external environment, is absolutely essential to the integrity of the organism and its constituent cells.

As you have already learned, a cell can exercise a substantial degree of control over its internal environment. In part, this regulation is the passive result of physical characteristics of the cell membrane. To a remarkable extent, however, control is exercised through energy-requiring active transport, in which the cell may exclude certain substances and acquire others against normal concentration gradients or at a rate greater than diffusion would allow. Cells themselves require a reasonably constant environment, and we are therefore not surprised to find that natural selection has equipped organisms with a variety of devices to regulate the chemistry of body fluids. Many of these devices exploit the ability of cells to carry out active transport.

In this laboratory you will examine, as representative regulatory devices, the excretory systems of the fetal pig and of a variety of other animals. Because the reproductive system of many vertebrates is rather closely associated with the excretory system, we shall use this opportunity to study the reproductive system of the fetal pig. In the following laboratory, we shall look experimentally at active transport in a living mammal.

One of the principal requirements of animals is a means whereby the by-products of metabolism can be removed from the body. The principal wastes are carbon dioxide, from oxidative metabolism, and ammonia, or some other form of nitrogenous waste, from protein deamination; both are toxic. You have already seen that carbon dioxide is discarded through diffusion from some type of gas exchange surface. In the case of nitrogenous waste, excretion is a more serious problem, since ammonia, the primary product, is highly toxic.

Some small animals can discard ammonia directly by diffusion from the body surface. Others must resort to excreting it in suitably dilute solution, by converting it into some less toxic form, such as urea or uric acid, or by combining these methods. In the great majority of animals, metabolic wastes are poured from individual cells directly into the body fluids; these fluids are then processed by special excretory organs, which release the wastes into the environment. Since most or all of the body fluids are processed by such devices, it is not surprising to find that these organs have often been endowed with the ability to retain selectively those materials useful to the organism, and thus to serve as fluid regulatory systems as well as excretory systems.

**PART I.
EXCRETORY
AND FLUID
REGULATORY
SYSTEMS OF
INVERTEBRATES**
A. *Contractile vacuoles
in protozoans*

These organelles, which you have already seen in several previous laboratories, function primarily to rid the animals of excess water, which enters by osmosis. By flushing water continuously from the cytoplasm, they may function incidentally in ridding the cell of ammonia, but it is doubtful that the contribution to nitrogenous excretion is significant. Carbon dioxide and ammonia produced by a protozoan cell can easily leave the animal by diffusion.

What would happen to the contraction rate of a vacuole if you added a bit of salt to the water in which an *Amoeba* was being kept? In what sort of environment could you expect to find some protozoans without contractile vacuoles?

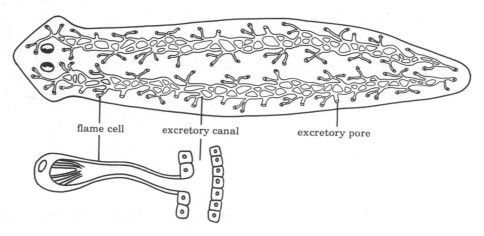

FIG. 10-1. Flame-cell system of planarian. (*Modified from R. Buchsbaum*, Animals Without Backbones, *University of Chicago Press, 1948.*)

B. *Flame-cell systems*

Several primitive animal phyla (Platyhelminthes [flatworms] and Nemertina [ribbon worms]) have excretory systems in which the tissue fluids are collected and driven out of the animal by **flame cells** and associated tubules. FIGURE 10-1 illustrates a flame cell; within it is a group of long flagella whose flickering motion resembles the motion of flames. The individual flame cells are located among the tissues of the worm's body and collect, by diffusion, tissue fluids with dissolved waste materials. Beating of the flagella then serves to drive the fluid into a system of tubules and thence to the exterior.

If worms are available, place one on a white card in a small amount of water; with a razor blade, cut off the posterior third, and make a wet mount in a drop of pond water. A small amount of pressure applied to the cover slip (a clean, dry cork is best for this purpose) will disperse the cells. Examine under low and high power and try to find flame cells; you will recognize them by the characteristic beating of flagella.

As water is constantly entering fresh-water species of these worms by osmosis, and tissue fluids are constantly being drained away by the flame-cell system, how may the animals avoid losing large quantities of solutes that are present in the body fluids?

C. Nephridia   The animals in several phyla with well-developed circulatory systems have evolved excretory organs known as **nephridia.** These remove wastes from both the tissue fluid and the blood, and thus exhibit some properties similar to flame cells and some properties similar to kidneys, which we shall examine later. We shall examine the earthworm as representative of an animal with nephridia.

1. The body of the earthworm is composed of a series of distinct segments, each of which is partitioned internally from adjoining segments by a membrane. Except for a few anterior segments, each compartment thus formed has its own pair of nephridia, which open independently to the outside.

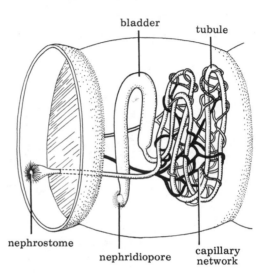

FIG. 10-2. Earthworm nephridium.

2. Refer to FIGURE 10-2, and notice that the nephridium consists of several parts. Fluids first enter a ciliated funnel, the **nephrostome;** the beating of cilia moves the fluids into a long **tubule.** A great many capillaries are located around the tubule, and it is here that exchanges occur between the blood and the nephridium. Wastes are removed from the capillaries and presumably there is some reabsorption of water and other substances from the tubule. The tubule eventually empties into a large distensible **bladder,** with muscular walls. This storage structure, in turn, empties to the outside via a ventrally located **pore.**

3. Note that the nephrostome is located in the segment immediately *anterior* to the one containing the rest of the nephridium. Each nephridium thus picks up tissue fluid from one segment and collects wastes from the blood in the next posterior segment.

4. If living specimens are available, insensitize one in 35% alcohol until it no longer responds to probing (Topic 8, p. 104); dissect as illustrated in FIGURE 8-6, p. 104, pinning the worm near the side of a dissecting pan so that it can be positioned under a dissecting microscope later on. Choose a section near the middle of the worm, arrange the pins so that the membranes between segments are stretched as little as possible, and identify the above-mentioned parts of the nephridium. Careful dissection under the microscope

will yield the whole nephridium (which can be wet-mounted on a slide and examined under the compound microscope) or its parts: the beating of flagella can be seen in the nephrostome, capillaries can be observed in association with the tubule, and parasitic roundworms can often be seen in the bladder.

D. Malpighian tubules   Examine demonstrations of insects with the digestive tract dissected out to show the Malpighian tubules, evaginations of the hindgut. These blind sacs are bathed directly by the blood in the open sinuses of the animal's body. Wastes move across the membranes of the cells composing the tubule walls, and are emptied into the hindgut. Label FIGURE 10-3, a dissection of the cockroach digestive system, taking this opportunity to review the cockroach digestive structures as well as the Malpighian tubules.

What is the primary nitrogenous excretory product of an insect? How is it adaptive?

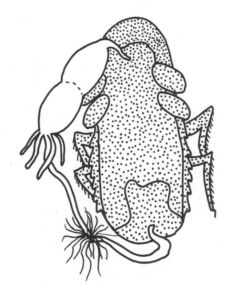

FIG. 10-3. Cockroach digestive system, showing arrangement of organs and placement of Malpighian tubules.

PART II.
EXCRETORY AND
REPRODUCTIVE
SYSTEMS OF
VERTEBRATES
A. Mammals

Since the excretory and reproductive systems often develop together and are otherwise closely associated in vertebrates, the structures are often collectively termed the **urogenital system.**

1. In your fetal pig, locate the large paired **kidneys,** reddish organs attached firmly to the dorsal wall of the peritoneal cavity. Clean the peritoneum away from one of the kidneys, and study it more closely. If possible, locate the **adrenal gland,** a narrow, light-colored body lying along the anteriomedial portion of the kidney; it is sometimes difficult to find in the fetal pig. This gland is part of the endocrine system (What is its function?), but its position makes it convenient to see at this time.

2. At the midpoint of the medial side of the kidney is a depression where blood vessels are attached. Locate the renal artery and renal vein, tracing the former to its origin in the aorta and the latter to its junction with the posterior vena cava. You will note, also, that the **ureter** leaves the kidney and runs posteriorly under the parietal peritoneum. Clean away the peri-

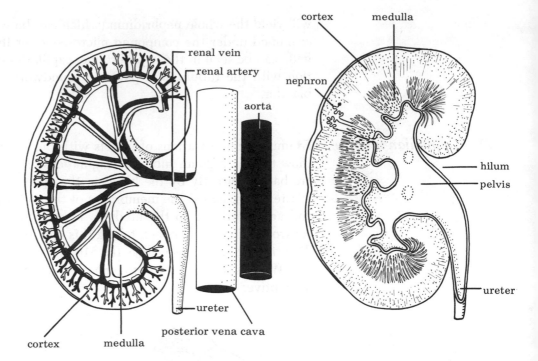

FIG. 10-4. Sections of the human kidney. Left: Kidney's blood supply. Right: Large renal pelvis into which numerous collecting tubules empty.

toneum, and follow the ureter to the **urinary bladder,** which normally lies in the posterioventral portion of the abdominal cavity. In your specimen it will be found on the inner surface of the flap of tissue to which the umbilical cord was attached.

3. As you recall, the vertebrate kidney is very closely associated with a closed circulatory system. Each functional unit or **nephron** of the kidney is composed of a **capsule** enclosing a mass of capillary loops, the **glomerulus,** and a long tubule divided into three relatively distinct regions: the proximal convoluted tubule, Henle's loop, and the distal convoluted tubule. A second capillary network is associated with the tubule. The distal tubules of different nephrons empty into common collecting ducts, which join, rejoin, and finally empty into the **renal pelvis,** a collecting area. The **cortex** of the kidney (the outer part, where the capsules are located), the **medulla** (the inner part through which tubules and collecting ducts pass), and the renal pelvis can all be seen clearly in a longitudinal section of the kidney. Make such a section yourself, halving the kidney, or examine one on demonstration. Compare it to FIGURE 10-4.

Because the nephron is closely associated with the circulatory system, the nephrons need not be distributed throughout the body (as is the case with the nephridia of earthworms, for example), but can be grouped together in the compact kidneys.

4. The **urethra,** which arises from the bladder medially between the ureters, runs posteriorly parallel to the rectum. In the adult, it carries urine to the exterior. Follow the urethra until it passes from view into the ring formed by the pelvic girdle.

5. We shall now examine the reproductive system. As your dissection proceeds, compare the structures in the fetal pig with the human reproduc-

tive organs, illustrated by figures in your text. Although we will not discuss the physiology of reproduction at this time, make certain that you understand the function of each structure studied, and review the hormonal control of reproduction in your text.

If your animal is a female, locate the reproductive structures described in this section, with reference to FIGURE 10-5. If your specimen is a male, examine another student's female specimen. Each student is responsible for knowing the reproductive structures of both sexes.

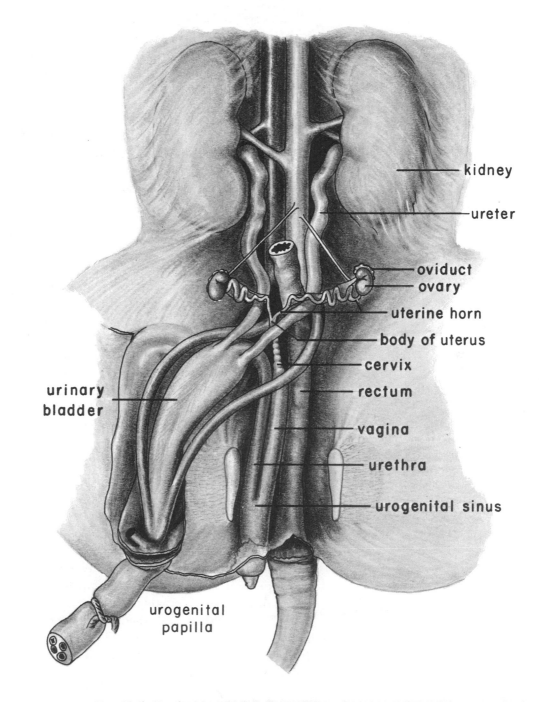

FIG. 10-5. Fetal pig, ventral dissection of female urogenital system.

a. Locate the paired female gonads, the **ovaries,** small bodies suspended from the peritoneal wall in mesenteries, posterior to the kidneys. Examine one ovary closely, and note the small short, coiled **oviduct,** sometimes called the Fallopian tube. The oviduct does not attach directly to the ovary, but ends in an open, funnel-shaped structure that partially encloses the ovary. It is visible only with the dissecting microscope. When ovulation takes place, the eggs are carried by ciliary currents into the mouth of the funnel, and pass down the oviduct to the uterus.

b. In your pig, most of the **uterus,** or womb, is divided into two sections called **uterine horns.** At its anterior end, each of the uterine horns is continuous with an oviduct; at their posterior ends the uterine horns join to form the smaller, median **body** of the uterus. Be careful not to confuse the uterine horns with the oviducts; the latter are much smaller, and are located very close to the ovaries. (In the human female, the uterus is a single, median structure.)

c. The body of the uterus is continuous posteriorly with the **vagina,** slightly larger in diameter than the uterus, and delimited from it by a slight constriction, the **cervix.** The vagina, lying between the rectum and the urethra, disappears from view into the ring of the pelvic girdle.

d. *After* you have identified the urethra, vagina, and rectum as they pass into the ring of the pelvic girdle, cut through the skin and muscle in the midventral line (between the legs; see FIGURE 5-5, p. 59), and then cut very carefully through the bone of the pelvic girdle. With your fingers, spread the cut edges apart and use blunt dissecting instruments to separate connective tissue from the vagina and urethra. Then follow these two ducts posteriorly to the point where they join to form the **urogenital canal;** this canal opens to the exterior through a **urogenital opening.** Bordering the urogenital opening are two small rounded lips or **labia.** Make an incision in the *side* of the urogenital canal, open the canal, and locate the **clitoris,** a small, rounded papilla arising from the inner ventral surface of the urogenital canal, very near the exterior. (*Note:* If your specimen is very small, a dissecting microscope will be necessary in order to see the clitoris clearly.) What is the function of the clitoris, and what is its homologue in a male mammal?

A urogenital canal is absent in adult female pigs and other mammals. As the fetus continues to develop, the urethra and vagina continue to separate posteriorly until finally the excretory and reproductive systems are completely independent of one another. Such independence of the two systems is relatively rare in vertebrates, and is distinctive of female mammals.

6. The structures described in this section refer only to the male; if your specimen is a female, examine another student's male specimen. Refer to FIGURE 10-6.

a. The paired male gonads, the **testes,** are best located by first finding the ducts that carry sperms from the testes to the urethra. In the

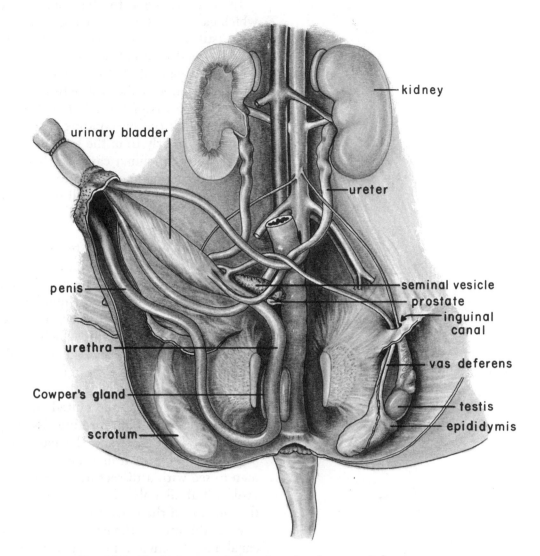

Fig. 10-6. Fetal pig, ventral dissection of male urogenital system.

space bounded ventrally by the urinary bladder, laterally by the ureters, and dorsally by the rectum, you will see two thin ducts that join medially and then attach to the dorsal side of the urethra. Trace these ducts, each called a **vas deferens,** from their medial junction, and note that they diverge and loop over the ureters. Follow one vas deferens posteriorly until it enters a small opening in the posterior peritoneal wall. The passageway into which this opening leads is called the **inguinal canal,** and it carries not only the vas deferens but a number of nerves and blood vessels as well.

b. Each testis first develops in the peritoneal cavity near the kidney, in a position similar to that occupied permanently by the ovaries in the female. As the fetus matures, however, the testes move posteriorly, or **descend,** each passing through the inguinal canal and into a pocket at the base of the hind leg. This pocket, called the **scrotum,** is lined by peritoneum and is actually formed by evagination of the peritoneal cavity.

In your specimen, the testis may be only partially descended, in which case it will be seen lying within the inguinal canal or, in rare cases, still in the peritoneal cavity proper. Whether the testis has fully descended or not, you should next carefully remove the skin at the base of one hind leg. Just beneath the skin, you will notice a white, elongated bulbous structure, covered with a transparent membrane. The testis of the adult will lie in this **scrotal sac.** Carefully clean away all the surrounding tissue (but do *not* cut any structures running anteriorly from the scrotal sac) and cut through the muscles ventral to the inguinal canal so that the full length of the scrotum, testis, and spermatic cord, a prominent cord extending from the anterior end of the scrotum well into the abdominal cavity, is exposed to view.

c. Located on or around the testis, find the **epididymis,** a very tightly coiled tubule, generally lighter in color than the testis itself. The coils of the epididymis serve for temporary storage of sperm cells, received from the testis through a large number of very tiny collecting tubules, too small to see in your dissection. Trace the epididymis to its junction with the vas deferens, which, as you have already seen, runs medially to join with the other vas deferens and then with the urethra.

d. At the point where the united vasa deferentia (the plural form of vas deferens) enter the urethra, locate a pair of small glands, the **seminal vesicles.** Between the bases of the seminal vesicles is the **prostate gland,** which may be developed enough in an older specimen to see with a dissecting microscope; in younger specimens it is visible, if at all, only as a slight swelling around the urethra. What is the function of these glands?

From this area to the exterior, the urethra functions as a urogenital canal, i.e., a common duct for urine and semen. During the passage of seminal fluid, of course, the duct does not also carry urine; it is cleansed by secretions that precede the semen, and any acid that remains is neutralized by substances in the semen.

e. Being very careful to leave intact the skin of the midventral flap, which contains the bladder, and the region posterior to the flap, cut carefully through the skin, muscle, and bone of the pelvic girdle, just as in the dissection of the female. Spread the cut edges, and locate the dorsal rectum and ventral urethra. Find the pair of large, elongate **Cowper's glands,** or bulbourethral glands, lying beside the urethra. These glands, along with the seminal vesicles and prostate, secrete fluids that make up part of the semen.

f. When it nears the posterior end of the body, the urethra turns rather abruptly anterioventrally and runs forward just under the skin of the midventral body wall. It may be felt in that region as a firm tube. Structures surrounding the urethra will undergo further development, and the complex forms the **penis.**

B. Amphibians You have seen that the excretory and reproductive systems of mammals share a common duct to the outside at least during development, and that this condition persists in the adult male. In many lower vertebrates, including the frog, the two are even more intimately united.

MALE

(FIGURE 10–7)

a. Under the dissecting microscope, carefully examine a dissection of a male leopard frog. (If you carry out the dissection yourself, refer to p. 75 for directions in opening the abdominal cavity.) Note the pair of reddish kidneys, each with a light-colored adrenal gland lying along its ventral surface. The ureter, running posteriorly from the kidney, is easily seen. Note that in the frog, the ureter does not empty directly into the urinary bladder as in mammals, but that it joins the dorsal surface of the **cloaca,** a posterior chamber of the digestive tract. Urine thus flows from the ureter into the cloaca, then into the bladder for storage, and finally back into the cloaca and out the opening leading from the cloaca to the exterior. What other types of vertebrates show a cloaca, rather than separate urinary, genital, and digestive ducts?

b. Find the yellowish testes, lying close to the kidneys. In lower vertebrates, the testes remain in the abdominal cavity, and no scrotum is present. You may be able to see a series of small ducts that carry sperms from each testis into the kidneys.

It is apparent that in the male frog the union of excretory and reproductive systems is much more complete than in the mammal. The duct we have called the ureter is actually a common urogenital duct, and often it is referred to by a special name: the **Wolffian duct.** From studying comparative anatomy and embryology of amphibians and mammals, we know that the mammalian ureter is a new duct, and that the old Wolffian duct, through evolution, has become the epididymis and vas deferens.

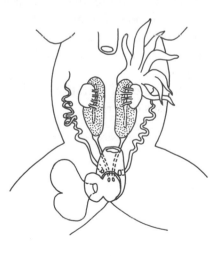

FIG. 10-7. Male frog, urogenital system.

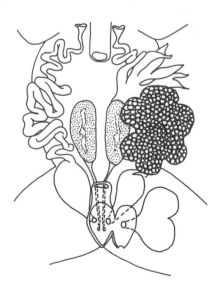

FIG. 10-8. Female frog, urogenital system.

c. Rudimentary **oviducts** can be seen in some male leopard frogs. Female organs are seldom so evident in adult male vertebrates of other species, but the structures of both sexes do usually begin to develop in all individuals during early embryology; one set continues to develop while the other retrogresses and disappears before or shortly after birth.

FEMALE
(FIGURE 10–8)

2. Examine a dissection of a female leopard frog, and quickly locate kidneys, adrenals, ureters, cloaca, and bladder, all of which are like those already seen in the male. Examine the large ovaries, each containing a very large number of eggs. Ripe eggs are released from the ovaries into the body cavity, and are then carried into the open ends of the oviducts, located far forward in the body cavity. The two oviducts empty directly into the cloaca. Apart from the cloaca, the excretory and reproductive systems of the female frog do not utilize the same ducts.

QUESTIONS FOR FURTHER THOUGHT

1. What is the function of testis descent in a mammal? Why should the testis not descend in a frog?

2. What happens to ammonia and other nitrogenous wastes after they are excreted by animals?

3. You have seen that in male mammals and amphibians the urinary and genital systems share common ducts along much of their length. What selective advantages would there be to having *separate* tracts in the females?

4. What structure in the pig is functionally comparable to the Wolffian duct in the frog?

5. What do kidneys and nephridia have in common? Why must the insect have an excretory system that operates on an entirely different principle?

6. Contrast osmoregulatory adaptations of fresh-water bony fish with those of marine bony fish and with those of sharks.

7. Discuss the process of urine formation in man, explaining
   a. the source of urea and where the urea content of the blood is highest
   b. the roles of filtration and reabsorption, and where each of these processes occurs.

Biological Science, pp. 59-61, 255-259.
Elements of Biological Science, pp. 44-46, 160-163.

Where diffusion would be too slow, or where substances must move against a concentration gradient, natural selection has exploited the ability of cells to carry out active transport. As you already know, energy-requiring transport processes are extremely widespread in organisms, ranging from mineral uptake in roots and selective reabsorption in the kidney tubules, to the "sodium pump" that maintains the electrochemical potential difference necessary for the proper functioning of nerves and muscle fibers. While the mechanism of active transport is not well understood, we do know that it always involves the expenditure of ATP, and that it often discriminates in favor of certain molecules. Amino acids, for example, exist in two configurations, differing only slightly in the spatial arrangement of atoms; the hamster intestine, in which you will presently study the transport of glucose, actively transports into the blood one of these forms (the L-configuration) but not the other (the D-configuration). Such specificity suggests the presence of some "carrier" molecule able to make the discrimination, but that is about all we know at present.

The absorptive surface of the hamster small intestine, like that of many other animals, is greatly augmented by large numbers of **villi,** fingerlike projections of the intestinal wall. Each villus, furthermore, is lined with cells that constitute the intestinal **mucosa;** each mucosal cell (FIGURE 11-1) further

FIG. 11-1. Electron micrograph of hamster intestine showing microvilli projecting from the border of several epithelial cells; notice the numerous mitochondria. (*Courtesy E. W. Strauss, Brown University.*)

increases the surface area for absorption with numerous tiny projections (**microvilli**) from its surface. Almost all the absorption of food materials, including glucose, amino acids, fatty acids, and glycerol, occurs across the membranes of these mucosal cells; behind the mucosa, there are numerous capillaries and lymph vessels. Food is carried in the blood to the liver, where it may be stored and released to the body as required.

If the transport of digestive products depended only upon diffusion, their level in the blood would never exceed their level in the small intestine, and much food would be lost. In fact, the concentration of glucose and other digestive products in the blood may substantially exceed the intestinal concentration. The difference is brought about and maintained by active transport.

The procedure you will use today was designed by T. H. Wilson of the Harvard Medical School and Dr. G. Wiseman of the University of Sheffield (England). It involves cutting the hamster intestine into sections of workable length, everting the sections so that the mucosal side is outermost, and then filling the sections with a solution of glucose in Ringer's solution. The sacs are then incubated in the same glucose-Ringer's solution; after incubation, the levels of glucose inside and outside the sacs are measured and compared. During the incubation interval, active transport will have moved glucose across the mucosa into the new "lumen" of the intestinal sacs. This basic procedure may be modified in various ways to demonstrate a requirement for ATP or for oxygen, or to demonstrate the selective transport of the L-amino acids.

Read through the entire experiment carefully before you begin work. In FIGURE 11-2, identify all the equipment as you read through the directions. During the incubation, the instructor may wish to discuss the experiment or show you a film. In the time you have left, make sure that the equipment for making glucose analyses is operating properly. If the instructor does not demonstrate the method of analysis, carry out several trial runs on samples of the glucose-Ringer's solution and on dilutions to make sure that you understand how the analyses work. Students should work in pairs throughout the experiment; in the first two sections, one student can operate while the other makes sure that instruments, solutions, and other equipment are made ready for the operator.

STEP 1. REMOVAL OF TISSUE FROM EXPERIMENTAL ANIMAL (FIGURE 11–2)

Each team will be provided with a freshly killed hamster. Open the abdomen with a midline incision, and extend it laterally to give yourself more working room. Quickly identify the small intestine and its upper and lower limits (the pylorus, and the region near the caecum).

Cut partway across the duodenum near the stomach; insert the cannula and tie it firmly in place with thread. Run warm 0.9% saline solution from the burette through the cannula into the small intestine. Watch the flow through the intestine, and when fluid appears in the caecum, stop the flow of saline. Then cut across the intestine just above the caecum, and allow more saline to rinse through the gut. Lifting the intestine gently near the caecal end, strip the gut of its mesentery up to the end of the duodenum; remove the intestine, and place it in a petri dish of *warm* saline. Wrap the

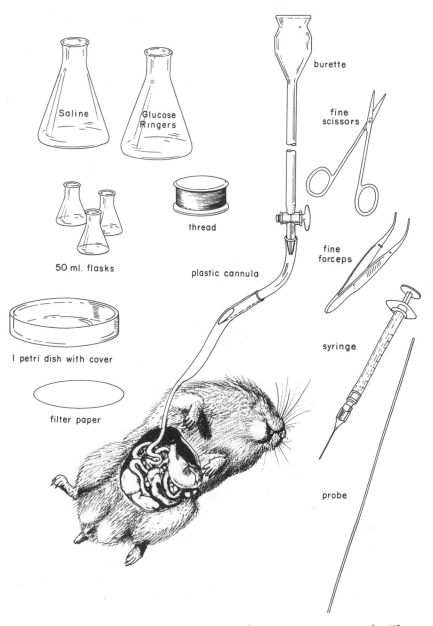

Fig. 11-2. Hamster dissection and equipment for the active-transport study. The intestine has been cut across the duodenum and a plastic cannula has been inserted.

animal in a paper towel and set it aside for a quick review of the internal anatomy when you have time later in the period.

STEP 2.
PREPARATION OF
INTESTINAL SACS

Your instructor will show you a film demonstrating the manner in which the intestinal sacs are prepared, or he may demonstrate the technique himself. After you have seen the demonstration, you will be ready to follow the outline below.

1. Blot the *lowest* centimeter of the intestine (that farthest from the stomach) on filter paper, and push it back through the lumen with your spe-

cial probe. When it appears at the other end, advance the tissue along the probe until the entire gut is everted. Remove the tissue by pulling it carefully off the probe, and place it at once in fresh, warm saline.

2. The gut is now to be cut into three or four sections, depending on the length of intestine available, and each section made into a sac. Tie a tight thread around the intestine about 3-5 cm from the duodenal end (about one-quarter the length of the whole gut); cut the tissue distal to the thread. Repeat this procedure until you have several such sections, each closed at one end by a thread ligature.

3. Each section will now be filled with the glucose-Ringer's solution and tied off; it is important that the sacs not be overdistended with solution, so follow the directions carefully.

Holding a short section of tissue by its thread, blot it gently on filter paper to remove adhering fluid, and place the tissue on an inverted petri dish (FIGURE 11-3). Fill a 1-ml syringe with glucose-Ringer's solution, and tie a thread loosely over the barrel. Holding the open end of the tissue with

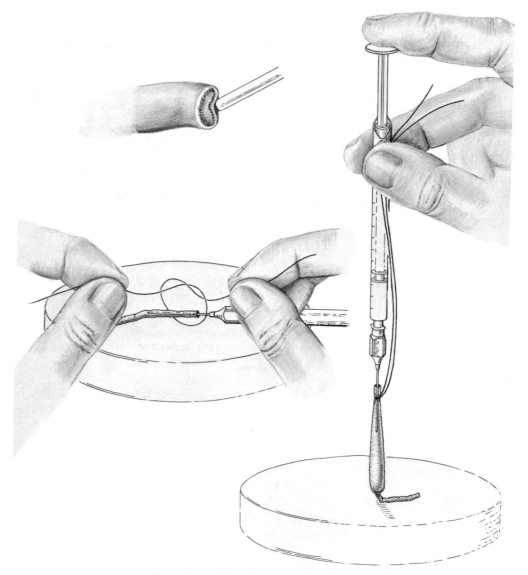

FIG. 11-3. Steps in preparation of intestinal sacs.

forceps, introduce the blunt needle into the sac. Rest the syringe on the edge of the petri dish, pull the thread over the tissue and needle, and draw tight.

With one hand, hold both the syringe barrel and the two ends of the ligature, and lift the sac off the dish, allowing the tissue to hang freely. Inject 0.5 ml - 1.0 ml of glucose-Ringer's solution until the sac is filled, but *not* distended. If overdistension is noted at this stage, withdraw a small amount of fluid into the syringe. Slide the sac off the needle and pull the thread tight. Trim any large "ends" of tissue or thread, and place the sac in a 50-ml Erlenmeyer flask, containing 5 ml of glucose-Ringer's solution.

Repeat this procedure until the other sacs have been prepared; each should be placed in a separate flask.

You have now prepared several sacs, each containing an isosmotic solution of salts and 10mM glucose. Each has been placed in a flask of the same solution. The osmotic concentration difference between the mucosal (outer) side of the sac and the serosal (inner) side of the sac is thus zero. If active transport occurs, this relationship will be altered, and can be detected by analyzing the glucose concentration on each side of the sac.

4. To test the requirement in active transport for ATP, add a drop of dinitrophenol, an inhibitor of cellular respiration, to one of the flasks, and make sure it is labeled so that it can be clearly differentiated from the others.

5. Each of the flasks should now be aerated and incubated according to one of the methods below.

STEP 3. AERATION AND INCUBATION
Option A.

Fit each flask with a two-hole stopper. Connect one of the openings to an air line, and bubble air through the flask at room temperature for one and a half hours. The air stream should provide some circulation of the flask fluid, but should be gentle enough so that the tissue will not be damaged.

Option B.

Fit each flask with a two-hole stopper, and run oxygen from a small tank into one of the holes for thirty seconds, under pressure to produce a flow rate of 2 liters per minute. Promptly stopper the flask, or plug the holes, and incubate at 37°C with shaking for one hour, or at room temperature with shaking for one and a half hours.

While the sacs are incubating, gather the materials for the glucose analyses, carry out trial runs, examine the hamster to review the internal anatomy of a mammal, and follow any other directions given by the instructor at the beginning of the period.

STEP 4. COLLECTION OF SAMPLES

After incubation, remove the stoppers from each flask and pour the solution from each one into a separate test tube, which you have previously labeled. With forceps, pick up each sac by one of its threads. Touch the dangling end *once* with filter paper to remove excess fluid, and hold the end of the sac just inside the lip of a small test tube. *Carefully* cut the end with scissors, and drain the sac contents for fifteen to thirty seconds. Keep the samples from each flask separate, and be especially careful to label those tubes used with the dinitrophenol-treated tissue.

STEP 5.
GLUCOSE ANALYSIS

Compare the glucose concentrations of sac contents and incubation solution by one of the following methods, as directed by your instructor.

Option A.

If a colorimeter is available, the glucose-oxidase method of analysis may be used. In the presence of this enzyme, glucose is oxidized to gluconic acid and hydrogen peroxide. The peroxide reacts with a hydrogen donor in the presence of peroxidase to give a yellow-orange product, the concentration of which may be estimated colorimetrically.

1. Pipette 0.10 ml of final serosal solution, final mucosal solution, and initial glucose-Ringer's solution into separate tubes.

2. Add 0.9 ml water to each tube. Use 1.0 ml water in a fourth tube as a "water blank." The initial solution, containing 10mM glucose, may be used as a standard.

3. Now, add 5.0 ml of the glucose oxidase reagent to each successive tube, making the additions one minute apart.

4. Exactly ten minutes after the reagent was added to the first tube, add 1 drop of 4 N hydrochloric acid, to stop the reaction. Repeat this procedure with the other tubes, adding 4 N HCl to each, at one-minute intervals, until all tubes have been completed.

5. After five minutes, the color may be read in a colorimeter at a wavelength of 400mμ. The colored product is stable for a few hours.

Option B.

If a colorimeter is not available, somewhat less accurate estimations of concentrations may be made by reacting a small amount of each solution to be tested with Benedict's reagent. The results are compared with known glucose standards.

In the Benedict's reaction, blue cupric ion (Cu^{++}) is reduced to red cuprous ion (Cu^{+}) by the aldehyde group of the sugar. The cuprous ions react with hydroxyl ions to yield cuprous oxide, a red product. The color obtained varies with the amount of glucose originally present: a clear blue solution indicates no glucose, a deep red indicates high glucose concentration.

To test a solution for glucose concentration, place three drops of the solution in a small test tube, add 2 ml of Benedict's reagent, and place the tube, properly labeled with masking tape, in a boiling water bath for five minutes. Then observe the color.

Follow this procedure for a sample of the solution inside each intestinal sac, a sample of the external solution for each, a sample of the initial external solution, and "standard" comparison tubes containing 0, 1, 2, and 3 drops of the standard glucose-Ringer's solution.

REPORTING THE
RESULTS

Organize the results into a table, and prepare diagrams, appropriately labeled to indicate the direction and extent of glucose transport in each sac.

ADDITIONAL
EXPERIMENTS

If time and materials are available, this experiment can be made more quantitative, or can be modified to demonstrate the hydrolysis of starch in the upper jejunum (vs the ileum) and to demonstrate selective transport of L-tyrosine. Consult the instructor's guide accompanying this book for details.

12 VERTEBRATE COORDINATION

Biological Science, Chapter 10.
Elements of Biological Science, Chapter 10.

From your previous work with the fetal pig, it should be clear that the vertebrate body is an association of extremely complex and intricate parts. Some of these parts, such as the skeleton and its associated muscles, expedite voluntary movements; others, such as the respiratory and circulatory systems, carry on their functions involuntarily. While each system of the body has a specific contribution to make, all must function together as a coordinated whole. Such integration depends upon the rapid exchange of information between widely separated regions of the body, and in this process the nervous system plays a vital role. In this unit you will examine the development and structure of the vertebrate nervous system, and conduct some experiments to determine the role of the central nervous system in communication within the organism.

PART I. STRUCTURE OF THE NERVOUS SYSTEM

For purposes of discussion, the nervous system is conveniently divided into three major sections. The **central nervous system,** or CNS, is composed of the brain and spinal cord; the **peripheral nervous system** constitutes all nervous connections outside the CNS, and the **autonomic nervous system,** a subdivision of the peripheral, regulates involuntary functions.

A. Central nervous system

Examine the skull and spinal column of the vertebrate skeleton on demonstration. The brain lies inside the cavity of the skull, and the spinal cord runs from the base of the brain down inside the spinal column. Paired **spinal nerves,** containing nerve fibers that bring information from and carry it to all parts of the body, leave the spinal cord at intervals through openings in the **vertebrae.**

DEVELOPMENT OF THE BRAIN

1. Examine on demonstration a two-day-old chick embryo, referring to FIGURE 12-1. The brain is its most prominent structure, and has three distinct parts, separated by constrictions. The **midbrain** is the most anterior because the brain has doubled back on itself during development. The **forebrain** can be identified posterior to the midbrain and adjacent to the developing eye, the optic cup. The **hindbrain,** consisting of several lobes, blends into the spinal cord, which can be traced posteriorly in the longitudinal axis of the body. These parts of the brain differ somewhat in anatomy and in function in different vertebrate groups, but early development is very much alike in all vertebrates. Which of the parts becomes dominant in man?

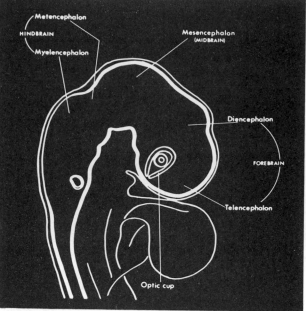

Metencephalon

HINDBRAIN

Myelencephalon

Mesencephalon
(MIDBRAIN)

Diencephalon

FOREBRAIN

Telencephalon

Optic cup

FIG. 12-1. Outline of 48-hour chick brain. (*Courtesy Carolina Biological Supply Company.*)

MAMMALIAN BRAIN

2. You will be provided with a mammalian brain in a container of water. Although it was previously fixed in 10% formalin, it has been thoroughly rinsed by your instructor. Be very careful in handling your specimen, since it is extremely delicate.

As you examine this mammalian brain, compare it with FIGURE 12-2, illustrating adult brains of other vertebrates, and with FIGURE 12-3, showing a longitudinal section through the mammalian brain.

Apart from tracts that conduct impulses through its various regions, the brain contains centers, each of which dominates regulation of one or more major activities in the body.

a. The most posterior part of the hindbrain, part of which may be missing in your specimen, is the **medulla,** which controls a number of important reflexes, such as coughing, sneezing, blinking, and swallowing, and contains centers for the parasympathetic control of respiration, heartbeat, visceral movement, and glandular secretion. Notice numerous nerves leaving the medulla. The **cerebellum** is located just anterior to the medulla on the dorsal side of the hindbrain; it maintains equilibrium of the body by correlating information from kinesthetic ("movement detector"), touch, pressure, and static receptors.

b. The small midbrain, front part of the brain "stem," is best seen in the longitudinal section, FIGURE 12-3; it processes information from the eye and ear.

c. The **pituitary gland,** which exercises control over most of the other glands in the body, hangs from a portion of the forebrain called the **hypothalamus;** the pituitary may have been damaged or removed when the brain was prepared for preservation. The largest and most anterior part of the forebrain is the **cerebrum,** consisting of halves that have almost completely overgrown the rest of the brain. Convolutions of the cerebrum provide a great surface area for neuron cell bodies, located in the outermost layer or **cortex** of the cerebrum.

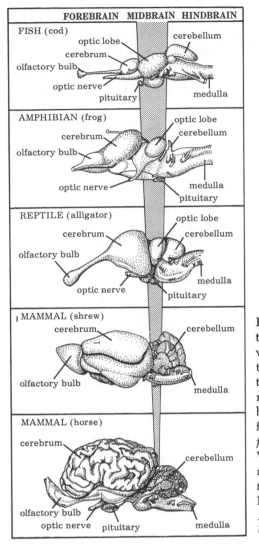

FORToREBRAIN MIDBRAIN HINDBRAIN

FISH (cod)
optic lobe
cerebrum
olfactory bulb
optic nerve
pituitary
cerebellum
medulla

AMPHIBIAN (frog)
cerebrum
olfactory bulb
optic nerve
optic lobe
cerebellum
medulla
pituitary

REPTILE (alligator)
cerebrum
olfactory bulb
optic nerve
optic lobe
cerebellum
medulla
pituitary

MAMMAL (shrew)
cerebrum
olfactory bulb
cerebellum
medulla

MAMMAL (horse)
cerebrum
olfactory bulb
optic nerve   pituitary
cerebellum
medulla

FIG. 12-2. Evolutionary change in relative size of midbrain and forebrain in vertebrates. In the sequence from fish through amphibian, reptile, and primitive mammal (shrew) to advanced mammal (horse), the relative size of the midbrain markedly decreases, while the forebrain expands enormously. (*Modified in part from A. S. Romer,* The Vertebrate Body, *Saunders, 1962; and in part from G. G. Simpson, C. S. Pittendrigh, and L. H. Tiffany,* Life: An Introduction to Biology, *copyright © 1957 by Harcourt, Brace & World, Inc. Used by permission of the publishers.*)

What is the function of the cerebrum in a mammal? Is it the same in other vertebrates? Beneath the front edge of the cerebrum, locate the **olfactory lobe.** What is its function?

SPINAL CORD   3. a. Examine a dissected spinal cord on demonstration. Note the shape

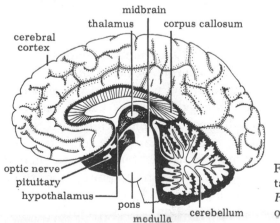

midbrain
thalamus   corpus callosum
cerebral
cortex

optic nerve
pituitary
hypothalamus
pons
medulla   cerebellum

FIG. 12-3. The human brain in sagittal section. (*Modified in part from S. W. Ranson and S. L. Clark,* The Anatomy of the Nervous System, *Saunders, 1959.*)

of the cord, and identify two slightly enlarged regions, the cervical and lumbar enlargements, reflecting the increase in neurons required to supply the fore- and hindlimb muscles.

b. Dissect the skin, muscle, and bone away from a portion of your fetal pig's spinal cord (by using a razor blade to shave away tissue down to the cord, and then by blunt dissection). Do your dissection under a dissecting microscope, and identify dorsal and ventral roots of a spinal nerve (see description in B below).

c. Also examine a cross section of the spinal cord. Locate the points of attachment of spinal nerves, and study one of them carefully, noting dorsal and ventral roots of the nerve. Which root carries afferent impulses, and which efferent impulses? The inner part of the cord is an H-shaped area containing gray matter, the nerve cell bodies; outside the gray matter is the white matter, composed of nerve fibers, extending lengthwise and also branching off into spinal nerves. In the center of the cord is a small **central canal,** which contains spinal fluid and is continuous anteriorly with the **ventricles,** fluid-filled cavities in the brain.

The dorsal part of the gray matter in the spinal cord contains cell bodies of association neurons, and the ventral part contains the cell bodies of motor neurons.

NEURONS IN THE CNS

4. Examine your slide of giant multipolar neurons. These slides are made by staining smears of cells from the ventral horn of the spinal cord of an ox. Locate one of the large bodies, and note the many processes running from it. Make a sketch of one of these neurons.

B. Peripheral nervous system
SPINAL NERVES

1. Twenty to thirty pairs (or more, depending upon the vertebrate in question) of spinal nerves leave the spinal cord between the vertebrae. Each of them has a **dorsal root** carrying sensory (afferent), and a **ventral root** carrying motor (efferent) impulses. These roots can be located on a slide of a cross section or may be seen in dissection, as noted previously. The bulge on the dorsal root, the dorsal root ganglion, contains the cell bodies of sensory neurons. Wherever nerve cell bodies aggregate outside the CNS in this manner, we speak of them as forming a **ganglion.**

OTHER NERVES

2. The specialized, impulse-conducting cells that form the nervous system are called **neurons** or nerve fibers. FIGURE 12-4 illustrates three major types of neurons, and will remind you that each consists of a **cell body** and variously branched processes called **axons** and **dendrites.** Dendrites are processes that conduct impulses toward the cell body, while axons usually conduct impulses away from the cell body. Neurons commonly have a single axon and many dendrites, but there are many exceptions to that rule.

Sensory neurons carry impulses from the organs of special sense to the CNS; motor neurons conduct impulses toward the muscles and other effector organs; and association neurons conduct impulses between the sensory and

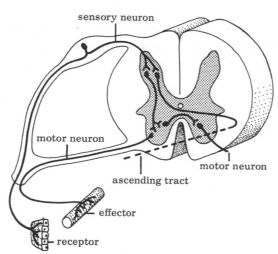

FIG. 12-4. Diagram of a reflex arc with association neurons. The sensory neuron synapses with several association neurons in the gray matter of the spinal cord. Some of these association neurons may synapse directly with motor neurons on the same side, but some cross to the other side of the cord and there synapse with the other motor neurons and with additional association neurons that run in ascending tracts through the cord to the brain.

motor fibers, usually within the CNS.

Nerves are collections of neurons into tracts of substantial size.

a. Examine a slide showing a cross section through a peripheral nerve. Note the many separate fibers (neurons) of which it is composed, and the dark myelin sheath around each fiber. What is the function of the myelin sheath?

b. Examine a slide showing the **neuromuscular junction,** that area where a neuron and the group of muscle fibers innervated by it connect. The axon of a motor neuron divides into a number of fine terminal branches, each ending as a **motor end plate** on the muscle fiber. Transmission of the impulse across the cleft of the neuromuscular junction is mediated by chemicals such as acetylcholine and adrenalin.

PART II. THE REFLEX ARC

Some stimuli cause rapid responses that come about involuntarily, and often without the intervention of the higher centers of the brain, although in many cases those centers are informed and may modify the response. These automatic responses are called **reflexes.** The sensory impulses that initiate a spinal reflex enter the spinal cord through the dorsal root; within the cord, they usually synapse with association neurons, which in turn transmit the impulses to motor neurons, and thence to some effector, such as a muscle. The whole pathway, from sense organ through the CNS to effector, is called a **reflex arc** (FIGURE 12-4).

As you examine some reflexes in a frog, do not forget that spinal reflexes, though they do not *require* the intervention of the brain, may nevertheless be modified by it, and that most reflexes are not nearly as simple as we have outlined above.

(*Note:* The following exercises require several frogs per student group. If frogs are in short supply, students can work in reverse order on the experiments, and can work in groups of up to four.)

Pith a frog in the usual manner, but only anteriorly, leaving the spinal cord intact. (Directions for pithing are in Appendix 5.) This procedure destroys the brain and separates it from the spinal cord; presumably, the frog has no sensations and can initiate no voluntary actions.

A. Coordination
reflexes
POSTURE

Place the frog on a table. Does it assume, or try to assume, a squatting posture? Do the hind legs position themselves under the body when pulled straight?

JUMPING

Try to stimulate the frog to jump. Do you have any success?

RIGHTING

Place the frog on its back. Does it make any attempt to right itself?

BALANCE

Place the frog on a board. Tilt the board. Does the frog respond at all?

SWIMMING

Does the frog swim when placed in deep water? Does it make any attempt to get out of the water?

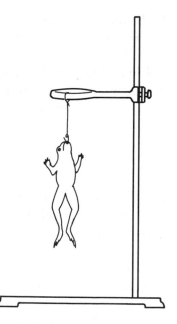

FIG. 12-5. Manner in which frog is hung from ringstand.

B. Response to
noxious stimuli

Hang the frog from a ringstand (FIGURE 12-5), so that the legs are allowed to hang freely. One way of accomplishing this is to put a safety pin through the frog's lower jaw, using a string to attach the pin to the ringstand clamp.

1. Use forceps to pinch one of the toes. What is the result? Try increasing the severity of the stimulus; what are the effects?

2. Pinch various parts of the body. Are the responses localized with reference to the stimulus? Are these responses predictable (i.e., Will the same stimulus always yield the same response)?

3. Apply some 10% acetic acid to one of the toes with a cotton swab. Are the responses the same as the ones from pinching? Try 25% acid, being

sure to rinse well by dipping the toe in water between renewed applications. What are the results?

4. Soak a small piece of filter paper (½ x ½ cm) with 25% acetic acid, and apply it middorsally to the skin of the pelvic region. What are the responses? Try the same technique on other parts of the body, washing off the acid with a cotton swab soaked in water, between applications. Can these movements be accurately predicted? What happens when one of the responses is interfered with (as, for example, by holding one of the legs)?

C. Control by the cerebellum and medulla

Place a refrigerated frog on crushed ice until it no longer moves about. Locate the tympanic membranes (the brown, circular membranes behind the eyes); what is their function? Using the anterior margin of the two membranes as a guideline, cut off the anterior portion of the skull with a pair of heavy scissors. Leave the lower jaw attached. You have removed the fore- and midbrains as shown in FIGURE 12-6.

After ten or fifteen minutes, introduce the frog to the reflex situations of A and B above, and record the results. What can you conclude about the site of control of coordination reflexes? Of the site of control of the reflexes in B above?

FIG. 12-6. Frog, showing cut to remove fore- and midbrain.

D. Functions of the midbrain and cerebrum

Place an unoperated frog in the same situations as the other two frogs (except part B above), and thereby determine the general additional functions of the midbrain and cerebral hemispheres. Are the responses always predictable?

PART III. SENSE ORGANS

Neurons can be stimulated to "fire" by almost any kind of treatment: electric shock, chemical agents, cutting, pinching, etc. Under normal conditions, sensory neurons are activated by the sense organ to which they are connected. A sense organ is itself activated by some type of environmental change in energy, a **stimulus,** and it somehow causes attached neurons to fire when it is so stimulated. The neurons then transmit the information to the CNS. The sense organ itself does not interpret stimuli as sensations, although it may subject the stimuli to filtering or other selective processes before information is passed on to the CNS. Interpretation takes place and response is initiated in the CNS. Interpretation is a function of the region of the CNS to which the impulses are transmitted, not of the type of stimulus or the type of receptor.

We shall investigate one important property of sense organs, and shall dissect and examine a representative receptor.

A. Adaptation

Under conditions of continuous stimulation, most types of receptors cease to become excited, or their rate of transmission of impulses to associated neurons declines. This phenomenon, which helps to reduce the transmission of useless information to the CNS, is called **adaptation.** Proprioceptive sense organs — those concerned with the transmission of information about muscle tension, posture, etc. — are generally slow-adapting; if this were not so, we would not be able to stand up or to balance ourselves without continuous conscious effort. Exteroceptive sense organs — those that supply information about events at the body surface, such as heat, cold, touch, etc. — generally adapt very rapidly; if they did not, impulses from all parts of the body, mostly useless information, would impinge continuously upon the CNS.

Test the adaptation of heat and cold receptors in the following manner:

1. Set up three beakers, one with cold water, one with tolerably hot water, and one with water at room temperature.

2. Place one index finger in the cold water and the other index finger in the hot water. Report your initial sensations, and sensations after several minutes. Before removing your fingers from the hot and cold water, read B below.

B. Identifying the stimulus required for "hot" and "cold" sensations

Move the two fingers together into the beaker of water at room temperature. Note that although the stimulus is superficially the same, the sensation in each finger is different. Are the stimuli *really* identical for each finger?

Receptors in the cold finger respond to heat gain, whereas receptors in the hot finger respond to heat loss. Does this experiment tell you whether the receptors for these stimuli are the same or different?

C. The eye: a representative sense organ

1. Obtain a sheep or cow eye in a dissecting pan; locate the brownish stubs of muscles that controlled the position of the eyeball and the white **optic nerve.** With scissors, cut the eyeball in approximate halves, in a plane parallel to the front surface.

2. Excluding fat on the surface, note that most of the wall of the eye is composed of three layers (FIGURE 12-7):

a. A white, tough outer **sclera,** specialized anteriorly as the transparent **cornea;** it may be milky because of preservation.

b. The black **choroid;** it contains numerous blood vessels supplying the retina. (What is the function of the black color of the choroid?)

c. The innermost, membranous gray **retina,** containing the light-sensitive cells.

3. Locate the biconvex **lens** lying in the interior of the eye toward the front. In life, the lens is transparent and acts as the focusing device. The cavity anterior to the lens is filled with **aqueous humor,** while the cavity between the lens and the retina is filled with the more viscous **vitreous**

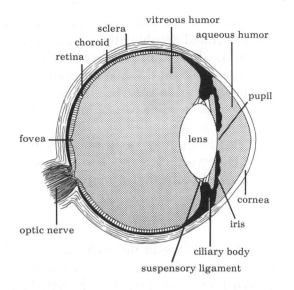

vitreous humor
sclera
aqueous humor
choroid
retina
pupil
fovea
lens
cornea
optic nerve
iris
ciliary body
suspensory ligament

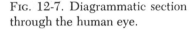

Fig. 12-7. Diagrammatic section through the human eye.

humor. What is the function of these fluids? Remove the lens and fluids from the eyeball. If you break the lens apart with your fingernail, you will find that it is composed of many fiberlike cells packed closely together.

4. Examine the choroid layer more closely. You will see that toward the front of the eyeball, near the lens, the choroid forms a thickened, ring-shaped, radially striated muscular area, the **ciliary body,** that curves inward to enclose the lens. The ciliary body is the point of attachment of the suspensory ligament, a thin membrane attached to the lens around its equator. During the process of accommodation, contraction of the muscle fibers in the ciliary body reduces the pull on the suspensory ligament and thus on the lens, thereby allowing the naturally elastic lens to assume a more rounded shape, which brings nearby objects into better focus. Conversely, when the muscle fibers in the ciliary body relax, they elongate, thus exerting great pull on the suspensory ligament and on the lens, and pulling the lens into flatter shape, better adapted to distance vision. In older individuals, the lens often loses its original elasticity, and accommodation diminishes. Older persons then need corrective lenses for near vision.

5. Continue to follow the choroid layer, past the ciliary body and forward. It continues as the **iris,** which turns inward, leaving the wall of the eye and covering part of the front of the lens. The iris is perforated immediately in front of the lens; the hole in the iris is called the **pupil.** The iris has muscle fibers, some circular and others radially arranged. The iris can thus act as a diaphragm regulating the size of the pupil and, in turn, the amount of light admitted to the retina.

6. Examine the rear half of the eye more carefully. Note that the pale, loose retina is attached to the wall of the eye at the **blind spot,** where the sensory fibers from the retina join the optic nerve, and where blood vessels pass into the eye. Beneath the rear portion of the retina, the choroid layer is covered by a shiny greenish-blue material. This is the **tapetum,** a reflecting area that sends incident light back through the retina, thus making more efficient use of the light. There is no tapetum in the human eye, but the device is common in nocturnal terrestrial vertebrates and in fishes that live far below the surface. It is the tapetum that causes the so-called "glowing" eyes of some animals at night.

PART IV.
ELECTRICAL
CHANGES IN
CONDUCTIVE
TISSUE

An oscilloscope is an instrument capable of picking up electrical stimuli from an object and graphing them on a screen that closely resembles a television screen. Changes in the electrical input to the instrument are viewed as deflections of a light trace on the graphlike screen. An oscilloscope can be a very valuable tool in studying the electrical changes that take place in a muscle or nerve or other part of an organism under specified experimental conditions.

The instructor may carry out a demonstration of electrical changes occurring in nerve, muscle, or other tissues, and he will explain the peculiarities of the oscilloscope that is used.

QUESTIONS FOR
FURTHER
THOUGHT

1. What are some of the evolutionary tendencies in the organization of the nervous system from protozoans to mammals?

2. What is the function of the hypothalamus?

3. Explain polarization reversal.

4. Define and explain the role of the sodium-potassium pump.

5. Explain how an impulse is transmitted along an entire reflex arc, mentioning all the components of the arc and all the processes occurring in transmission.

6. Mention the different sense receptors in man, and indicate the type of stimulus to which each responds.

7. Describe the structure of the eye and its method of accommodation, using diagrams where necessary.

TOPIC 13   EFFECTORS

Biological Science, Chapter 11.
Elements of Biological Science, Chapter 11.

One of the chief fascinations most of us find in animals is their endlessly varied capacity for movement. Animals in action in fight or flight are the most obvious and dramatic examples of movement, but almost all of an animal's activities, from chasing, chewing, and digesting food to communication with prospective sexual partners, are impelled by movements of one sort or another. In most animals, movements are brought about by muscles acting against some form of skeleton.

Muscles, of course, are **effectors,** biological devices that do things in response to stimuli, usually from the nervous system. There are many other kinds of effectors, such as nematocysts (in which phylum?), cilia, and glands, but muscles are the most common and intensively studied form. Some effectors may be "independent" (not subject to any control other than that built into themselves) or may be under the control of something other than the nervous system, but skeletal, or voluntary muscles, which you will examine today, are almost exclusively under nervous control.

In this unit, you will investigate several different kinds of skeletons, their associated muscular effectors, and the physiology and biochemistry of muscular contraction.

PART I. SKELETAL SYSTEMS

One can immediately appreciate the importance of structural support in terrestrial organisms, but it is less obvious that a skeletal system is essential for rapid and efficient movement in any habitat. Arthropods and vertebrates — the two phyla of animals that have evolved the most effective locomotion — are also the two that have well-developed jointed skeletal systems. We shall examine representatives of these phyla and shall look briefly at one other type of skeleton, the hydrostatic skeleton.

A. Hydrostatic skeleton: the earthworm

Although one may not ordinarily think of the earthworm as having a skeleton, the action of body-wall muscles on the incompressible fluids within brings about very effective movement. We therefore describe this arrangement as a hydrostatic skeleton.

1. Examine a cross section of the earthworm (*Lumbricus*) and locate in the body wall two sets of muscles (FIGURE 13-1): an inner, longitudinally-oriented set, arranged in featherlike bundles, and an outer, circularly-oriented set. The longitudinal muscle fibers are oriented parallel to the long axis of the body, and you will therefore see them in cross section; the circular fibers will be seen in longitudinal section, since they run in the same

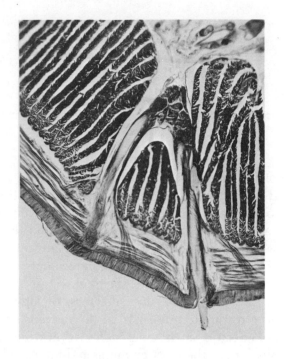

FIG. 13-1. Cross section of earthworm outer body wall. This photomicrograph shows longitudinal and circular muscles. (*Courtesy Ward's Natural Science Establishment, Inc., Rochester, New York.*)

plane as the cross section of the body. Label FIGURE 13-1.

What effect would contraction of the longitudinal muscles have on the shape of the worm? What effect would contraction of the circular muscles have on the shape?

Take this opportunity to review other structures in the cross section: intestine with typhlosole (what is its function?), dorsal blood vessel, ventral nerve cord, setae, and perhaps parts of nephridia.

2. Since the body is divided by membranes (septa) into compartments, and since each compartment (segment) contains relatively incompressible fluid, a compartment that contracts in diameter (by the action of which muscle set?) must simultaneously increase in length. Following this basic principle, the earthworm's fully segmented body is capable of a great variety of movements. Which movements occur depends upon the activity of the longitudinal and circular muscles in different parts of the body.

Obtain a pair of living earthworms. Place one on a wet paper towel in a pan and the other on a wet plate of glass. Observe carefully the movements of the body and the effectiveness of locomotion on these two substrates.

Forward movement is accomplished by a wave of contraction of the circular muscles, accompanied by relaxation of the longitudinal muscles, and therefore elongation of the body. Which way along the body does the contraction wave pass? Following the push forward, there is a contraction of longitudinal muscles, accompanied by a relaxation of circular ones; this expands the anterior region of the worm against the soil, and the animal is anchored by setae, bristles in the body wall that can be felt by running a finger posteriorly down the worm. Notice that the animal requires a traction-producing surface in order to move effectively; of the surfaces you provided, which affords the best traction?

3. Follow the waves of muscular activity in forward and backward movement, and in anchoring. What modifications must be made in the symmetrical contractions if the worm is to make turning movements?

4. Try making a small cut in the ventral side of the worm, to section the nerve cord; notice the effect on movement, and interpret your observations.

B. Exoskeletons

Examine representatives of the phylum Arthropoda, preferably alive. These animals have a rigid exoskeleton, composed principally of the polysaccharide **chitin,** covering most of the body. The exoskeleton is secreted by the underlying epidermis.

The exoskeleton functions to prevent excessive water loss and to protect the internal organs from mechanical injury; in addition, it provides a point of attachment for muscles and forms lever systems between movable parts of the skeleton.

If a crayfish or fiddler crab is available, watch its movement carefully. Rapid locomotion and fine gradations of movement are possible for animals with jointed skeletons. Notice the joints, and the manner in which they are manipulated during movement.

What are some of the problems that result from having an exoskeleton?

C. Endoskeletons

Vertebrate animals have endoskeletons, internal skeletons composed of bone and/or cartilage, with the muscles and other soft parts of the body outside. The relatively large amount of muscle tissue supportable by an internal skeleton, along with the remarkable internal and external structural adaptability of bone, make possible a wide variety of different types of locomotion in vertebrates, despite the serious problems of support inherent in the heavier vertebrate body.

Examine the vertebrate skeletons on demonstration. This outline refers to the human skeleton, but it is broadly applicable to most other vertebrates, and reference will often be made to other commonly available skeletons (e.g., frog, cat). You will find it much easier to appreciate skeletal adaptations and to learn the names of a few parts of the skeleton if you try to think about how a particular skeletal structure is organized to carry out its special functions. One of the easiest ways to understand function is to compare as many skeletons as possible in the time available. It will help you to remember parts if you label FIGURES 13-2 through 13-5 as your study proceeds.

The vertebrate skeleton is arbitrarily divided into **axial** and **appendicular** components. The axial skeleton is the main longitudinal portion and the appendicular skeleton includes the bones of the appendages and their supportive pectoral and pelvic (shoulder and hip) girdles.

AXIAL SKELETON

1. The skull is composed of many small bones fused together. Note particularly:

> a. the **cranium,** a bony case enclosing the brain and major sense organs (Which sense organs?)
> b. the **facial bones**
> c. the **mandible** (lower jaw), the only bone of the skull not immovably fused to the rest

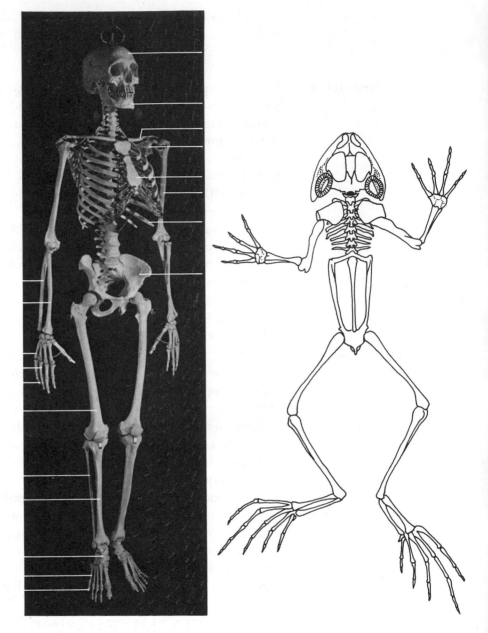

FIG. 13-2. Human skeleton. (*Courtesy Ward's Natural Science Establishment, Inc., Rochester, New York.*)

FIG. 13-3. Frog skeleton.

Locate the **hard palate,** the roof of the mouth, in the human skull. Note that it separates the nasal passages from the mouth. You will recall from your study of this region in the fetal pig that the separation is continued posteriorly by the soft palate, so that the air passages open into the pharynx quite far back in the mouth. Such an arrangement makes possible simultaneous breathing and mastication of food (try it). It would be difficult for a mammal to interrupt breathing every time it had to feed, since its metabolic rate is high enough to require regular and constant breathing. Examine the skull of a frog and notice that the hard (secondary) palate is lacking; the nasal passages open near the front of the mouth. Remember that a frog's metabolic rate is lower than that of a mammal, and that it has

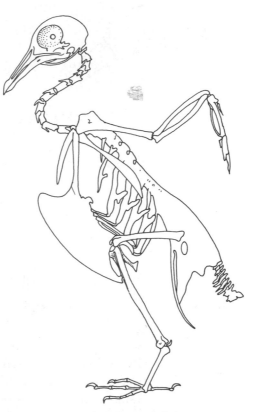

Fig. 13-4. Pigeon skeleton.

alternative sites for gas exchange, whereas a mammal has only its lungs.

2. The **vertebral column** serves for support and for housing the spinal cord, and is composed of many vertebrae, separated from one another in life by intervertebral **discs**. The column is customarily divided into five series:

    a. (7) cervical vertebrae (forming the neck region)
    b. (12) thoracic vertebrae (with which the ribs articulate)
    c. (5) lumbar vertebrae (in the abdominal region)
    d. (5) fused sacral vertebrae (enclosed by the pelvic girdle)
    e. several caudal vertebrae (forming the tail, and fused in man to form the **coccyx**)

The number of vertebrae in each region differs from species to species, and the variation is particularly marked in the caudal vertebrae. The numbers in parentheses refer to man.

Notice the stout, dorsal-projecting processes (neural spines) on the vertebrae, and how they differ in length. Compare the length of these processes on a human skeleton with those in a cat skeleton. The difference in relative length is particularly marked in the cervical and thoracic regions. Why?

3. Locate the ribs, forming a bony cage that with its associated muscles forms a support for the chest wall. The ribs articulate dorsally with thoracic vertebrae, and some are also attached by cartilage directly or indirectly to the **sternum.** Those ribs without ventral attachment are called floating ribs.

The frog has a small sternum (partly bone and partly cartilage), but it

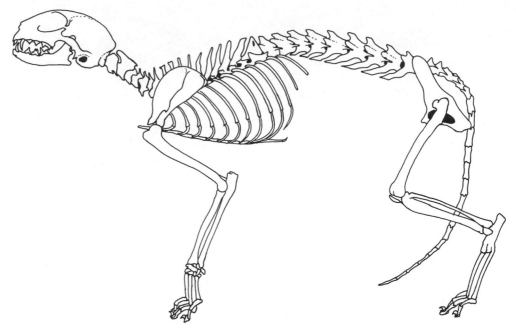

Fig. 13-5. Cat skeleton.

lacks ribs. What functions in mammals are performed by the ribs? How are these functions performed in a frog?

Examine the skeleton of a bird, if available, and note the remarkable relative difference in size between the sternum of a bird and that of a mammal. The large area of the bird sternum provides a surface for the attachment of wing muscles, a function that the mammalian sternum does not have to perform, although the latter does serve as a point of attachment for muscles that support and flex the arm toward the body axis.

APPENDICULAR SKELETON

4. The **pectoral girdle**, supporting the forelimbs, is composed of two pairs of bones:

    a. **clavicles** (collar bones)
    b. **scapulas** (shoulder bones)

Note the prominent clavicle in man, and the very tiny one in the cat (in some cat skeletal preparations, it may be absent). The clavicles provide pivot points in animals that can move their forelimbs in many different directions, but interefere with movements of these limbs during running (how?). The clavicles also provide support for animals that hang or lift with the forelimb.

5. The forelimbs are composed of loosely articulated bones, as follows:

    a. **humerus,** the large long bone of the upper portion of the limb
    b. **radius,** the long bone of the lower arm, with a ball-and-socket joint at the elbow, allowing rotational motion
    c. **ulna,** the other long bone of the lower arm, with a hinge joint at the elbow, allowing motion in only one plane
    d. **carpals,** a group of small bones forming the wrist
    e. **metacarpals,** slender bones forming the palm

f. **phalanges,** the bones of the fingers

Compare the radius and ulna of the human or cat skeleton, with those of the frog, in which rotational motion is neither necessary nor desirable.

The position of the ulna makes it useful as a point of attachment for muscles, but it bears very little weight relative to the radius; in some animals, the ulna is therefore reduced in size or partially fused with the radius. If the forelimb of a horse is available, note the fusion of radius and ulna, with only the upper portion of the ulna still clearly defined.

Notice the difference between carpels, metacarpels, and phalanges in cat and human skeletons. The cat stands on its fingers and toes, an arrangement that gives it additional leverage for running.

6. The **pelvic girdle** forms the basal support for the hindlimbs, and is composed of three pairs of tightly fused bones.

7. The hindlimbs are composed of a series of loosely articulated bones, as follows:

a. **femur,** the long bone of the thigh
b. **tibia,** the larger of the two long bones of the shank
c. **fibula,** the smaller of the two long bones of the shank
d. **tarsals,** a group of small bones forming the ankle
e. **metatarsals,** slender bones of the foot
f. **phalanges,** the bones of the toes

The fibula is comparable to the ulna, in that it bears very little weight relative to the tibia, and may therefore sometimes undergo reduction or fusion with the larger bone.

D. Bone and cartilage

Most elements of the endoskeleton are preformed in cartilage and later replaced by bone, but some parts remain permanently cartilaginous. Cartilage and bone are two types of *connective tissue*, in which the cellular elements secrete a **matrix** (fibers and a polysaccharide gel in cartilage or calcium salts in bone).

Examine slides of hyaline cartilage and compact bone in cross section; study the demonstration comparing longitudinal sections through long bones of birds and mammals if available.

1. In the slide of **hyaline cartilage,** note the clear, noncellular gel matrix (the fibers will not be visible) and the cavities containing cartilage cells. Depending upon the area from which the tissue for your slide was taken, you may also find some fibrous connective tissue and fat associated with the cartilage.

2. Bones have a complex internal structure. In a long bone, such as the femur, the outer and stronger parts usually consist of compact bone, and the inner region, often hollow and traversed by bars of bony material, is spongy bone. Crossbars in the spongy region support the whole bone in much the same manner that the superstructure of an airplane is supported by steel struts. The exact organization of spongy bone depends upon the mechanical stresses to which the bone is subject. Marrow, which fills out the spaces in spongy bone, serves as a region of blood-cell manufacture and as an area of fat storage.

Fɪɢ. 13-6. Photomicrograph of cross section of bone, showing Haversian systems. *(Courtesy Thomas Eisner, Cornell University.)*

Birds, in which skeletal lightness is advantageous, lack spongy bone and marrow although crossbars are present to provide strength. The interior of many bird bones contains air sacs which contribute to lightness.

3. Examine a cross section of compact bone. Notice the many **Haversian systems,** units of concentric deposition of the mineral salt matrix of bone. The lumen of each Haversian system, called an Haversian canal, carries blood vessels and nerves. Outside the lumen, the matrix of the bone is deposited in concentric layers, perforated by dark structures (regions where the bone cells lived) and by tiny radially-oriented channels, called canaliculi. These channels connect the bone cells to one another and to the Haversian canal. In a mature Haversian system, the bone cells are isolated from one another, and the exchange of materials with the vascular system occurs by means of diffusion from individual cells to the canal through fluid in the canaliculi.

The complicated internal structure of compact bone is the result of changes that occur as the bone develops, and as different stresses are placed upon it. The mineral salts (calcium phosphate and calcium carbonate) of which the bone matrix is composed are constantly undergoing resorption in some areas to form small, tubular channels. New matrix material is deposited on the walls of these channels in concentric layers. This continual reformation and reworking of bone accounts for the appearance in cross section of the Haversian systems, each of which was once an area of resorption.

Label the photomicrograph of cross section of compact bone (Fɪɢᴜʀᴇ

13-6). Why would it be adaptive for compact bone to be reworked and organized in this complicated manner? In what parts of the body would you expect to find bony structures that did *not* show the development of Haversian systems?

PART II. THE ARRANGEMENT AND ACTION OF SKELETAL MUSCLE

A. Muscle arrangement

Working in groups as directed by your instructor, dissect and study the arrangement of muscles in the hind leg of a frog.

Cut all the way around the loose skin next to the hip at the base of one of the rear legs of a doubly pithed frog. Grasp the skin firmly and pull it back, exposing the muscles of the thigh and shank.

Locate the **gastrocnemius,** a large muscle on the back of the shank. Observe the white connective tissue (fascia) surrounding the fleshy belly of the muscle. The fascia continues beyond the muscle at either extremity to form strong white bands, the **tendons,** that attach the muscle to bones (Figure 13-7).

When a muscle contracts, one of its extremities (usually the proximal one) remains essentially stationary; this point of attachment is called the **origin** of the muscle. The movable attachment is called the muscle's **inser-**

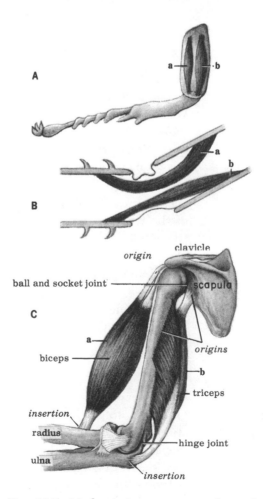

Fig. 13-7. Mechanical arrangement of muscle and skeleton in human arm and insect legs.

tion. The gastrocnemius has two proximal origins. The larger one is on the posterior portion of the distal end of the femur, near the knee; the smaller forms a small tendon, attached to a larger, sheetlike tendon covering much of the anterior portion of the knee. The large tendon attaches directly to the femur and is actually the insertion of one of the large thigh muscles, the triceps. At its distal end, the gastrocnemius inserts on the ankle and foot by way of the long **Achilles tendon,** which extends over the heel.

B. Muscle action

The type of action resulting from contraction of a particular muscle depends upon the exact position of the muscles' origin and insertion, the type of joint between them, and the action of other muscles—both **antagonists,** arranged to oppose the action of the muscle in question, and **synergists,** which guide and limit the principal movement, and thereby make possible finer gradations of motion.

There are several terms commonly applied to describe the action of muscles:

1. flexor   — causes jointed parts to move toward each other
2. extensor — causes jointed parts to move away from each other
3. adductor — causes movement of a part toward a vertical plane running through the longitudinal midline of the body
4. abductor — causes movement of a part away from a vertical plane running through the longitudinal midline of the body

Hold the thigh rigid with one hand and grasp the belly of the gastrocnemius with the other. Gently pull the gastrocnemius toward its origin. What is the action of the muscle under these conditions? Next hold the shank rigid and again pull the gastrocnemius toward its origin. Now what is the action of the gastrocnemius? Decide by examination which muscles, acting as antagonists, determine for any given instance which of these two major actions the gastrocnemius performs.

Locate the **triceps femoris,** a very large muscle on the anterior surface of the thigh. It is best seen in dorsal view. This muscle has two origins on the pelvic girdle, and inserts on the femur by way of the broad white tendon already seen passing over the anterior surface of the knee. Hold the thigh firmly and pull on the triceps. Is this muscle an extensor or a flexor of the shank?

The **biceps femoris** is a narrow muscle best seen in dorsal view just posterior to the triceps. Its origin is on the pelvic girdle. Where does it insert? What is its action?

PART III. MUSCLE PHYSIOLOGY

One way to study the characteristics of muscle contraction is simply to stimulate a muscle and watch it. This procedure will tell you that muscles, like nerves, respond to many kinds of stimulation (touch, chemicals, heat, electricity, drying, etc.), and it can give you a crude idea of the relationship between the amount of stimulation and the extent to which the muscle responds.

Much more precise information can be obtained by using the kymo-

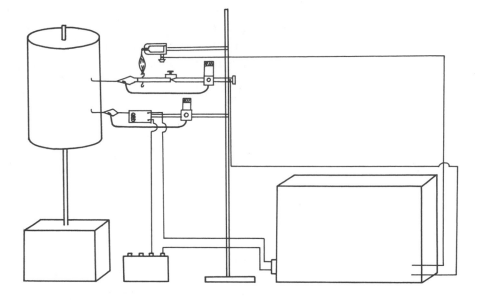

Fig. 13-8. Kymograph apparatus, for studying muscle contraction; fill in and label the controls present on the instrument used in your laboratory.

graph apparatus, one variation of which is illustrated in Figure 13-8. Using this arrangement, an electrically stimulated muscle is attached to a lever that amplifies the contraction and records it by writing on paper attached to a moving drum. Simultaneously, an electrical device, the signal marker, records the application of the stimulus, so that the relationship between the response of the muscle and the application of the stimulus can be ascertained. This arrangement has many advantages over *in situ* stimulation of muscles. For one thing, very small responses can be detected because of amplification through the lever system; in addition, the moving drum allows a detailed examination of the different phases of contraction and relaxation. The application and measurement of stimuli are much more readily controlled than *in situ*.

Electrical stimuli are used because they are easily applied and controlled for intensity, frequency, and duration, because they do not damage the muscle, and because they are most like the muscle's natural stimulation in the body.

Your instructor may have you perform these experiments in groups, or they may be done as a demonstration. In either case, learn the parts and operation of the apparatus *before* the experiments are done, and find out if and how your apparatus differs from that illustrated. As you identify the parts, label them in Figure 13-8, adding any part not already in the diagram.

A. Parts of the apparatus

1. The **stimulator** is a device used to generate electrical stimuli of controlled duration, frequency, and intensity (voltage). Check the following controls, adding to Figure 13-8 those which are present on your stimulator:

a. Signal marker switch — set at "on" position. This is a device that gives a "click" when your stimulator delivers a shock to the muscle. You therefore have an audible and a visual (see 4 following) check on the delivery of stimuli.

b. Frequency dial — set at lowest frequency. This adjustment controls the number of times per second that stimuli are delivered to the muscle. Most stimulators have a "Multiply By" dial adjacent to the frequency control dial; this usually permits an increase in the frequency by multiples of ten. The total frequency in any instance is the value shown on the "Multiply By" dial times that shown on the frequency dial. Set the "Multiply By" dial at 0.1, or its lowest setting.

c. Duration dial — set as specified by your instructor. This adjustment controls the length of time over which the stimulus is delivered, and the setting is usually very small. Set the "Multiply By" dial, adjacent to the duration dial, at 0.1, or its lowest setting.

d. Voltage dial — set at lowest value. This adjustment permits you to control the intensity of the stimulus, measured in volts. As for the frequency and duration adjustments, total intensity is calculated by multiplying the value shown on the "Multiply By" dial by the value shown on the voltage dial. Set the "Multiply By" dial at its lowest value, or 0.1.

e. Mode switch — set on "off." This switch permits the operator to deliver single stimuli at his own rate (usually the "down" position of the switch, but check the label carefully) or to deliver continuous or "repeat" stimuli at a selected frequency.

f. Stimulus selector — set on "regular."

g. Output — set on "mono."

h. Polarity — set on "normal."

i. Power switch — set "on."

Check to see that the stimulator is properly plugged in. The wiring leaving the stimulator and running to the rest of the apparatus should not be changed from one laboratory section to the next. If you find free-flying wires, consult your instructor before proceeding.

2. The **kymograph** consists of an electrical or spring-driven motor that powers a revolving drum mounted on a vertical post. Your instructor will explain how to control its speed, and will tell you how to mount paper upon the drum. The drum is covered with a glossy paper, which may be plain if the writing is to be done in ink, or may be smoked with carbon particles, if the writing is to be done by scraping off the carbon. Your apparatus may have an additional vertical post and drum, to give a longer paper. Mount the drums as high as possible on the posts, and follow your instructor's directions for further handling.

3. The **femur clamp** holds the bone to which the experimental muscle is attached.

4. The **writing lever** is attached to the bottom of the muscle, and serves to record its movements on the kymograph paper.

5. The **signal marker** is a device that writes on the kymograph paper to indicate the points at which stimuli are applied to the muscle.

The femur clamp, writing lever, and signal marker should all be arranged so that the muscle hangs vertically, after it has been attached to the

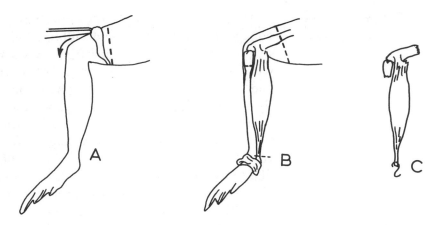

Fig. 13-9. Steps in preparation of frog muscle.

apparatus, and so that the two writing points contact the drum with approximately equal pressure. Adjust the position of the levers *after* you have attached the muscle. If pressure on the drum is too great, the levers will lag in their movements due to friction against the paper; if the pressure is too slight, the lever will come off the drum at some points in its movement. Either situation will produce an uneven record. If you are using ink-writing apparatus and it is not operating properly, call your instructor.

Read through *all* the experiments before you prepare the muscle.

B. Preparation of the muscle (Figure 13–9)

1. Cut around the skin at the base of the intact leg, just as you did in examining the muscles earlier (Figure 13-9A).

2. Skin the leg as if you were removing a glove from a hand (Figure 13-9B).

3. Cut off all muscle from the thigh bone (femur), *except* the gastrocnemius origin, which should be left intact.

4. Insert the scissors-tip between the calf muscle and the shank bone (tibiofibula) and carefully separate the muscle from the bone. Then insert the scissors-tip between the Achilles tendon and the heel. Cut the tendon, leaving as much as possible of it attached to the gastrocnemius. Completely separate the gastrocnemius with its tendon from the remainder of the lower leg, up to the knee. Cut away all muscle and bone *except* the gastrocnemius, the Achilles tendon, and the femur to which the gastrocnemius is attached. The resulting preparation should look like Figure 13-9C.

5. Moisten the muscle with frog Ringer's solution. Insert the femur into the femur clamp, and fix it firmly therein. Carefully hook a small S-shaped pin through the Achilles tendon, and hook it to the writing lever. Put a small amount of tension on the muscle by moving the femur clamp on its vertical post. Adjust the position of the levers relative to the drum (as explained in A5). The writing levers should be horizontal and should have some tension on them so that when the muscle contracts slightly the lever will record the movement.

6. One person on each team should be assigned responsibility for keeping the muscle moist with frog Ringer's solution throughout the experiments. It may be dropped over the muscle with a medicine dropper, but be careful *not* to spill any on the apparatus.

C. *Experiments*

THRESHOLD

1. For any given stimulus duration and intensity, there is a minimum voltage, called the **threshold voltage,** below which the muscle will not respond. With the duration and frequency set on their lowest values (or otherwise, as specified by your instructor), start the voltage at its lowest value and stimulate the muscle with single shocks a few seconds apart. Increase the voltage slightly, moving the paper about 1 cm each time, until you see a response. Record this voltage — along with the frequency and the set duration value — as the threshold voltage. (*Note:* If no response occurs, even if you have advanced the voltage dial to its *highest* value, decrease the voltage dial to its *lowest* value and turn the "Multiply By" dial up one factor — usually from 0.1 to 1.0. Then begin turning up the voltage dial again, and testing with single shocks until a response occurs.)

MAXIMAL STIMULUS

2. After reaching the threshold, continue to increase the voltage, stepwise, turning the drum each time. Note that as the voltage is increased, the extent of the response increases as well. This is explained by the fact that the muscle contains many fibers, not all of which have the same threshold. As the stimulus intensity is increased, more and more fibers contract, and the muscle shows a larger and larger movement. Eventually you will reach a **maximal** value, beyond which further increase in voltage will not produce an increase in contraction. Any stimulus applied after the maximal stimulus is called a **supramaximal** stimulus.

Individual muscle fibers respond like nerve fibers, in an all-or-none fashion. The muscle as a whole, however, can respond incrementally to increasing stimulus intensity because the individual fibers have different thresholds.

SINGLE MUSCLE TWITCH

3. Arrange the muscle for a single contraction and relaxation, a single "twitch," with the drum turning at full speed. Label the curve with contraction period and relaxation period. The **latent period** of the muscle, that interval between the arrival of the stimulus and the event of contraction, is too small to be measured without additional special equipment. Refer to your text for details.

SUMMATION OF RESPONSE, AND TETANUS

4. Set the voltage at the level discovered as maximal stimulus for your muscle, and set the kymograph speed on high, without yet turning it on. Set the mode switch to repeat, and, with the drum turning, gradually increase the stimulus frequency from two to twenty per second. You will notice during the increase in frequency that the muscle does not fully relax after each contraction, but is hit by another stimulus during its relaxation period, and begins to contract again, stronger than before. As the frequency increases, this effect, called **summation** of response, becomes more and more marked. How might you explain this greater contraction than that elicited by a maximal single stimulus? Finally there is *no* relaxation at all, but only a smooth curve; this condition is called **tetanus.** Normal contraction of skeletal muscles is tetanic contraction, brought about by the delivery of very frequent stimuli from the central nervous system. If skeletal muscles did not contract in this manner, muscular movements would be jerky and uncoordinated.

FATIGUE

5. With a frequency setting just below that needed to obtain summation, and with the voltage setting required to get maximal response from the muscle, allow the apparatus to run rather slowly until the muscle fatigues and begins to relax, even though stimuli continue to impinge on it. What does the fatigue curve look like? Why does the contraction not continue to

be full? What biochemical events take place during fatigue?

6. When you have finished the exercises, set the stimulator dials at the lowest levels, turn off the power, unplug the stimulator, clean up the apparatus and your desk area, and turn in your fully labeled record to the instructor.

PART IV. THE BIOCHEMISTRY OF MUSCLE CONTRACTION

While we know much about the mechanics and physiology of muscle activity, the mechanism by which a muscle fiber shortens is still imperfectly understood. The most widely accepted hypothesis, which accounts for the difference in ultrastructural appearance of relaxed and contracted muscle cells, suggests that contraction involves two muscle proteins, actin and myosin. The molecules of these proteins are believed to slide just past one another during the contraction, from an extended position to an apposed, shortened position. The sliding will not occur without ATP and certain mineral ions.

In this exercise, you will have an opportunity to observe directly the influence of ATP and of the mineral ions on the contraction process, and to obtain a quantitative measure of contraction. The material used in these experiments is psoas muscle, one of the skeletal muscles lying along the backbone. It was removed from a freshly dead rabbit, tied on a stick at approximately its natural body tension, placed in a solution of 50% glycerol, and stored in a freezer. The glycerol removes almost everything except the contractile proteins, but the muscle retains its striated appearance and ability to contract.

Each group of 2–4 students will be provided with a 2 cm section of psoas muscle in a dish of cold glycerol. Using glass needles or *clean* stainless steel forceps, one member of the group should separate the muscle into bundles, one for each student. Each bundle should be placed in a Syracuse dish of glycerol solution; from this point on, each student should follow the procedure on his own.

1. Using 7–10X magnification if possible, and using glass needles or clean stainless steel forceps (preferably the latter), gently tease the muscle into very thin groups of myofibers, the muscle cells. Single cells or thin groups are best; strands thicker than a silk thread will curl up when they contract and must not be used.

2. Mount one of the strands in a drop of glycerol on a slide, with a cover slip. A Pasteur pipette or dropper is best for transferring the fiber. Examine under low and high powers; note the striations, and compare your observations with photograph 2 in FIGURE 13-10. Notice that each cell has many nuclei.

3. Transfer three or more of the thinnest strands to a drop of glycerol on a second slide. The strands should be the same length, and a cover slip should *not* be added. Using a dissecting microscope and a millimeter ruler held beneath the slide, measure the length of the fibers. If there is more than a small drop of glycerol on the slide, soak up the excess on a piece of lens paper held at the edge of the glycerol farthest from the fibers.

4. Flood the fibers with several drops of a solution containing both ATP and the required mineral ions. Observe the reaction of the fibers and, after

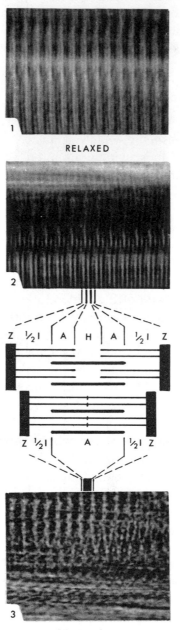

FIG. 13-10. Glycerinated muscle fibers, phase contrast 210X. (1, 2) Relaxed muscle. (3) Contracted muscle. Drawing between 2 and 3 shows banding difference between relaxed and contracted muscle. (*Courtesy Carolina Biological Supply Company.*)

thirty seconds or more, remeasure the length of the fibers. Calculate the degree of contraction.

5. Repeat the experiment twice, using clean slides and pipettes, and being especially sure to use *clean* teasing needles or forceps. In one of the repetitions, use ATP alone; in the other, add metal ions alone. Record and explain the results. (*Note:* If you contaminate either of the solutions with the other, however indirectly, you will obtain ambiguous results. If your results are ambiguous or otherwise unexpected, compare your procedure with that used by your neighbors, and try to explain your results.)

6. Observe one of the contracted fibers under low and high powers of a compound microscope, and look for differences in appearance between the muscle in a contracted and in a relaxed state. If the differences are not ap-

parent in your preparation, note them in Figure 13-10, along with a diagram that relates the banding pattern seen in the muscle to the hypothetical sliding of filaments.

It appears that the shortening of a muscle fiber is due to the shortening of its constituent sarcomeres. This, in turn, results from a progressional sliding of the thin filaments inward between the thicker filaments, pulling the dense "Z" membranes nearer together. The free ends of opposing thin filaments eventually may meet one another at the midline of the sarcomere. Striations in contracted glycerinated muscle are thus distinguished from those in relaxed muscle by an obliteration of the "H" discs and a decrease in length of the "I" bands. Figure 13-11 compares relaxed and contracted muscle in electron micrographs, and should make these differences clear.

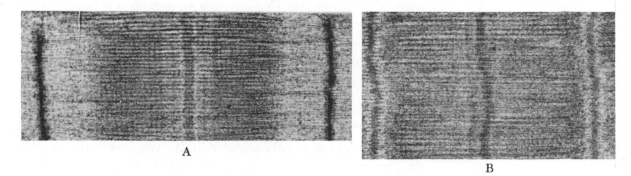

A   B

Fig. 13-11. Electron micrographs of skeletal muscle. (A) Relaxed. (B) Contracted. Compare with Fig. 13-10. (*Courtesy H. E. Huxley, Cambridge University, England.*)

QUESTIONS FOR FURTHER THOUGHT

1. Write a concise but thorough description of the physiology of muscle contraction at the nonmolecular level. Be sure to include an explanation of each of the following: all or none, threshold, maximal stimulus, summation of response, tetanus, latent period, fatigue.

2. Explain the theory of the molecular physiology of skeletal muscle contraction proposed by H. E. Huxley, and give some of the evidence on which this theory is based.

3. Explain the skeletal and muscular arrangements and functions that make it possible to use the same body part to perform many different motions or to perform the same motion at many different speeds. Include mention of: origin, insertion, antagonist, synergist, joint, etc.

4. What functions, apart from support, does the endoskeleton perform?

5. In a single sentence, explain the operation of the hydrostatic skeleton in an earthworm. How would the repertoire of movements differ if the body was not segmented?

TOPIC 14 BEHAVIOR

Biological Science, pp. 261-266, 403-407.
Elements of Biological Science, pp. 165-170, 249-251.

One of the basic characteristics of life is its capacity for response to environmental change. An organism reacts to a specific change in its environment, a **stimulus,** by carrying out some activity that we call a **response.** The repertoire of responses that characterizes an organism is called its **behavior.** Living organisms have evolved a remarkable diversity of behaviors, which, for convenience of discussion, we may group into two categories: learned behavior, that array of responses acquired by an organism in the course of its experience, and inherited or innate behavior, which is a part of the organism's genetic endowment. In this laboratory, we shall study a few examples of inherited behavior.

Like the nervous and effector systems from which it is molded, behavior is subject to evolutionary adaptation. As you watch different behaviors today, try to think about them as adaptive mechanisms for the organisms in question, and ask yourself how they may be useful to the organism in its natural situation. Think also about how the stimuli you are using may differ from those to which the organism is normally subject. Discussion with your neighbors and comparison of results is encouraged.

Several shorthand terms will be helpful to you in describing the behaviors you see in this laboratory, and also in thinking about their adaptive value. Locomotor behavior, that involving movement of an organism from one place to another, is described by two terms, **taxis** and **kinesis.** A taxis is an oriented movement, toward or away from the stimulus that brought about the movement. A kinesis, on the other hand, is random movement, caused by a stimulus but not necessarily oriented by it. Some insects, for example, select a dark habitat in preference to a lighted one. If we can demonstrate that an animal actually avoids light by moving directly away from it, we can describe its behavior as a taxis (in this case, a phototaxis, i.e., one caused by light). It is perfectly possible, however, that light, or the "absence of darkness," may merely initiate random movements which eventually chance to carry the animal into the darkness, where the movement "turns off." This sort of behavior would be called a kinesis. The terms taxis and kinesis are most often used to describe animal behavior, but they apply equally well to any organism that shows locomotion, for example, the motile algae.

Two other terms we shall find useful are **nastic movement** and **tropism;** we shall use them to describe nonlocomotor behavorial movements in plants. A nastic movement is a change in the posture of a plant or plant part, and it is the same no matter what the stimulus direction. It does not orient the plant toward or away from the stimulus, as could the locomotor behaviors we discussed above. A tropism, in sharp contrast, is a growth curvature, a postural change oriented by the direction of the stimulus.

We may add a prefix, as shown in the following list, to any of the terms discussed above, to describe the nature of the stimulus. We may add the word positive or negative to describe the direction of the movement relative to the stimulus. For example, the bending of a sunflower toward the sun would be called a **positive phototropism.**

photo-	reaction to light
thermo-	reaction to temperature
geo-	reaction to gravity
hydro-	reaction to water
hygro-	reaction to moisture or humidity
thigmo-	reaction to contact
chemo-	reaction to chemicals
galvano-	reaction to electricity
opto-	reaction to change in the visual field
seismo-	reaction to shock

Do the following experiments in groups, as directed by your instructor. Your group may be asked to give a report on one of the experiments. In the report, you are expected to pool the results obtained by the rest of the class and to summarize information regarding the adaptive value of the behavior you discuss.

PART I. PLANT MOVEMENTS
A. Mimosa

Obtain a potted specimen of the sensitive plant, *Mimosa*, and *without touching* the leaves, move it carefully to your desk.

1. Stimulate one of the end leaflets by gently touching it; if nothing happens, stimulate again by touching with a little more force. What is the plant's response? How would you describe this response, using the terminology in the introduction? What benefit might the plant derive from this behavior? Note the approximate time when the response was given.

FIG. 14-1. Photograph of *Oxalis*. On the right, notice the *Mimosa*.

Fig. 14-2. Angle of *Oxalis* leaves before experiment.

The movement you saw was due to a rapid change in the turgor of cells at the base of the leaflet. Can you locate in this region any structure that might help to transmit the pressure of your touch to these sensitive cells?

2. Time the recovery of the leaflets to their normal posture. Design and carry out an experiment to test the effect of increased severity of stimulus, or try using other stimuli. You might wish to try to find out whether the plant is particularly sensitive to touch in certain regions of the leaflets. While you are waiting for the plant to recover, go on to the next part.

B. Photonasty in Oxalis

Obtain an *Oxalis* plant (FIGURE 14-1) that has been kept in sunlight, and examine the leaves and stalk. In the space provided (FIGURE 14-2), sketch the angle of the leaves relative to the stalk.

1. Place a cardboard box over the plant, and leave it for about an hour. In the meantime, go on to other parts of the laboratory. At the end of that time, remove the box just long enough (less than a minute) to glance at the plant so that you can again sketch the angle of the leaves relative to the stalk. Immediately replace the cardboard box, and sketch the angle in FIGURE 14-3.

2. With the box still covering the plant, design and carry out an experiment to prove that this behavior is a nastic movement (that its direction is not determined by the stimulus).

C. Phototaxis in flagellated algae

You will be supplied with a petri dish containing *Euglena*, a flagellated unicellular alga. FIGURE 14-4 diagrams the structure of *Euglena*; make a wet mount for comparison.

Fig. 14-3. Angle of *Oxalis* leaves after experiment.

Fig. 14-4. *Euglena*, diagram of structures seen under oil immersion.

1. Stir the water in the dish gently, until the algae are equally dispersed throughout the container. Cover the dish completely with aluminum foil, making certain that no light can enter. After half an hour, remove the foil and record the distribution of the algae in a sketch.

2. Cut a small circle (5 mm diameter) in the aluminum foil cover, and put it back on the container. Illuminate the *Euglena* through this hole, using a microscope lamp or other source positioned to deliver a vertical beam of light into the container, and set up far enough away so that the water will not become heated. After half an hour, remove the foil cover, and record the distribution of the algae. What would you call this response? How would you design an experiment or observations to determine whether these were taxes or kineses?

3. Design and carry out an experiment to demonstrate negative taxis. How could you design an experiment that would demonstrate negative *pho*-*to*taxis? Eliminate any possible effects of a temperature stimulus.

D. Demonstration of plant tropisms

Examine the demonstration of plant tropisms; each setup will be labeled with an explanatory card. Distinguish between positive and negative photo-tropism and geotropism.

If there is a film of plant tropisms, be certain to see it sometime during the period; list the kinds of tropisms seen in the film.

PART II. ANIMAL MOVEMENTS
A. Reactions to moisture and to light

We will use animals known as **isopods.** These terrestrial crustaceans are sometimes called pill bugs, sow bugs, or volkswagens. While most crustaceans are aquatic, such as lobsters, crayfishes, shrimps, etc., the isopods have also established themselves on land. Their survival on land depends on a very moist habitat, and terrestrial isopods are usually found only in very moist places, such as under stones, rotting logs, leaf mold, etc.

1. Prepare two glass bowls by putting a three-inch square of moistened (*not* soaking) paper towel into one, and an identical unmoistened square into the other. Obtain ten isopods from the stock container, and introduce five of them into each bowl. Cover each bowl with a box to exclude the possibility of light acting as a stimulus to activity.

After ten minutes, remove the boxes and *immediately* observe the degree of activity in each bowl. How would you describe the rate of locomotion in each bowl? Is there any directedness to movement in either of the bowls? Using the terms previously introduced, how would you describe this

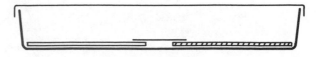

Fig. 14-5. Isopod chamber with waxed paper dividing wet and dry paper toweling.

reaction? Can you tell, from this experiment, whether the stimulus is presence or absence of water? Can you be certain that the stimulus has anything to do with direct sensing of liquid water? Could it be a response to presence or absence of *moisture*? How could you set up experiments to test the answers to these questions?

2. Obtain a piece of waxed paper and a pan from the supply desk. Line the bottom of the pan with two pieces of paper toweling, separated in the middle and overlapped by a two-inch strip of waxed paper (Figure 14-5). Secure the waxed paper with tape. The paper should fit the pan snugly. With distilled water, moisten one of the pieces of paper towel (*not* dripping wet) and leave the other dry. Introduce ten isopods to the waxed-paper strip, and cover the pan with aluminum foil. After ten minutes, remove the foil and observe the position and activity of the isopods. How many are on the wet paper? How many are on the dry paper? What is the state of activity of the isopods on the wet paper? On the dry paper? Repeat the experiment several times. What conclusions can be drawn?

3. Design and carry out some experiments using the pan, paper toweling, waxed paper, and foil to show the effect of light alone, or of light and moisture together, as stimuli. When you finish, return all animals to the stock container at the supply desk.

In interpreting these experiments, it is important to recognize the existence of problems in experimental design that may greatly affect the outcome of your experimental procedures. One such problem is the matter of *where* the animals are introduced in the pan. If your introduction chances to direct more of them toward the moistened area, more will naturally be counted there; and it is possible that they might never have been counted there if the introduction of animals had been altogether random. The use of the waxed-paper strip is a step toward the attainment of random introduction, but there are many better methods of solving the problem. Discuss the experiment with others in your group, and be prepared to suggest other ways in which the problem of attaining randomness in the introduction of animals could be solved. The simpler your solutions, the better. Your instructor may be able to provide you with some equipment to carry out your improved experiments, if there is time.

B. Reaction to temperature

Obtain a piece of glass tubing (bent into a 90° angle at each end), two wooden stoppers for the glass tube, two beakers, and a thermometer.

1. Fill the tube with distilled water, and add one or two drops of a rich *Paramecium* culture from the supply desk. Stopper the ends of the tube.

2. Fill one small beaker with ice and water, and another with water at about 36°C, from the hot-water tap. Arrange the beakers and glass tube as

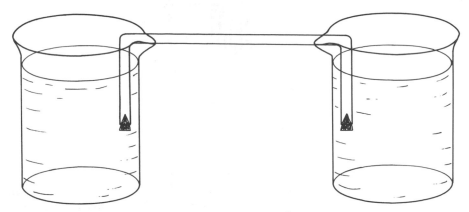

Fɪɢ. 14-6. Temperature-gradient setup.

in Fɪɢᴜʀᴇ 14-6, placing each end of the tube in a different beaker. After fifteen minutes, record the temperature of the water in each beaker, and, under the dissecting microscope, observe the position of the animals. Are they distributed uniformly throughout the tube, or have they become concentrated in a specific region? In order to describe this behavior, you should make observations under the dissecting microscope. Can you determine whether the behavior is a taxis or a kinesis?

3. Repeat the experiment, using one beaker with water at 20°C and the other with water at 55°C. Record the temperature of the two beakers at the beginning and at the end of the experiment. After fifteen minutes, observe and record the distribution of the animals, and explain your results.

Assuming that the temperature within each of the tube-ends is the same as that in the beaker in which it is immersed, and assuming that a uniform gradient of temperature has been established within the tube as a whole (probably *not* a valid assumption), prepare a diagram of the tube showing the degree of temperature change per unit of length. Observing the region of the tube where most of the animals have congregated, determine the optimum temperature for *Paramecium*.

C. Reaction to gravity

Obtain a turtle from the stock container. Hold it parallel to the table and about a foot away, quietly, until the animal extends its head. *Slowly* lower the posterior part of the turtle. How does the head move? Then slowly raise the posterior part of the turtle, and note the accompanying head movement.

Repeat the experiment with increased speed, and note the response. Try tilting the animal around its longitudinal axis: what is the result? What sensory devices mediate these responses? Where are they located?

D. Reaction to change in the visual field

Using the kymograph drum and motor, you will change the visual environment of a fish. Fɪɢᴜʀᴇ 14-7 shows the setup: a cylinder of paper, striped on the inside with tape, has been attached to the top of the drum. When the drum is turned on, the paper and its stripes move with the drum. Also notice that a beaker is hung from a ringstand down into the paper cylinder.

1. Remove the beaker. Fill it half full of pond water, and add a small

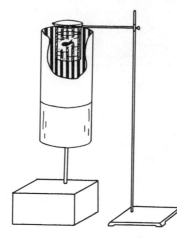

Fig. 14-7. Optomotor apparatus — using a kymograph to change an animal's visual environment.

fish. Return the beaker to its position on the ringstand, and slowly turn the kymograph to full speed. What does the fish do? How would you describe this reaction? Of what benefit would this response be to the fish in its natural environment?

2. Repeat the experiment, using the turtle. How do you explain the results?

Look up the Phylum or Division and Class of each of the animals and plants you have studied in this period, and enter each of them, if they are not already included, in the appropriate space in Appendix 2.

TOPIC **15**

CHEMORECEPTION
AND BEHAVIOR IN
THE BLOWFLY

Dethier, V. G., 1962. *To Know a Fly,*
Holden-Day, San Francisco.

Human beings depend very little on their chemical senses for survival, but they are unusual in this respect. For most animals, scent is among the most crucial of the senses, subserving a remarkably complex repertoire of activities. Natural selection has exploited the possibilities of chemical substances as information transmitters and the properties of sensory cells as chemoreceptors in a variety of different ways. Chemicals are widely used in communication between members of the same species, to relay information relating to the recognition of territory, the locating of potential mates, and many other types of activity. Nor is chemical communication confined to intraspecific interactions; many animals use chemicals as defense mechanisms, in locating potential enemies, or in locating specific habitats or foods. The ability of insects to make fine odor discriminations not only makes possible intraspecific communication and even "tribe" identification within the same species, but has also been seized upon by evolution in the adaptations of many flowering plants. A great many such plants utilize odors to attract pollinating insects; without the insects the plants would often be incapable of sexual reproduction. One European orchid species even mimics the odor of a particular species of female bee, in order to attract males of the same species, which pollinate the blossom when they alight and attempt to mate with it.

Chemicals are very important as environmental cues that trigger specific behaviors, particularly in animals with relatively stereotyped behavior patterns. In this laboratory, we shall explore in some detail a feeding response given when the presence of sugar is detected by chemoreceptors, one of many chemically controlled behaviors in the common blowfly *Sarcophaga bullata*. From this study, you will gain some insights into insect behavior, and learn about the nature of the chemoreceptors under consideration. The study is also a good example of how a relatively simple analysis of behavior patterns is carried out.

Before beginning any of the experiments, read the directions carefully and plan responsibilities for each member of the group with which you are to work. Before each experiment, discuss and clarify the hypotheses upon which the experiment is based (often purposely not clarified in the directions), and think very carefully about what each experimental approach can show. Understand the significance of the *controls* for each experiment. Prepare reports as directed by your instructor.

PART I. HANDLING THE FLIES

Sarcophaga bullata is much like the housefly, though it is larger. In nature, it feeds upon carrion and plant sugars; in the laboratory, it can be raised on liver. The females usually lay their eggs in the carcasses of dead animals. The larvae (maggots) feed on this material, grow in size through a series of molts, and eventually pupate, forming a cocoon much like that of a moth, except that it consists of the larval epidermis and not of woven material. FIGURE 17-2 on p. 193 diagrams the life cycle: at room temperature, the egg hatches into a larva in about two days; the larva feeds for eight days and then pupates; pupation lasts for about four days, after which an adult fly hatches out of the pupa. Adults live about two months.

You will be provided with a screen-top glass jar containing enough flies for your experiments. Follow the directions below very carefully to prepare the animals for use in your experiments.

1. Push the jar with the flies into a container of ice until all but the top is covered. Also cool a culture dish, to receive the flies after they have been immobilized by the cold. Leave these containers in the ice for ten minutes or more, during which time the poikilothermic flies will cool, become slowly immobilized, and drop to the bottom of the jar. While they cool, go on to 2.

2. Prepare half as many small wooden sticks as you have flies, by placing a dab of sticky wax (such as Tackiwax) at one end of the stick or on a cork fastened to the stick. Place the prepared sticks in a holder. Before proceeding, be sure to wipe any condensation from the culture dish that is cooling in the ice container.

3. Place some crushed ice in a dissecting pan, remove the cooled culture dish, and place it on the ice in the dissecting pan, in a convenient spot for the following manipulations.

4. When the flies are immobilized, place two of them in the culture dish (the bottom of which should be dry), and promptly replace the screened container in the ice. A camel's-hair brush is useful for making transfers. Using forceps to push the wings into the wax, attach the flies to the sticks, as shown in FIGURE 15-1. One pair of flies can be attached to each stick, one to each side. Place the sticks back in their holder.

Repeat this procedure, using a few more flies each time until all have been attached to the sticks. (Be *very careful* to release only a few flies at a time from the stock container, especially when you are handling them for the first time. They warm up quickly, and will escape to annoy other persons.)

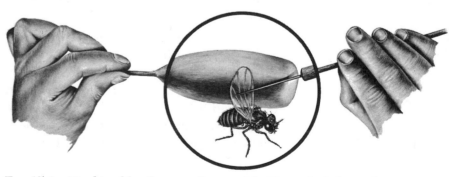

FIG. 15-1. Attaching blowfly to applicator stick. The end of the applicator (or a small cork mounted on it — see FIG. 15-4) is coated with Tackiwax or paraffin. After being anesthetized with CO_2 or cold, a fly is picked up gently with forceps, and its wings are pressed into the wax. (*Taken from* BSCS Biological Science: Interaction of Experiments and Ideas, *Prentice-Hall, 1965; used by permission of the Biological Sciences Curriculum Study.*)

5. Label each stick with a small numbered piece of masking tape, and keep a record of the treatment each fly has undergone. Careful records of this kind will eliminate any question of whether a fly is suitable for a particular experimental series, by virtue of its previous treatment.

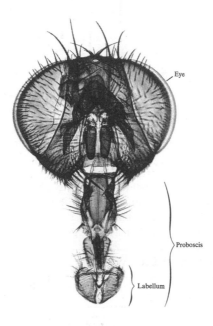

Fig. 15-2. Head of housefly showing mouthparts. (*Courtesy Carolina Biological Supply Company.*)

**PART II.
OLFACTORY
REACTIONS**

Although we shall distinguish **olfaction** (smell) from **gustation** (taste), it is well to recognize that they are only variations of the same type of sense. Both depend upon contact between chemicals and chemoreceptors, and the distinction between them is based only upon the medium in which the stimulus is carried — gaseous medium in smell, fluid medium in taste. In some situations it is difficult to make this distinction.

Figure 15-2 shows the front view of the head of a housefly, much like *Sarcophaga*. A combination of mouthparts forms the **proboscis,** a strawlike organ used in sucking up juices the fly feeds on. The proboscis is extended in Figure 15-2. The flattened, aboral end of the proboscis is called the **labellum.** In the following experiments, a positive response to some chemical is characterized by extension of the proboscis, while a negative response is characterized by failure to extend the proboscis.

1. Obtain a rack of labeled test tubes containing a variety of substances with which flies often come into contact. For example, use water, concentrated sugar solution, fresh meat juice, and spoiled meat juice.

Using *separate* wooden sticks for each substance, dip one end of the stick into the solution and then hold it close to the head of each of four flies, being *sure* that it is safely out of reach of the proboscis and all legs. Record positive and negative results (+ or −) in Table 1 below.

2. Put some drops of water on a clean glass slide or in a petri dish. Allow the forelegs of each fly to touch the water, and let each one drink its fill (Figure 15-3). Repeat the above tests for olfactory responses with each fly, and record the results in Table 2. Does water satiation have any effect on the olfactory response? If so, how might the differences in result between Tables 1 and 2 be explained?

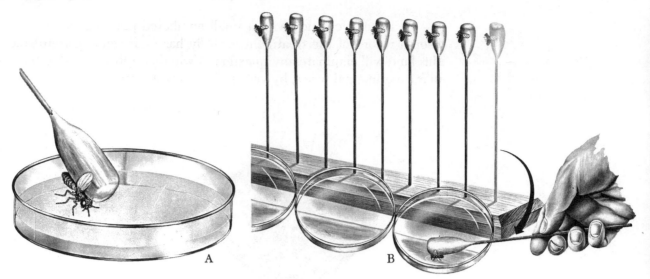

FIG. 15-3. (A) Feeding or watering a blowfly. Note the extended proboscis. (B) Holder with blowflies awaiting experimental tests. Each stick should be numbered so that flies can be distinguished. (*Taken from* BSCS Biological Science: Interaction of Experiments and Ideas, *Prentice-Hall, 1965; used by permission of the Biological Sciences Curriculum Study.*)

FLY NUMBER TEST SUBSTANCE	1	2	3	4
A				
B				
C				
D				
E				
F				

TABLE 1. OLFACTORY REACTIONS BEFORE DRINKING

FLY NUMBER TEST SUBSTANCE	1	2	3	4
A				
B				
C				
D				
E				
F				

TABLE 2. OLFACTORY REACTIONS AFTER DRINKING

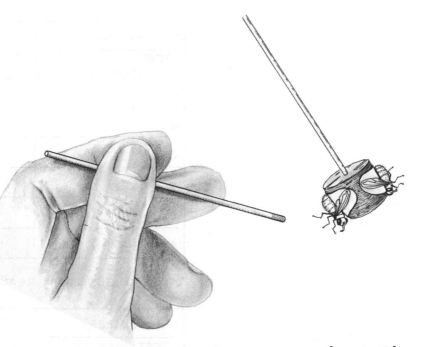

FIG. 15-4. Testing olfactory responses. Be *sure* to present the potential food materials from *behind* the fly.

**PART III.
GUSTATORY
RESPONSES**

1. With four fresh flies not previously used for experiments, test for a taste response to each of the substances used in II. Allow the forelegs to contact drops of each substance placed on a slide or in a petri dish. Make sure that you present the materials from *behind* the fly (FIGURE 15-4) so that no part of the head comes into contact with the test material. *Do not* let the flies feed at this point; simply watch for a reaction, and remove the fly promptly after it has occurred, or after a few seconds, in the case of no response. Record results in Table 3.

2. Repeat the tests in 1, after allowing each of the flies to drink its fill of distilled water. Rinse the forelegs of each fly by passing them through several drops of distilled water, before beginning the experiments and again between each test. Record the results in Table 4.

3. Compare the olfactory and gustatory responses under conditions of thirst and water satiation, and suggest hypotheses to account for the differences observed. Is it possible to be certain, in the gustatory tests, that olfactory responses are not contributing to the results? Is it possible to separate these responses completely?

**PART IV.
THRESHOLD OF
RESPONSE TO
SUGAR**

As you should have been able to guess already, taste receptors in adult flies are located on the legs and labellum. When a sugar chemoreceptor in a fly is stimulated, a very specific response occurs: the proboscis is extended and the sugary liquid is imbibed. Some sugars are more effective than others in stimulating the receptors. There is some evidence that the stimulatory effectiveness of a compound may in part depend upon its molecular shape, and this principle may extend to human chemoreceptors, although there is now conflicting evidence on this point. Some substances,

FLY NUMBER / TEST SUBSTANCE	5	6	7	8
A				
B				
C				
D				
E				
F				

TABLE 3. GUSTATORY REACTIONS BEFORE DRINKING

FLY NUMBER / TEST SUBSTANCE	5	6	7	8
A				
B				
C				
D				
E				
F				

TABLE 4. GUSTATORY REACTIONS AFTER DRINKING

such as salts and alcohols, stimulate different neurons in the sensory hairs, and will cause rejection of a previously acceptable sugar solution.

The lowest stimulus effective in producing a response is called the **threshold stimulus.** Thresholds of chemoreceptors may be altered by many factors, such as age, sex, temperature, humidity, and the nutritional state of the fly. In the following experiments, we will determine the threshold for several different sugars, the rejection threshold for salt, and the effect of the nutritional state of the animal on the sugar threshold. If you have time, you may wish to design and carry out other experiments, but you should first obtain your instructor's permission.

A. Sugar thresholds

The following dilutions of sugars will be available:

Galactose	0.005	0.01	0.05	0.1	0.2	0.5	1.0 MOLAR
Glucose	"	"	"	"	"	"	"
Sucrose	"	"	"	"	"	"	"

GALACTOSE:
GLUCOSE:
SUCROSE:

TABLE 5-1. HUMAN SUGAR THRESHOLDS

FLY NUMBER TEST SUGAR									
GALACTOSE									
GLUCOSE									
SUCROSE									

TABLE 5-2. FLY SUGAR THRESHOLDS (BEFORE FEEDING)

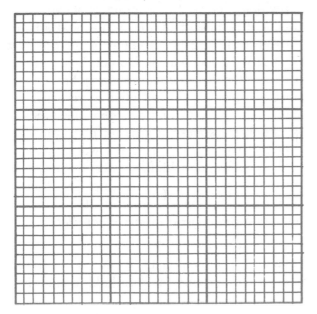

GRAPH A. FLY SUGAR THRESHOLDS

(INDIVIDUAL DATA)

GRAPH B. FLY SUGAR THRESHOLDS

(CLASS DATA)

1. Find your own threshold to each of the sugars, by dipping a clean finger into each solution (from least concentrated to most concentrated) and tasting until you first detect the sugar. Enter the results in Table 5–1.

2. Use depression slides or other small containers to hold the sugars for testing the flies, and make sure that each container is properly labeled. With so many concentrations it is very easy to get them confused. Allow each of the experimental flies (which should have been starved) to drink its fill of distilled water, and then test it by bringing the forelegs into contact with each of the solutions, beginning with the least concentrated one. Record as each fly's threshold that concentration at which the fly first gives a positive response (i.e., proboscis extension). Do *not* allow the flies to feed,

and make sure that each gives several negative responses to water before you begin tests with it. (Why are these precautions necessary?)

Enter all the results in Table 5–2. How much variability is there between individual flies? Plot your results on the graph, using the horizontal axis to express sugar concentration and the vertical axis to record the number of flies that responded to each concentration. Graph A should carry three plots, one for each sugar. Graph B can be used to plot the results polled from the entire class.

Would you say that your own flies represent an adequate statistical sample? Do the class data form a better sample? Why?

Compare sensitivity of human and fly to the different sugars, considering the threshold to each sugar as the lowest concentration to which at least 50% of the flies (or humans) respond.

B. Repellents

The threshold concentrations of substances that flies find repellent (such as salt, NaCl) may be determined by combining such substances with a solution known to be acceptable and above threshold. Failure to extend the proboscis to a solution in which sucrose, for example, is above threshold, is an indication that the repellent is present in above-threshold amount.

Select one of the above-threshold concentrations of sucrose to which a fresh fly consistently responds with acceptance. Do not, however, allow the fly to feed; remove it at once after the positive responses have taken place.

Mix an equal number of drops of the selected sucrose solution with drops of the repellent solution (1.0 M NaCl or some other substance supplied by your instructor). Test the fly's response to this mixture, and record the results in Table 6. Repeat the experiment, using a lower concentration of NaCl* if the fly rejected the mixture, or a higher concentration of repellent (e.g., 2 M NaCl) if the fly did not reject the first mixture. Try the first repellent solution with higher concentrations of sucrose. Does the rejection threshold depend entirely upon the concentration of repellent?

C. Effect of nutrition on the sucrose threshold

Select several flies, and allow each to feed on 1.0 M sucrose for ten to fifteen seconds. In the usual manner, redetermine the threshold of each fly to each of the test sugars; record the data in Table 7 and compare it with your original data for threshold in starved flies. Explain the results, and suggest hypotheses for the differences observed.

PART V. THE EFFECTS OF FOOD ON LOCOMOTION

You will be supplied with several flies from which the wings have been removed, or your instructor may ask you to remove the wings yourself, with fine scissors. These flies have been fed 0.1 M sucrose, a minimal diet, and have been allowed to get used to their condition overnight.

1. Place one of the flies on the table, and watch its pattern of walking. Does it walk in straight lines, or does it deviate from straight lines?

2. Place a drop of above-threshold (1.0 M) sucrose in contact with one of the flies, or bring the fly briefly into contact with the drop. Remove the

---

* The concentration may be halved, for example, by mixing a drop of 1M NaCl with a drop of water.

FLY NUMBER REPELLENT SOLUTION										

TABLE 6. REJECTION THRESHOLDS

FLY NUMBER REPELLENT SOLUTION										

TABLE 7. FLY SUGAR THRESHOLDS (AFTER FEEDING)

animal immediately from the drop, and compare its walking pattern with the walking pattern of the first test. If no difference is noticed, return the fly to the stock container and try the experiment with another animal.

Can you suggest an explanation for this difference in behavior? How might it be of advantage to the animal in nature?

3. With a fine brush, draw thin, concentric circles of distilled water and sugar solution on white paper as shown in FIGURE 15-5. Draw one pattern at a time.

FIG. 15-5. Circles used for study of effects of food on locomotory behavior.

a. Place a wingless fly in the innermost circle of the first set (A), and watch its behavior continuously until it walks out beyond the outermost circle. Record the walking with a line penciled into FIGURE 15-5.

b. Repeat the procedure in 1 with sets (B) and (C) (FIGURE 15-5), using a fresh fly each time. Record the walking as before.

Explain the behavior in each case.

4. On a piece of white paper, draw a "Y" about half an inch high in 1.0 M sucrose. Place a fresh fly at the stem of the "Y" and let it walk. Carefully record its movements, and watch especially what happens when the fly reaches the branching point of the "Y." What determines the direction the fly chooses at the branching point? Repeat the experiment several times until you have the answer.

5. On a piece of white paper, draw an "M" in 1.0 M sucrose. Put a wingless fly at one end of the "M" and let it walk. Can the fly follow the trail? If so, have you trained it, or has it "learned" to behave in this fashion?

TOPIC 16 CELLULAR REPRODUCTION

Biological Science, Chapter 13.
Elements of Biological Science, Chapter 13.

Though individual cells and organisms die, the continuity and increase of life are guaranteed by two types of nuclear division: **mitosis** and **meiosis,** processes that differ principally in the behavior of the chromosomes.

Coupled with cytoplasmic division processes, mitosis is a mechanism for transmitting unaltered genetic constituents from one generation of cells to another. This is ordinarily accomplished through a duplication of chromosomes, so that two complete sets are present in the parent cell; after the duplication, the sets are separated in such a manner that each daughter cell inherits a complete chromosomal complement, identical to that of the parent (FIGURE 16-1A). Mitotic cell divisions are important in growing and regenerating regions of organisms; in plants, they are confined mainly to the meri-

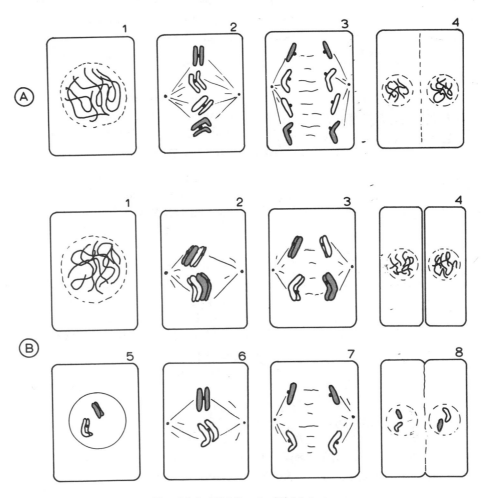

FIG. 16-1. (A) Mitosis. (B) Meiosis.

177

stems, such as buds, root tips, and cambium. In animals, mitotic divisions are found in cells at the base of epithelia, in the bone marrow, and in many other regions where tissues are growing or cells are being replaced.

Meiosis, in contrast, is a mechanism for reducing the number of chromosomes by half. In animals, this takes place in cells that will give rise to gametes, but it may occur at other points in a sexual reproductive cycle, as indeed it usually does in plants. To understand meiosis, you need first to recall that chromosomes commonly occur in pairs; members of a pair carry genetic factors for the same traits and are called **homologous chromosomes.** During *mitosis*, homologous chromosomes behave *independently*, and a copy of each is transmitted to each daughter cell. During *meiosis*, however, the homologues associate or **synapse;** synapsis is followed by separation of the members of each homologous pair into *different* daughter cells (FIGURE 16-1B). This fundamental difference in chromosomal behavior — independence of homologous chromosomes in mitosis and association and separation of homologous chromosomes in meiosis — is the key to understanding the differences between these two otherwise mechanically similar processes.

In this unit, you will make and study your own slide of dividing cells in the root-tip meristem of broad bean, *Vicia faba*. This study will give you an opportunity to review your textbook understanding of cell division and a chance to observe at firsthand the behavior of chromosomes in mitosis. A clear understanding of chromosomal behavior in both mitosis and meiosis is essential to your later work in genetics.

PART I. TECHNIQUE FOR MAKING SLIDES

Prior to the laboratory period, seeds of the broad bean were germinated and the root tips, which contain many actively dividing cells, were cut off, killed, and hardened in an acid solution. Preparation of this material for microscopy consists of three steps:

1. The root tip is prepared for staining; it is treated with hydrochloric acid, a procedure that exposes chemical groups called **aldehydes,** present in the chromosomal DNA, to which the stain may then attach.

2. The root tip is stained with **Feulgen reagent,** a dye mixture that adheres to the aldehyde groups exposed in the first step, and thereby colors the chromosomes a deep magenta.

3. The cells are dispersed into a single layer on a slide, and the slide prepared for microscopic observation. The slide may be studied at once, or it may be run through several other steps to make it permanent.

Making slides is not a haphazard process; it requires a great deal of care if the resultant preparation is to be worthwhile for study. Since you will have only one opportunity to make a good slide, read the directions carefully before you begin and follow them to the letter, noting any modifications your instructor may suggest at the beginning of the period.

A. Directions for temporary mounts

1. From your instructor, obtain a root tip in a vial of fixative. Discard the fixative into the waste beaker at your desk. Add distilled water to the vial, and leave for one minute. Discard the water, and add 6 N hydrochloric acid to half the capacity of the vial. Note the time. Leave the acid on the

roots for three minutes and then, with a dropper, remove it at once and discard into waste beaker. Add distilled water to the vial to rinse the roots. Agitate gently for one minute and discard the water.

2. With forceps, place the roots in a second vial, containing Feulgen reagent. Note the time, and remove the roots after 20 minutes. While they are staining, clean the first vial, and partially fill it with 45% acetic acid. Put the roots in the acid when they are brought out of the stain.

(*Note:* If you wish to make a permanent slide, omit step 3 and follow instead the steps in B. Read all the steps in that procedure before you try it.)

Fig. 16-2. Technique for making tissue smash. Tease the tissue into small pieces; add a drop or two of liquid, a cover slip, and a cork; apply vertical pressure.

3. With insect pins or a razor blade, tease or chop the tissue into very small bits in a drop of 45% acetic acid on a clean slide. This operation damages a few cells, but makes the majority much easier to smear later on.

Adding a drop of acetic acid if necessary, gently put a cover slip in place, taking care to eliminate air bubbles. Put a small clean cork upside down on the cover slip and press down firmly, in a vertical direction (Figure 16-2). Properly performed, this operation will smear the cells into a very thin layer. If the pressure is vertical, it is almost impossible to break the cover slip; it takes a great deal of pressure to make a proper smear, so don't hesitate to apply it!

B. Additional directions for permanent mounts

1. Clean a slide and cover slip with 95% alcohol, and place both these items on a clean paper towel at your desk. Put one drop of Mayer's albumen on the slide. With a clean finger, wipe almost all of it off, so that only a thin film remains. It is *essential* that a *clean thin* film covers the middle third of the slide; if this doesn't work the first time, clean the slide and try again.

2. Follow the directions in A3 above for making the smear.

3. Place the slide on a piece of dry ice for five to ten minutes. At the end of this time, while the slide is still on the dry ice, insert a razor blade at the edge of the cover slip and flip it off. Immerse the slide *at once* in a Coplin jar of 95% alcohol, where it should remain for five minutes.

4. Run the slide through three changes of 100% alcohol, each lasting five minutes, and one change of xylene, also five minutes.

5. Remove the slide from the last solvent change, let the excess solvent run off the slide onto a towel (but *don't* let the slide dry), and promptly place a drop of Euparal, Damar, or other mounting medium over the cells. Using your best technique to avoid air bubbles, mount the clean cover slip over the

Euparal or Damar. If there are bubbles in the medium before the cover slip is applied, they may be removed by poking with a heated needle.

PART II.
IDENTIFICATION
OF MITOTIC
PHASES

Although cell division is commonly separated into "phases" for convenience of discussion, it is well to remember that such separation is arbitrary, and that the process is in fact continuous. Keep in mind as you examine your slide that the stain you have used colors only DNA; other structures will show up indistinctly, if at all. The directions in A refer to the broad bean, in which you will see *only* the chromosomal events in cell division; other events, especially those involved in cytoplasmic division (**cytokinesis**) are much more conveniently studied in prepared slides, described in section B.

Study your slide under both low power and high power of the compound microscope, with reference to FIGURE 16-3.

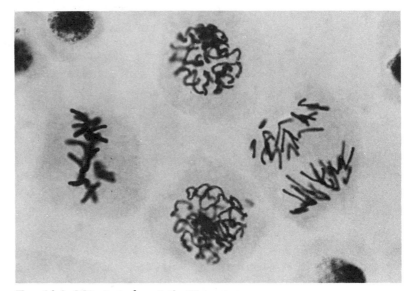

FIG. 16-3. Mitosis in hyacinth. This photograph was prepared from a squash mount similar to that you will make from *Vicia faba* root tip. (*Courtesy Carolina Biological Supply Company.*)

A. Broad bean

1. A majority of the cells on your slide will be in **interphase.** The stained granular nucleus of such cells is clearly delineated in the cytoplasm, though individual chromosomes are not visible. Often two or three nucleoli will be visible. Are they stained? What events take place during interphase? What evidence is there that they occur?

2. During **prophase,** the chromosomes — now doubled so that each consists of a pair of identical **chromatids** — shorten and thicken. Although they were not visible as discrete bodies during interphase, the chromosomes now appear as tangled threads. Later, as the threads become more distinct, the nuclear membrane disappears. Compare the appearance of the chromosomes in early and late stages of prophase. Note that in the former stages the chromosomes are clearly enclosed in a spherical area, the nucleus. Later, after the nuclear membrane has disappeared, the chromosomes spread out.

The pair of chromatids composing each prophase chromosome are attached at one point, the **centromere.** The centromere, in turn, apparently is the point of attachment of the chromosome to the **spindle.** The spindle (not visible on your slide, but easily seen in prepared slides) is formed in early prophase and consists of protein fibrils that probably play a key role in moving the chromosomes during the division.

By the end of prophase, the chromosomes are very short and thick and have migrated to the center of the cell.

3. During the brief stage termed **metaphase,** the chromosomes are arranged along the equatorial plate* of the spindle and appear to form a line across the middle of the cell. Each chromosome is attached by its centromere to a fibril of the spindle. (The centromere is visible in some metaphase chromosomes, under high power, as a clear area in the chromosomes.) It is actually the centromeres that are lined up precisely along the equatorial plate, while the "arms" of the chromosomes may extend toward the poles in various directions, giving the chromosomes a "frayed" appearance. At the end of metaphase, each centromere divides; each of the former chromatids now has its *own centromere* and is designated a separate chromosome.

4. At the beginning of **anaphase,** the two new single-stranded chromosomes move away from one another, one going toward one pole of the spindle and the other to the opposite pole. You can recognize a cell in early anaphase by observing that the chromosomes are in two groups, a short distance apart. By late anaphase, each chromosome group has almost reached its respective pole.

Under high power (or with an oil-immersion objective, and following the directions in Appendix 6, if your instructor allows) compare the appearance of metaphase and anaphase chromosomes. If possible, locate these two phases close together on your slide. Make sure that your lenses are very clean and that the light is properly adjusted. Under favorable conditions, you will be able to see that a metaphase chromosome consists of two chromatids, connected by the unstained centromere, while an anaphase chromosome is single-stranded. If you cannot actually see the double structure of the metaphase chromosome, a comparison of the *width* of an anaphase and metaphase chromosome in adjacent cells should demonstrate clearly that metaphase and anaphase chromosomes consist of different amounts of material.

5. In early **telophase,** the chromosome clusters have reached the poles, but the "arms" of the chromosomes are still oriented along the spindle. In mid-telophase, the clusters are more compact, and individual chromosomes are no longer oriented along the spindle. By late telophase, new nuclear membranes have formed around the clusters, and individual chromosomes are indistinct. Telophase ends when the nucleus assumes the characteristics of interphase, and the full mitotic cycle is completed.

When you feel you have a clear idea of chromosomal behavior in mitosis, find examples on your slide that *differ* from the illustrations and make a large drawing of the nucleus and/or chromosomes of each phase. It may be necessary to make several different sketches for some of the phases that change in appearance as they progress. Label the drawings carefully.

---

* An imaginary plane running through the middle of the cell at right angles to the long axis of the spindle.

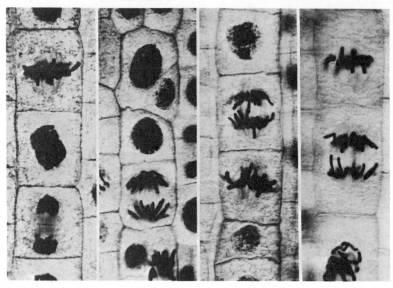

FIG. 16-4. Mitosis in onion root tip — longitudinal section. (*Courtesy Carolina Biological Supply Company.*)

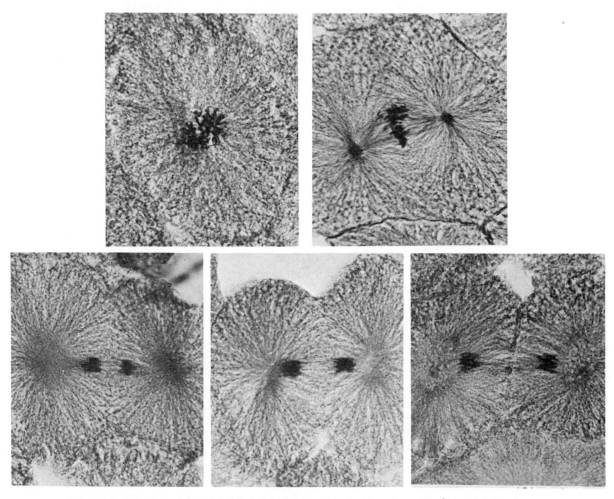

FIG. 16-5. Mitosis in whitefish blastula (early embryo). (*Courtesy Carolina Biological Supply Company.*)

B. Onion and whitefish (FIGURES 16–4 and 16–5)

1. Locate the five stages of mitosis in a prepared slide of onion (*Allium*) root tip (longitudinal section). Nucleoli will be clear in interphase cells, and the spindle is usually seen easily in anaphase cells. Cytokinesis begins during early telophase and is generally complete before telophase has finished. As in most plants, cytokinesis occurs through the formation of a **cell plate** (which eventually becomes the middle lamella of a new cell wall) in the region of the equatorial plate.

2. Locate the five stages of mitosis in a prepared slide of whitefish blastula (*Leucichthys*). A blastula is an early embryonic stage, produced by successive early divisions of the zygote. A word of caution: When you study these slides, remember that the cells of the blastula are not oriented in the regular longitudinal fashion typical of a plant root. Therefore, a section through the whole embryo cuts cells in many different ways. Try to locate longitudinally sectioned cells to study mitosis.

Animal cells contain a **centriole,** a structure associated with spindle formation that is generally absent in plant cells. The centriole is too small to be seen without the electron microscope, but it is located in the centrosome, a clear area that can be easily seen. Some of the fibrils that radiate from the centriolar region are fibrils of the spindle; others, radiating toward the poles, form two starlike **asters,** also absent in plant cells.

Cytokinesis occurs as a constriction, or **cleavage furrow,** which forms in the equatorial region. This furrow deepens progressively until it cuts through the cell and its spindle, producing two new cells.

Make sketches to show the differences between the plant and animal cells in mitosis and cytokinesis, or label FIGURES 16-4 and 16-5.

PART III. THE TIME SEQUENCE IN MITOSIS

In your slide examination, you probably noticed that some stages were found much more often than others. Which phase did you see most frequently? There are differences in the frequency of appearance of the phases because each one takes a different time to occur. Considering only what you know about the events of each phase, and *without* consulting your slide, make a list of the five phases in order of *decreasing* frequency in which you would expect to find them.

1.
2.
3.
4.
5.

Now examine the slide and, in the following manner, determine the relative frequency of each of the five stages in mitosis:

1. Select a region of the root tip that seems to have dividing cells scattered regularly throughout it, and is one cell thick.

2. In this field, count the number of cells in each phase, entering that number in the chart below and in the chart for data from the entire class. Add the numbers in your column to find the total number of cells counted.

3. Divide the number of cells in each stage by the total number counted, and multiply by 100 to get the percentage of cells in each stage.

	NUMBER OF CELLS	PERCENTAGE OF CELLS	TIME
Interphase			
Prophase			
Metaphase			
Anaphase			
Telophase			
Your Total			
Class Total			

4. Assuming that the difference in time for completion of each stage accounts for the difference in observed frequency of stages, and that mitosis in broad bean normally takes about 16 hours from the beginning of interphase until the end of telophase, calculate from your percentage data the approximate time (in minutes) it takes for the completion of each phase of mitosis. Enter these calculations in the appropriate column of your table.

Which stage is the longest? Is this what you predicted?

Which stage is shortest? Does this comply with what you know occurs during this stage?

Compare the frequencies of stages in the class data with your own. If there are discrepancies, how could you explain them?

Why do you think it is better to draw conclusions from the sum of data submitted by the class than from your data alone?

QUESTIONS FOR FURTHER THOUGHT

1. In a single sentence, how would you explain the basic significance of mitosis?

2. Explain the difference between the terms chromosome and chromatid.

3. In which stage of mitosis does DNA duplication occur?

4. Suggest an hypothesis to explain the adaptive value of having genetic material packaged in chromosomes.

5. What evidence is there in anaphase cells to indicate that chromatids are mechanically moved apart during anaphase? What structures are probably responsible for the movement?

6. Which of the phases do you think would require the greatest expenditure of energy? How would you design an experiment to find out?

7. If you know the time it takes to complete a mitotic cycle, and if your information regarding the time for completion of each phase is correct, what other information would you need in order to calculate the *rate of movement* of chromosomes in anaphase?

8. You inferred the length of mitotic phases from information about the total time for a mitotic cycle and the relative frequency of each stage. What technique would you need to get the same information directly?

9. In a single sentence, state the fundamental difference in chromosomal behavior between mitosis and meiosis.

10. Without looking at your notes or your laboratory manual, take a sheet of paper and make large diagrams of mitosis and meiosis in an organism with three pairs of chromosomes. Make a list of the most important differences between mitosis and meiosis.

11. Define the term or, where appropriate, describe its role in cell division:

prophase	cell plate
metaphase	nucleolus
anaphase	centromere
telophase	aster
interphase	homologous chromosomes
cytokinesis	synapsis

GENETICS AND THE ANALYSIS OF DATA

Biological Science, Chapter 14.
Elements of Biological Science,
Chapter 14.

The chromosome theory of heredity is based largely upon experiments that have involved the crossing, or mating, of diploid organisms, those that contain pairs of alleles for each trait. In heterozygotes, the presence of one kind of allele often totally masks or dominates the other, and the actual genetic makeup of the organisms can be ascertained only through breeding experiments, where the alleles are *segregated* into gametes and then *recombined* in different ways. In this laboratory, you will study the results of some crosses very similar to those performed by Mendel in his classical experiments on heredity in peas. In part, this study will enable you to determine what kind of genetic information can be inferred from such experiments; in addition, you will learn how the data resulting from such experiments can be subjected to statistical tests of validity. Finally, you will perform some crosses yourself, and will examine a few genetically controlled traits in human beings.

PART I. MONOHYBRID AND DIHYBRID CROSSES
A. Monohybrid cross

A monohybrid cross is one in which the pattern of inheritance of one pair of contrasting alleles is being studied. The production of chlorophyll in tobacco is controlled by a dominant allele, which we may arbitrarily designate C; the alternative recessive allele c, when present on both chromosomes, gives rise to albino plants. You will be provided with a dish of agar upon which about 100 tobacco seeds have been germinated. These seeds were parented by a pair of heterozygous (Cc) tobacco plants. (Alternatively, your instructor may ask you to plant the seeds yourself one week prior to the laboratory.)

Classify and count the seedlings, and record the numbers in Table 1-A below, ignoring for the moment Table 1-B labeled "Analysis."

| | NUMBER OF SEEDLINGS | |
	GREEN	ALBINO
Individual		
Class		

TABLE 1-A.

Had you not been given the information about dominance of the allele for chlorophyll production, how would you have been able to determine that information *from your data*?

What crosses would you have made to produce the hybrid parents of the seeds you germinated? What would have been the genotype of each of the parent plants?

ANALYSIS	CLASS 1	CLASS 2
Observed values		
Expected values (e)		
Deviation (d)		
Deviation Squared (d^2)		
d^2/e		
$\chi^2 = \Sigma\,(d^2/e)$		

TABLE 1-B.

What is the genotype of albino plants?

What are the possible genotypes of your green plants? If you had to obtain more hybrid plants in order to get seeds for next year's class, how would you go about isolating the hybrids *from your own plants*?

B. Statistical analysis of data

On the basis of your knowledge of segregation and recombination of alleles, you would have predicted a *phenotypic* ratio of three greens to one albino, as shown in the Punnett square below:

	$\underline{C}$	$\underline{c}$
$\underline{C}$	$\underline{CC}$	$\underline{Cc}$
$\underline{c}$	$\underline{Cc}$	$\underline{cc}$

The ratio you actually found when you counted the seedlings was probably not exactly 3 : 1, just as flipping a coin a number of times will not yield exactly equal numbers of "heads" and "tails" even though the predicted ratio is 1 : 1. If the *deviation* (difference between your actual and expected results) is small, you would not be very disturbed; you intuitively, and correctly, expect some deviation in any random process, such as coin tossing or — in your tobacco experiment — the segregation of alleles at meiosis and their recombination at fertilization. But what if the deviation were large? Was something wrong with the way the experiment was run? Was something wrong with your hypothesis (what you expected)? How large must a deviation become before you will decide to reject your hypothesis and form a new one?

The problem of formulating mathematically consistent and objective answers to such questions faces almost every scientist and enters almost every realm of research. The use of statistics offers a reasonable and standardized solution. One statistical test of deviation significance, widely applied to genetic ratios, is the so-called **chi-square** (χ^2) **test.** The formula for this test, derived from mathematical calculations that we cannot discuss here, is as follows:

$$\chi^2 = \Sigma\,(d^2/e)$$

In this formula, d is the deviation from the predicted value, e is the predicted (hypothetical) value, and Σ means "the sum of." The expression (d^2/e)

alone gives a chi-square value for each "class" or group of data in which deviation can occur. For example, in our tobacco experiment, there are two classes of data, green and albino. Chi-square values for each group of data are added, and the total value is compared with a table that gives information about the significance of the deviation, as we shall see shortly.

We shall illustrate the use of this test by considering two hypothetical tobacco seedling experiments. Suppose that in Experiment A, we had obtained 85 green seedlings and 35 albinos. The total sample would then have been 120, and the expected ratio would have been 3 : 1 or 90 greens to 30 albinos. In hypothetical Experiment B, suppose that we had counted a total of 32 seedlings, of which 19 were green and 13 were albino. The data and chi-square calculations for Experiment A are in Table 2 below. In Table 3, you can fill in the corresponding information and calculations for Experiment B.

	CLASS 1	CLASS 2
Observed values	85	35
Expected values (e)	90	30
Deviation (d)	-5	$+5$
Deviation squared (d^2)	25	25
d^2/e	$25/90$	$25/30$
$\chi^2 = \Sigma\,(d^2/e) =$	$0.28 + 0.83 = 1.11$	

TABLE 2.

	CLASS 1	CLASS 2
Observed values		
Expected values (e)		
Deviation (d)		
Deviation squared (d^2)		
d^2/e		
$\chi^2 = \Sigma\,(d^2/e) =$		

TABLE 3.

Notice the difference in total chi-square value: about 1.1 for Experiment A and 4.1 for Experiment B. We shall want to consider three very important factors that influence the size of these numbers.

From the formula for the chi-square test, it should be clear that the larger the value of d, the larger the chi-square value for that class of data; second, the larger the value of e (expected), the smaller the chi-square value for that class of data. Finally, the more classes of data under consideration, the larger we should expect the total value to be. For example, if we had only two classes of data (as in our tobacco experiments), we expect less opportunity for deviation (and a smaller chi-square total) than if we had considered an experiment in which there were ten different data classes, and hence ten opportunities for deviation. Since the d-value and the number of classes of data were the same in both our hypothetical experiments, only the expected values varied. These, of course, reflect the size of the sample. In a sample as small as that in Experiment B, we should not be very surprised to find a large chi-square value.

Chi-square is therefore a reflection of three factors: deviation size, sample size, and the number of opportunities in which deviation can occur. It can help us to answer two questions:

1. How probable is it that our x^2 values would have occurred simply by chance?
2. Should we reject our hypothesis, or not?

We can answer the first of these questions by comparing our chi-square values with those shown in Table 4.

	NO REASON TO DOUBT HYPOTHESIS							REASON TO DOUBT HYPOTHESIS		
	DEVIATIONS INSIGNIFICANT							DEVIATIONS SIGNIFICANT		
p / C-1	.99	.95	.80	.50	.30	.20	.10	.05	.02	.01
1	.00016	.0039	.064	.455	1.074	1.642	2.706	3.841	5.412	6.635
2	.0201	.103	.446	1.386	2.408	3.219	4.605	5.991	7.824	9.210
3	.115	.352	1.005	2.366	3.665	4.642	6.251	7.815	9.837	11.341
4	.297	.711	1.649	3.357	4.878	5.989	7.779	9.488	11.668	13.277
5	.554	1.145	2.343	4.351	6.064	7.289	9.236	11.070	13.388	15.086

TABLE 4. VALUES OF CHI-SQUARE.

In brief, Table 4 shows the *maximum* chi-square values allowable if the deviations actually found are to be considered due to chance alone. The vertical column C-1 ("classes minus one") shows the number of classes of data under consideration minus one, a number often called the degrees of freedom. For both of our hypothetical experiments A and B, this number is 1. The p-value, on the other hand, expresses the *probability* that deviations yielding the indicated value are due to chance alone. Comparing our value of 1.1 (for Experiment A) with Table 4, we find that the corresponding p-value for one degree of freedom is between 0.20 and 0.30. This means simply that we could expect a chance deviation as large or larger than the one we actually obtained in 20 percent to 30 percent of the cases considered, if we were to repeat the experiment again and again.

To answer our second question, we must first decide on a reasonable p-value, and then find where our chi-square total lies with respect to that p-value. A high chi-square value usually indicates that the experimental results deviate greatly from the predicted results. Clearly, the more the results deviate from the predictions, the more likely it is that the predictions were based on an incorrect hypothesis. It follows, therefore, that a high chi-square value is a more significant indicator of the possible need for revision of the hypothesis than is a low chi-square value. By convention, we accept as statistically significant (and therefore as cause to seriously doubt our hypothesis) any deviation so great that the probability of its having occurred by chance alone is 0.05 or less. In other words, any chi-square value large enough, or larger, to yield a p-value as low as 0.05 is said to reflect a statistically significant deviation — a deviation that warrants re-examination of the hypothesis.

A chi-square value large enough to yield a *p*-value of 0.01 is said to be highly significant.

On this basis, our deviation in Experiment A was insignificant; that for Experiment B was significant. Had we actually obtained experimental values like those in hypothetical Experiment B, we might repeat the experiment with a larger sample, recognizing that the chi-square test is particularly sensitive to sample size; if we continued to get significant *p*-values with a larger sample, we would have to revise the hypothesis from which our expected values were derived, or we would have to look carefully at the way we ran our experiment. If we are quite sure of the hypothesis (as we would be in this case) we may have failed to count the seedlings accurately, someone may have pulled some of them up before we got a chance to examine them, or perhaps some physiological factor has caused differential germination. Only after ruling out such possibilities would we begin to look for a genetic explanation of our significant deviation.

Carry out the chi-square test on your results from the tobacco seedling experiment; a space is provided in Table 1-B. Are the deviations statistically significant?

C. Dihybrid cross

You will be given an ear of corn that represents the F_2 generation of a cross between plants differing in *two* characteristics. The F_1s were genotypic hybrids for two characters, kernel *color* and kernel *texture*. One pair of alleles controls color (purple or yellow) and the other pair controls texture (smooth or rough, often referred to respectively as "starchy" or "sugary"). The manner in which such crosses are made is shown by the photographs in FIGURE 17-1.

Classify and count the F_2 progeny (the kernels) and record the data in Table 5.

	CLASS 1	CLASS 2	CLASS 3	CLASS 4
Observed values				
Expected values (e)				
Deviation (d)				
Deviation squared (d^2)				
d^2/e				
$x^2 = \Sigma\,(d^2/e)$				

TABLE 5.

Which of the alleles for color is dominant? Which of the alleles for texture is dominant? How do you know?

How do you know that the color and texture alleles are located on different chromosomes, and are therefore segregated and inherited independently?

Determine chi-square values for your own data and then again for data collected from the whole class. Are the deviations significant? If anyone in the class *does* have significant deviations, try to determine their cause.

FIG. 17-1. Controlled pollination of corn. (A) Bag is secured over tassel, where pollen is produced. (B) Bag is placed over ear while it matures. (C) Bag is removed from mature ear, exposing silk brush. (D) Pollen is poured from tassel bag over silk brush. The silks are elongated styles (part of the female flower; see Topic 24 and your text); the pollen lands on the silks and eventually grows down through the silk into the ovary, within each kernel of the ear. (*Courtesy Carolina Biological Supply Company.*)

PART II.
SEX-LINKAGE IN
DROSOPHILA

The fruit fly, *Drosophila melanogaster*, has been a very important experimental organism for genetic studies. Like the other classical subjects of genetic research, it has many easily distinguishable mutants. Its main attractions, however, are small size, short life cycle, and ease of culture in the laboratory, where it can be readily induced to form large numbers of offspring in a relatively short time.

As in many other animals, one pair of chromosomes controls the expression of sex in *Drosophila*. These chromosomes are arbitrarily designated *X* and *Y*. A female is genotypically *XX*, and a male is *XY*. Certain traits, such as eye color in *Drosophila*, are carried only on the *X* chromosome; such traits are spoken of as being sex-linked. Sex-linkage was discovered with *Drosophila* in 1910, and we shall repeat the classical experiments to determine the phenotypic ratios obtained when alleles (in our case, for eye color) are sex-linked.

The entire experiment spans four weeks, according to the following schedule. Your instructor will explain modifications in the schedule, which parts you are to follow, and when you are to do them.

Week 1. Remove parental generation; learn how to handle and sex flies, and observe stages in the life cycle (this may also be done in Week 2 or 3).

Week 2. Remove and count F_1 (hybrid) generation; prepare a second cross using the F_1 hybrids as parents.

Week 3. Remove F_1 parents.

Week 4. Count F_2 generation; analyze data.

Week 1

During this laboratory, your instructor will return your culture so that you can remove the parent flies, and also learn about the life cycle and handling of the flies. Alternatively, he may remove the parents for you and return the culture the following week.

DROSOPHILA
LIFE CYCLE AND
NATURAL HISTORY

1. *Drosophila* occurs naturally on rotting fruit; both the adults and the larvae feed on plant sugars and on the wild yeasts that grow in rotting fruit. The female flies lay eggs on the same materials. After a day or two the eggs hatch into small larvae, which feed and grow for about eight days, depending upon the temperature. During the period of growth they molt twice, and there are therefore three larval periods of growth between molts, called **instars.** When fully grown, third-instar larvae cease feeding, climb onto the walls of the culture vessel, and pupate. During pupation, which lasts about four days, the larval tissues are reorganized to form those of the adult. The life cycle is summarized in FIGURE 17-2.

All the stages should be visible either in the medium or on the walls of the culture vial. Use a dissecting microscope to find them.

HANDLING THE FLIES

2. To facilitate handling and sexing, flies are usually anesthetized with ether. It is important to avoid too much or repeated anesthesia because it easily kills or sterilizes newly emerged flies and has deleterious effects on older individuals. The following directions apply to the anesthetizer made by

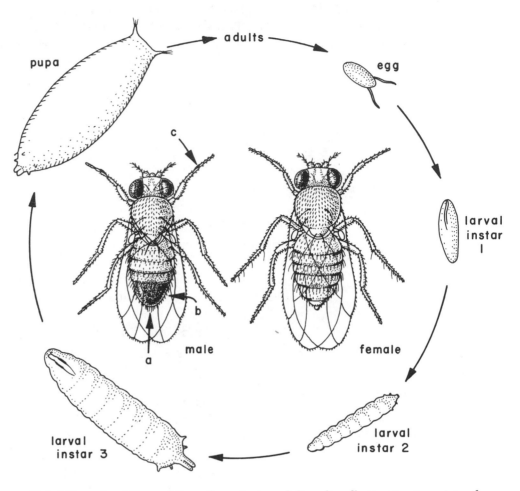

FIG. 17-2. Life cycle of *Drosophila*. a, b, and c on adult male call attention to points of sexual difference mentioned in text.

the Carolina Biological Supply Company, but your instructor may prefer to have you use some other method.

a. Remove the stopper from the top of the anesthetizer. Remove the cap from the bottom. Fill a third of the hollow stopper with ether for anesthesia. Pour the ether on the foam pad in the bottom of the etherizer. Immediately replace the cap and put the hollow stopper back in the top of the etherizer.

Keep burners and cigarettes **away** *from ether; it is highly flammable.*

b. When you are ready to etherize the flies, remove the stopper from the top of the etherizer. Tap the bottom of the culture vessel against the palm of your hand to knock the flies down. Remove the plug from the culture. Invert the culture over the etherizer and tap the flies into the chamber. After all the adults have been tapped into the chamber, quickly right the culture vessel so that its base covers the top of the etherizer. Plug the culture vessel. Tap the base of the etherizer on the table, remove the culture vessel, and plug the etherizer with its own stopper. Watch the behavior of the flies in the chamber. About twenty seconds after they stop moving, they may be dumped onto a white card for examination. (*Note:* If it is necessary to re-anesthetize flies, the anesthetizer may be reinverted over them with the hollow stopper removed. Avoid doing this repeatedly, as noted above.)

Flies are best observed and sorted on a white card, using a camel's-hair brush as manipulator.

c. Remove the parent flies from your culture, and purposely kill them by over-etherizing. They may then be observed more closely (for sexual characteristics) without fear of their escape.

SEXING DROSOPHILA

3. It is very important for you to be able to distinguish easily the sex of your flies. There are several sharp anatomical differences that make this a simple matter after you have had a little practice. In FIGURE 17-2 some of the sexual differences are indicated by arrows on those parts of the male specimen that differ from corresponding regions on the female specimen.

The male is generally smaller than the female, but this may be confusing if both animals are immature. One good indicator of sex is the presence of black bristles (a) about the male genitalia; these are absent in the female. Another good indication (b) is the dark-tipped male abdomen, but this again may be confusing on immature specimens. The sex combs (c), groups of black bristles on the male forelegs, are a good indicator of sex. A final criterion is the distance between the eyes, which is less in the male than in the female; you can make this comparison if the animals are seen together.

You will notice that some of the flies have red eyes, and some white eyes. These characters are controlled by a pair of sex-linked alleles. By making reciprocal crosses (red male × white female and white male × red female) we can learn how these alleles are inherited, and what the dominance relation is between them. In Table 6, in the box labeled "parents," note which culture you received and the number of adults of each sex and eye color that started the culture. This information already will have been recorded on the culture vial.

Discard the dead parent flies, and any other flies you subsequently wish to remove permanently, into a jar of alcohol marked "morgue" on the supply desk.

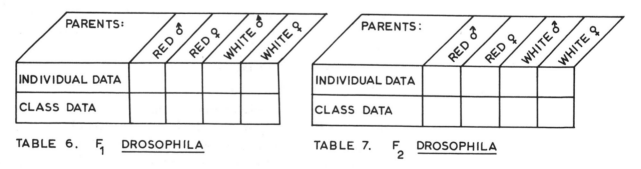

TABLE 6. F₁ DROSOPHILA   TABLE 7. F₂ DROSOPHILA

SALIVARY GLANDS

4. In *Drosophila* and some other flies, larval growth occurs principally as a result of cell enlargement. In some species the larval salivary glands consist of large cells containing nuclei with extremely large chromosomes. Such chromosomes have been widely used in studies relating chemical activ-

ity in the larva to the morphology (particularly the banding pattern) of the chromosomes.

To see the giant chromosomes, select from your culture a third-instar larva, crawling on the sides of the culture vessel, preparing to pupate. Place the larva in a drop of saline solution on a microscope slide, and examine with the dissecting microscope. Orient the larva with dissecting needles, so that the anterior end, which bears small, black toothlike structures, points away from you. Hold the larva by placing a needle across its posterior end, and insert another needle just behind the mouthparts. Pull the anterior needle away from you while keeping the posterior one in place. In the resulting debris, you will find the salivary glands — two saclike structures. They are almost transparent. Attached to them are small pieces of yellowish, opaque fat tissue. In most cases the glands will have been detached from their common duct.

When you have found the glands, remove the extraneous material, and take up excess saline with a piece of paper towel. Be careful not to soak up the glands as well. Add two drops of aceto-orcein stain, and leave for ten minutes, being *sure* the stain does not dry out. Add another drop of stain during the staining period if necessary. When staining is complete, remove most of the stain with a paper towel. Place a clean cover slip over the glands, a cork over the cover slip, and squash straight down with the cork. Pressure must be great enough to break the nucleus but not so great as to damage chromosomes. Examine the slide quickly, and if it is worthwhile examining at greater length, seal the edges of the cover slip with vaseline, to prevent drying out.

Week 2 In this laboratory, you will prepare a second mating, using as parents the F_1 hybrids that have now emerged in the first culture. In addition, you will count and score (for eye color and sex) all the F_1 individuals.

1. Prepare a culture vial and label it appropriately.

2. Etherize the F_1 flies lightly, until they have become motionless. *Do not kill them.* Select four males and four females, being *certain* of sex; do not use an animal if you are uncertain of its sex. Put the flies into the new culture vial. Make sure they are normal again (flying about, not caught in the medium) and that the vial is properly labeled before you return it to the supply desk.

3. Kill the remaining flies by over-etherizing, and classify them by sex and eye color. Record these data in Table 6.
(*Important note:* If your original culture was started from red-eyed females and white-eyed males, and this week you find progeny with white eyes, there has been a mistake in the preparation of your original culture. If your original culture was the reciprocal, and you find no white-eyed males, there was an error in its preparation. If either of the above applies to your culture, follow the remaining instructions using flies from another student's culture.)

4. Thoroughly wash and rinse the old culture vial, and return it upside down to the supply desk, along with its stopper.

Week 3   Remove the F_1 hybrid parents and discard them; make sure these parents are the same as indicated on the culture vial.

Week 4   Classify the F_2 flies by sex and eye color, and record the data in Table 7.

1. Which of the alleles is dominant? How do you know this from your data?

2. Knowing which allele is dominant and knowing the nature of sex-linkage, make predictions of the genotypic and phenotypic ratios for the F_1 and F_2.

3. Subject your data to chi-square analysis. Are the deviations statistically significant?

4. The instructor will lead a discussion of results and perform a chi-square analysis on data summed from the entire class.

PART III. MULTIPLE ALLELISM   Many, indeed most, gene loci in many species of organisms may be represented by more than two alleles. Generally all the alleles in a given series affect the same character (in a quantitative way) but they may control different aspects of a character. Most genes also have **pleiotropic effects** (i.e., they affect more than one character). Where multiple alleles have recognizable pleiotropic effects, the effects of members of a series on different characters may or may not run in parallel.

We shall consider only three alleles in an important series of multiple alleles in man that determines blood types in the **A-B-O** series. This case is a good example of how multiple alleles work, and will give you some knowledge of an important character in human beings. By typing your own blood, you will learn how blood groups are determined, and thus be introduced to some concepts from the fields of immunology and serology. After you have determined your own blood type, the genetic basis for these blood groups will be considered.

A. Blood typing   The red blood cells (erythrocytes) of human beings may contain proteins (**antigens**) designated A and B, or there may be no antigens present. Blood plasma contains other proteins (**antibodies**) designated anti-A and anti-B, which, when in the presence of particular corresponding antigens, will cause those antigens to adhere to one another. If blood from one individual is mixed with blood or plasma containing corresponding antibodies, the red blood cells clump together, or agglutinate, due to the adhesion of their antigens. Obviously a person's blood cannot contain the antibodies corresponding to his own blood cells, but his plasma normally contains antibodies for all other antigens in the series. When blood transfusions are to be given, then, it is always best to obtain a donor of the same blood type as the patient. If such a donor is not available, blood of another type may be used provided that the plasma of the patient is compatible with the red blood cells of the donor. Donor antibodies are diluted during the transfusion and will not cause serious difficulty unless the transfusion is very extensive.

An individual's blood type could be determined by mixing a drop of his blood with sera from persons of known blood types. It is more convenient, however, to use commercially prepared sera.

1. Using a heavy wax pencil, divide a perfectly clean slide into three compartments by drawing heavy lines and labeling as indicated:

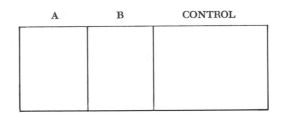

2. Add a drop of serum containing anti-A antibody to the first compartment on your slide and a drop of serum containing anti-B antibody to the second compartment.

3. Puncture your finger with one of the sterile disposable lancets provided. Do *not* use a lancet previously used by another student. Add a drop of blood to each compartment on your slide, being careful to allow the blood to fall into the drops of sera or to transfer it with a *clean* toothpick, so that there is no chance of transferring any of the sera from one compartment to another. Mix the material in the first compartment with the tip of a clean toothpick, and then discard the toothpick. With another clean toothpick, mix the material in the second compartment, and discard the toothpick. Stir the third drop with a third toothpick and allow it to stand as a control. What is the value of a control in this experiment? (*Caution:* Do not use a single toothpick for more than one antiserum. Why?)

4. After about half a minute, observe each drop on your slide. If agglutination has occurred, the erythrocytes will be seen sticking together in small clumps. If no agglutination has occurred, they will be scattered uniformly throughout the liquid. (Changes in appearance after two or more minutes are normal clotting and should not be regarded as agglutination. If you are uncertain of your results, use your dissecting microscope.) Which condition should be observed in the control? Record your findings in the table below, using the symbols given.

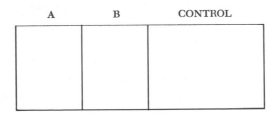

+ = agglutination
− = no agglutination

5. The left side of the table below shows the antigen and antibody content of blood of each of the four types in the A-B-O series. From these data you should be able to determine which types will give positive results when tested with the two types of sera; you are to place plus or minus signs in the appropriate places in the two columns on the right side of the table.

GROUP	BLOOD CONTAINS		REACTION WITH ANTIBODIES	
	Cellular antigens	*Plasma antibodies*	*anti-A*	*anti-B*
O	None	anti-A and anti-B		
A	A	anti-B		
B	B	anti-A		
AB	A and B	None		

6. Using the information in the above table, determine the answers to the following questions:

What is your blood group?

What type (or types) of blood could be safely transfused into your circulation?

Into what type (or types) of recipient(s) could your blood be safely transfused?

Explain your answers to each of the last two questions.

B. Genetics of the A-B-O blood groups

1. The inheritance of the A-B-O groups is best described by a theory of three alleles, which are designated I^A, I^B, and i (or sometimes L^A, L^B, and L^O). Both I^A and I^B are dominant over i, but neither I^A nor I^B is dominant over the other when both are present in the heterozygote. Accordingly, the four blood groups, or phenotypes, correspond to the genotypes indicated in the table below:

BLOOD GROUP	GENOTYPE
O	ii
A	I^AI^A or I^Ai
B	I^BI^B or I^Bi
AB	I^AI^B

On this basis, what is your genotype? _____ If you know the blood groups to which your siblings or parents belong, you may be able to specify your genotype more exactly.

2. The table below gives the approximate frequency of each phenotype in the United States population. The frequencies are often different in the populations of different nationalities and geographic regions and, as a result, can be used in studying the origins and relationships of different peoples. As you can readily guess, there is much variation in the United States population because of its diverse origins. Record the blood group frequencies for the entire class in the table. Using a chi-square test, compare your results with those indicated.

| FREQUENCY IN POPULATION | | PHENOTYPE | NUMBER IN | FREQUENCY IN |
(U.S. White)	(U.S. Negro)	GROUP	CLASS	CLASS (%)
45%	47%	O		
41%	28%	A		
10%	20%	B		
4%	5%	AB		
100%	100%		Total	100%

3. Blood typing is often used as evidence in paternity cases in courts. In a series of disputed paternity cases, the mother and child in each had the blood types listed in the table below. For each, indicate the blood types which, if found, would exonerate an accused man.

MOTHER	CHILD	MAN EXONERATED IF HE BELONGS TO GROUP
A	O	
B	AB	
O	A	
AB	A	
O	O	
B	B	
A	B	
AB	AB	
A	A	
A	AB	
B	A	
B	O	
AB	B	

PART IV. MULTIPLE GENE COMPLEXES

Often, several different genes all affect the same character. The ways in which such genes interact vary greatly. Some have additive effects, some are antagonists, etc. It is also common for several independent genes to affect closely related but different characters; the characters then interact in their effect on the life of the organism. Here you will study an example of the latter type of genetic complex.

You learned earlier that the sense of taste in human beings is commonly analyzed in terms of four basic tastes — sweet, sour, bitter, and salty. The sensitivity of each of these taste senses to various substances is controlled by many different genes. Thus one gene may determine whether substance A is tasted as sweet, a second gene may determine whether substance B is tasted as sweet, a third may do the same for substance C, etc. Each of these genes controls a separate character, but the characters all interact to affect the way in which a person tastes a meal. You will test yourself for several such genetic taste factors in this section.

1. Obtain a piece of test paper that has been treated with the chemical phenylthiocarbamide (PTC). Making sure that your tongue is moist, touch the paper lightly to your tongue and then record whether or not you taste it; if positive, record whether the sensation is a sweet, sour, bitter, or salty one. If you did not taste the PTC, try again by chewing a small piece of the paper, recording results as above.

The ability to taste PTC is inherited, and is apparently determined by a pair of alleles. Normally, the allele for tasting is dominant over that for non-tasting. In the United States, about 70 percent of the people are tasters (the great majority taste the chemical as bitter), while about 30 percent are non-tasters. The percentages vary greatly between different human stocks, and are used in racial research, as are those of blood types.

It should be emphasized that the ability to taste PTC is not always clear-cut; there is much variation in the sensitivity threshold. It is also interesting to note that the chemical can be tasted only if it is dissolved in the taster's own saliva; even a sensitive taster cannot detect the PTC if it is dissolved in water or in someone else's saliva and then placed on his dry tongue.

2. Obtain a piece of thiourea test paper and, following the procedures used above, determine whether you are a taster or non-taster. This substance is closely related chemically to PTC but the ability to taste it is inherited independently. Thus, although most people can taste thiourea (as in the case of PTC), the taster and non-taster groups for the two substances need not be the same.

3. Obtain a piece of sodium benzoate test paper and determine your ability to taste it. If you are a taster, determine whether the sensation is a sweet, sour, bitter, or salty one. The ability to taste sodium benzoate is inherited independently of the sensitivity to PTC, but the two taste characters apparently interact markedly in their effect on a person's reaction to various foods. For example, people who taste PTC as bitter and sodium benzoate as salty tend to like such foods as sauerkraut, buttermilk, turnips, spinach, etc., better than average, while people who fall in the bitter-bitter group for the two test chemicals like them less than average. Giving PTC first and sodium benzoate second, the most common taste group seems to be bitter-salty, followed by bitter-sweet, bitter-bitter, and tasteless-salty. Most other combinations also occur.

QUESTIONS FOR FURTHER THOUGHT

1. Construct a diagram showing independent assortment of alleles during meiosis.

2. How does meiosis in sperm formation differ from that in egg formation?

3. What steps would a plant breeder take to obtain seeds like those you

used in your tobacco-seedling experiments? How could the user of such seeds be absolutely certain that the seeds represented genuine offspring from a monohybrid cross? How could a 3 : 1 ratio be "faked?"

4. What is a sex-linked allele?

5. What kinds of characteristics make an organism suitable for breeding experiments? What groups of organisms would make especially useful subjects for such experiments? Would human beings be useful for genetic studies?

6. In a single sentence, carefully explain the significance of the chi-square test. What does the phrase "statistically significant" mean?

7. Type-A blood contains both A-antigen and anti-B antibody. Type-O blood contains neither antigen but both anti-A and anti-B antibodies. Why is it safe to transfuse moderate amounts of O blood into an A individual?

FUNGAL REPRODUCTION AND GENETICS

Biological Science, pp. 511-515, 738-743.
Elements of Biological Science, pp. 325-327, 469-473.

Saprophytic fungi, like many bacteria, are agents of decay and are therefore of great importance in degrading the complex organic compounds of once-living organisms to simpler compounds that can be utilized by other forms of life in the construction of new living matter. Many of the fungi are also useful in genetic research, for they grow rapidly, are easy to maintain in the laboratory, and have short life cycles. Another valuable feature is that the vegetative mycelia are haploid, so that the phenomenon of dominance does not interfere with the determination of genotypes after mating.

In the laboratory today, we shall examine briefly the structure and reproduction of the three major groups of fungi, and carry out some experiments that demonstrate the usefulness of fungi as tools for genetic research.

PART I. THE TRUE FUNGI, DIVISION EUMYCOPHYTA

The body of a fungus is either unicellular or filamentous. Individual filaments are called **hyphae,** and the mass of hyphae that ordinarily forms a fungal body is called a **mycelium.** The hyphal walls sometimes contain cellulose, but more often their most important component is chitin. Fungi lack photosynthetic pigments, and obtain nutrition as parasites or saprophytes, or by forming symbiotic relationships with other organisms.

The vegetative form of a fungus is often a continuous mass of multinucleate protoplasm, not organized into individual cells. Septa (cross walls in the filaments) occur in some forms, but they are usually perforated, so that fully independent cells are rarely formed except during reproduction.

The Division Eumycophyta is customarily divided into three classes, Phycomycetes, Ascomycetes, and Basidiomycetes, each of which is given full divisional status in some classification schemes. Most of the characteristics that distinguish the three groups are related to their sexual reproduction.

A. Phycomycetes (the algal fungi)

The vegetative hyphae of Phycomycetes characteristically lack cross walls, although septa do appear during the formation of reproductive structures. You will be provided with a culture of the black bread mold *Rhizopus stolonifer.* Through the covered dish, make the following observations with a dissecting microscope.

1. Examine the hyphae that form a mycelium on the surface of the agar.

2. Find **rhizoids,** which extend from the surface mycelium into the agar as rootlike threads.

3. Aerial hyphae called **sporangiophores** grow upright from the surface

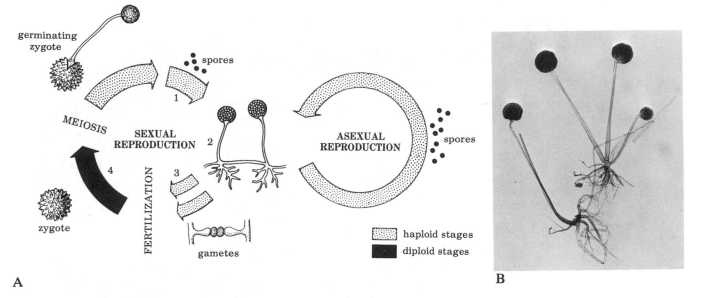

Fig. 18-1. *Rhizopus.* (A) Life cycle. (B) Hyphae with sporangia. (*B: Courtesy Carolina Biological Supply Company.*)

and bear globular **sporangia** at their tips. Each sporangium produces many hundreds of asexual spores. By what cell division process would such spores be formed?

4. *Rhizopus* also reproduces sexually. Short branches from two different hyphae (which must be of different mating types or "sexes") contact each other at their tips. Cross walls soon form just back of the tips in each hypha, thus delimiting gamete cells, which then fuse to form a zygote. Examine on demonstration a petri dish inoculated with both plus and minus strains of *R. stolonifer*. Notice the line of zygotes formed where the hyphae came into contact. Figure 18-1 diagrams the life cycle of *Rhizopus*. How would cells produced by the zygote differ genetically from those produced in the asexual sporangia?

5. In water-inhabiting species of Phycomycetes, the spores usually bear cilia or flagella that propel them through the water. Such spores can survive for a long time before finding a suitable substrate upon which to grow. The instructor has put some split hemp seeds into a dish of pond water. Examine the molds that have attached to each seed. For closer observation, borrow a contaminated seed and examine it under the microscope.

B. Basidiomycetes (the club fungi)

Many of the largest and most conspicuous fungi — puffballs, mushrooms, toadstools, and bracket fungi — are Basidiomycetes. Though the above-ground portion of these plants looks like a solid mass of tissue and in some forms *is* differentiated into a stalk and a prominent cap, it is nevertheless composed of hyphae, as are all the true fungi. Basidiomycete hyphae have perforated septa. It should be emphasized that the prominent mushroom, toadstool, or bracket — the reproductive structure of the organism — is only a small part of the whole plant; most of the plant is beneath the ground or ramifying through whatever other substrate the fungus is using.

As in the Phycomycetes, sexual reproduction is initiated by contact between vegetative hyphae of different mating strains. Compatible hyphae

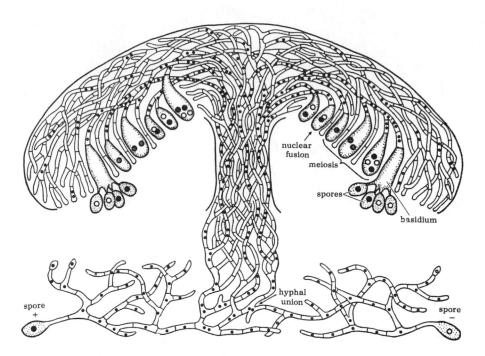

FIG. 18-2. Diagram of section through a mushroom. The entire stalk and cap are composed of hyphae packed tightly together. Spores are produced by basidia on the lower surface of the cap. (*From L. W. Sharp,* Fundamentals of Cytology, *McGraw-Hill Book Co., 1943. Used by permission.*)

fuse, and a new generation of binucleate cells, formed in the region of fusion, eventually grows into the large reproductive structure, or **fruiting body** (FIGURE 18-2).

Basidiomycetes are distinguished by their club-shaped reproductive cells, **basidia**. FIGURE 18-2 shows the relationship of the basidium to the fruiting body as a whole, and FIGURE 18-3 shows several single basidia. The cells of the hyphae that produce basidia are binucleate, containing one haploid "male" nucleus (from one of the parent mycelia) and one haploid "female" nucleus (from the other parent mycelium). Terminal cells of these hyphae become zygotes when their nuclei fuse in fertilization. The zygote then develops into a basidium. Its diploid nucleus divides by meiosis, producing four haploid nuclei. Each nucleus migrates into one of four small

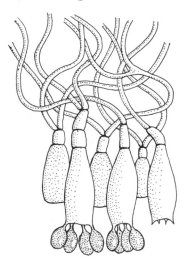

FIG. 18-3. Reproductive structures of Basidiomycetes. The large club-shaped cells at the bottom are basidia. A mature basidium bears four round spores at its end. The three shorter basidia in the figure are immature. The one on the far right is an old basidium that has shed its spores.

protuberances developed on the end of the basidium. The tip of each protuberance then becomes walled off as a **basidiospore,** which falls or is ejected from the basidium. Each spore may give rise to a new mycelium, if it lands on the proper substrate.

1. Examine on demonstration the fruiting bodies of various Basidiomycetes. Among them are shelf fungi, puffballs, and mushrooms.

2. Under your dissecting microscope, examine an edible mushroom and notice the many gills composing the umbrella-shaped cap. Basidia line the gills.

3. Examine a slide of a cross section through the cap of a mature *Coprinus*. The small, dark structures on the edges of the gills are the basidiospores.

C. Ascomycetes (the sac fungi)

The members of this large group include diverse forms: unicellular yeasts, powdery mildews, cottony molds, and complex cup fungi. However, they are unified by a common characteristic, the possession of a reproductive

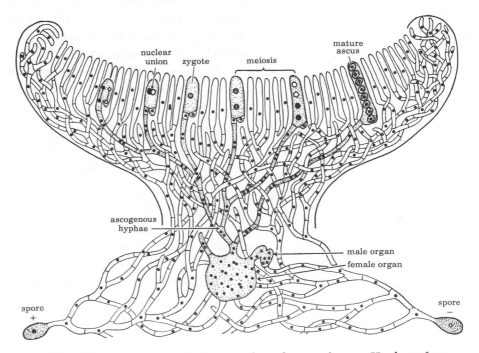

FIG. 18-4. Diagram of a magnified section through a cup fungus. Hyphae of two mycelia, one from a plus (female) spore and the other from a minus (male) spore participate in forming the cup structure and in producing the spores. The plus mycelium bears a female organ, and the minus mycelium bears a male organ. A tube grows from the female organ to the male organ, and then minus nuclei (small circles) acting as male nuclei, move into the female organ and become associated with the female nuclei (black dots). Next, hyphae (gray) grow from the female organ; each cell in these hyphae contains two associated nuclei, one minus and one plus. The terminal cells of these hyphae eventually become elongate, and their nuclei unite, forming a zygote nucleus. The zygote nucleus promptly divides meiotically, and each of the four haploid nuclei thus formed then divides mitotically, producing a total of eight small spore cells, which are still contained within the wall of the old zygote cell, now called an ascus. When the mature ascus ruptures, the spores are released. (*From L. W. Sharp,* Fundamentals of Cytology, *McGraw-Hill Book Co., 1943. Used by permission.*)

body called an **ascus**, a saclike structure within which haploid spores (usually eight) are produced. Like a basidium, an ascus develops as the result of a sexual process (FIGURE 18-4).

The vegetative hyphae of Ascomycetes have perforated septa.

PEZIZA

1. a. Examine a preserved specimen of *Peziza*, a cup fungus. The cup contains many developing and mature asci. Raindrops fall into the cup and splash out the spores. The cup is composed of many hyphae packed tightly together.

b. Study a slide of *Peziza*, sectioned through the cup. Notice the arrangement of asci; find some asci with eight spores intact.

NEUROSPORA

2. a. Observe the fluffy mycelium of the *Neurospora* (pink bread mold) culture on demonstration. *Do not open this culture.*

b. With a wire loop provided, obtain one of the black **perithecia** (singular, perithecium) from the *Neurospora* culture (marked "A") on the supply desk. Make a wet mount of it in a drop of water. The perithecium in *Neurospora* is comparable to the cup in *Peziza*; it is the structure within which the asci form. Place a cover slip atop the mount, and crush the perithecium with a cork applied to the cover slip.

c. Find intact asci, with all eight spores. Is there any difference in color of the ascospores? Are there variations in the arrangement of ascospores in different asci?

d. Ascomycetes can also reproduce asexually by means of **conidiospores,** produced in chains at the end of special hyphae called conidiophores, but not within sporangia. Each conidiospore (also called a conidium) can grow into a new plant. Make a wet mount of conidia from the culture marked "B" at the supply desk, or from a suspension of conidia provided by your instructor. Make a sketch below for your own records.

PART II. SEXUAL REPRODUCTION IN NEUROSPORA: SINGLE FACTOR CROSS

Sexual reproduction, as we have noted repeatedly, brings into existence new genetic variants, and thereby provides a greater reservoir of raw materials upon which natural selection can act. So important, indeed, is sexual reproduction, that organisms have evolved a multitude of remarkable adaptations to prevent self-fertilization. The male and female parts of flowers often ripen at different times, so that only cross-pollination is possible. Many insects are adapted to remove pollen from one flower and deposit it neatly on the next blossom visited. The existence of separate sexes or mating types is further insurance that variation will be introduced continually into populations of organisms through the mating of slightly different types of individuals.

In diploid organisms, certain cells undergo meiosis to form haploid gametes, which unite at fertilization to form a zygote. The zygote usually undergoes development to form a new diploid adult organism. While in one sense diploid cells are adaptive and useful as reservoirs of potentially beneficial recessive alleles, diploidy is cumbersome in some types of genetic studies because the genotypes of the offspring often cannot be determined directly. Because of dominance relationships between the members of some allelic pairs, the presence of one member of a pair may be completely masked by the other.

The use of haploid fungi as experimental organisms eliminates these difficulties. When such organisms undergo sexual reproduction, two haploid cells of different "mating types" fuse to form a zygote. The zygote does *not* develop into a multicellular diploid organism, but promptly undergoes meiosis to form new haploid cells. The products of the cross thus become available immediately, and since all the products are haploid, they are genotypically distinguishable from one another with respect to traits visible in the spores or the vegetative colonies. In the laboratory today, we will work with *Neurospora*, an ascomycete that has been widely used in genetics.

Neurospora occurs in two different mating "types," designated A and *a*, which are controlled by one pair of alleles. Fertilization occurs after cells of opposite mating types unite. The conidium of one strain fuses with the female reproductive structure of the other strain. This is followed by nuclear fusions, producing numerous zygotes, the only diploid stage in the life cycle. Within each zygote, which becomes an ascus, the diploid nucleus soon undergoes meiosis, producing four haploid nuclei. Each of these nuclei undergoes mitosis, so that the end result is eight nuclei. Because the ascus is so narrow, the nuclei do not slip past one another at the divisions, and so are arranged linearly within the sac. With the addition of a heavy spore wall, each haploid nucleus with surrounding cytoplasm becomes an ascospore. Starting at either end of the ascus, you can count off four pairs of spores, each pair representing one of the products of meiosis. Each spore, when liberated from the ascus, can germinate into a new mycelium. The life cycle is diagrammed in FIGURE 18-5.

Before the ascus shoots, the ordered ascospores form a record of what happened during meiosis. The products of a single meiosis are arranged in order within the same sac; by separating them and germinating each spore by itself, you can determine directly the results of allelic segregation. From this kind of procedure it is relatively easy to determine whether a particular trait is controlled by one pair of alleles, or whether perhaps more than one pair is involved. We know that the mating types A and *a* are controlled by one allelic pair through just such an analysis.

The usual procedure is to isolate and germinate individual ascospores. Since this is technically difficult, we shall use a somewhat different procedure that is the same in principle.

A wild type (pink) *Neurospora* of one mating type was grown with an albino mutant of the opposite mating type under suitable mating conditions. The asci formed subsequent to this mating were the same ones you observed earlier in their perithecium. When the asci matured, they burst and liberated spores on the lid of the petri dish. The spores were then collected and suspended in enough sterile water to yield a spore concentration of 200/cc.

You will take 200 spores, add them to melted agar, and allow them to remain at 60°C for thirty minutes. This procedure activates ascospore ger-

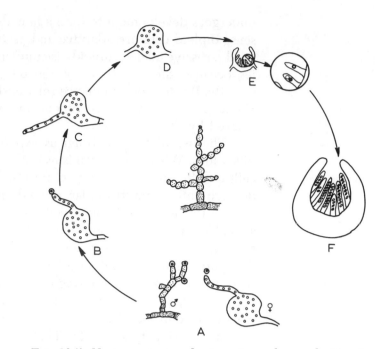

Fig. 18-5. *Neurospora* reproduction. Asexual reproduction is by conidia, formed on aerial hyphae; one such hypha is illustrated in the center of the life-cycle diagram. Sexual reproduction, probably initiated by adverse environmental conditions, follows the same basic pattern as that shown for the cup fungus in Figure 18-4. The hyphae first form male and female sexual structures (A); the male structure forms specialized conidia, which drift to the female structure (the protoperithecium) and land on one of its outgrowths (B). Nuclear fusion occurs, and the male and female nuclei mingle in the protoperithecium (C). Each undergoes mitosis several times (D); eventually male and female nuclei pair off and fuse (E) to form zygote nuclei. Each zygote becomes an ascus (F); when the protoperithecium ripens, it forms a perithecium, from which ascospores are later forcibly discharged. Although it is not shown in this figure, a ripe perithecium (one suitable for squashing to see asci and ascospores) can be identified by a small nipple at its top.

mination and kills conidiospores, which may have been mixed with the ascospores. The melted agar suspension of ascospores will then be poured into a petri dish containing a sorbose agar. The ascospores will germinate within the agar and, because the sorbose causes the hyphae to branch profusely, discrete colonies will form; this is in contrast to the usual spreading habit of *Neurospora*. By the following week, some of the hyphae in each colony will break through to the surface of the agar where either white or pink conidiospores will be formed, according to the genetic makeup of the colony. You will thus be able to score the results of the sexual cross by counting the number of colonies of each type.

A. Procedure   Before beginning the experiment (and preferably before coming to class), read IIB (pour-plate method) in Appendix 7; be sure to follow the sterile technique demonstrated by your instructor. A stock container of ascospore suspension in water will be provided for each group of eight students. Each student will be provided with a sterile pipette, sterile petri dish of sorbose agar, and a tube in the water bath with 5 ml of sorbose agar.

1. Label the petri dish of sorbose agar with your name and laboratory section.

2. Briefly remove the test tube of melted agar from the water bath. Swirl the container of ascospore suspension to be sure the spores are evenly distributed, and pipette 1 cc of suspension into the tube of agar. Label the tube with masking tape, *well above the water level,* so that it will not come off during the incubation. Let the tube remain in the water bath for thirty minutes.

3. At the end of the thirty minutes, mix the suspension thoroughly by tightly grasping the tube near the top and briskly tapping the tube bottom with the forefinger of your strong hand. Pour the spore-containing agar *promptly* into the sorbose-agar plate (FIGURE A7-3, p. 374), tilting the plate gently from side to side to spread the suspension uniformly over the surface of the sorbose agar.

4. When the agar has solidified and cooled, place the plate in the storage container provided. The instructor will incubate and return the plates for your examination during the next period.

5. Clean the test tube and pipette thoroughly with detergent and hot water, using a brush to make sure that all the agar is removed from the tube. Rinse at least five times in warm or hot water and place in the designated containers on the supply table to dry.

B. Results

Your data can be used to determine whether the albino trait is controlled by a single allele.

Count and record the number of white and pink colonies, and enter the data in the table below. Often one strain will mature faster than the other, and often the color of the wild type cannot be detected until it has produced conidiospores. Therefore, read the plates only when they are mature. Confluency will be resolved when the colonies reach maturity and form small tufts of conidiating hyphae.

NUMBER OF COLONIES

	White	Pink
Individual Data		
Class Data		

What does this ratio illustrate? Does it differ from the ratio you would have expected if the albino traits were controlled by a single gene? If it does differ, is the difference significant?

If *all* the products of meiosis had been separated and germinated by removing each spore from an ascus, the resulting ratios would be determined mechanically rather than statistically. In our case, however, many factors such as the differential release of ascospores from the ascus, differential removal of spores from the lid of the petri dish, and differential germination of the spores, may affect the colony ratio, so our results should be subjected

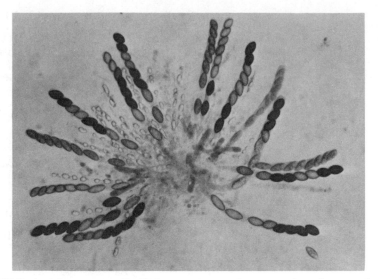

FIG. 18-6. Asci from a squashed *Neurospora* perithecium. This photograph illustrates five of the six possible spore patterns that may result from the segregation of a single gene pair. (*Courtesy R. L. Phillips, University of Minnesota.*)

to statistical analysis. Apply the chi-square test to the class results.

In the asci you observed earlier, could you have distinguished wild type from mutant spores? Make another perithecium crush, as in 2b, p. 206. In your slide and in the photomicrograph (FIGURE 18-6) locate as many different spore arrangements as you can, and diagram each one. Which of them is due to crossing over? Diagram the chromosomes to show the segregation of wild type and albino factors, and crossing over, that gave rise to each different spore arrangement.

PART III. HETEROKARYOSIS IN NEUROSPORA

As we have already seen, sexual reproduction in *Neurospora* involves the fusion of mycelia of two different mating types. The result of this fusion is an association of genotypically different nuclei in a common cytoplasm. Later, the nuclei fuse, meiosis occurs, and haploid spores are formed. Nuclear fusion may occur promptly after hyphal fusion, as in many Phycomycetes, or it may be postponed while the nuclei undergo mitosis, often to form a complex fruiting body.

It is also common for two hyphae to fuse, and for their nuclei to associate, *without* sexual reproduction. The multinucleate mycelium derived from such fusion is called a **heterokaryon.** The association of nuclei in a heterokaryon may be short-lived or persistent, depending on the conditions under which the fungus is growing. If conditions are unsuitable for the growth of the original hyphae, heterokaryosis provides an important mechanism whereby the species may become capable of growth and asexual reproduction, thus improving its chances of eventually reaching areas more suitable for growth. If the heterokaryon is capable of growing on a medium where neither of the original mycelia could grow, heterokaryosis gives a species an important adaptive advantage.

In the experiment we shall perform, one of the mycelia lacks the gene that controls the formation of an enzyme involved in the synthesis of panto-

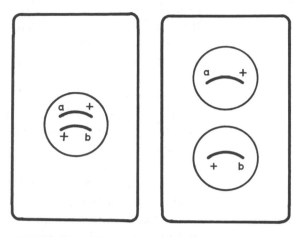

HETEROZYGOTE   HETEROKARYON

FIG. 18-7. Comparison of heterozygote and heterokaryon cells. The plus designates "wild type" or normal alleles, while "a" or "b" designates corresponding
mutants. Each chromosome in the heterozygote nucleus has one mutant locus, but its homologue is normal at the same locus. If the wild type allele is
dominant, it makes up for any deficiency the mutant
might exhibit. The same chromosomal condition exists
in a heterokaryon, except that the two chromosomes
are located in different nuclei.

thenic acid, a necessary vitamin; the other mycelium cannot synthesize
nicotinic acid, another important vitamin. Operating together in a heterokaryon, each nucleus makes up for the other's deficiency; this phenomenon
is called **complementation.** A heterokaryon is in some ways similar to a heterozygous diploid nucleus; in a diploid nucleus, of course, the two alleles for
a particular trait are located in the *same* nucleus (FIGURE 18-7), while in a
heterokaryon, they are located in *different* nuclei but in the same cell.

While both sexual reproduction and heterokaryosis involve hyphal
fusion, they are significant for the organism in different ways. Sexual reproduction is a source of genetic variants; heterokaryosis, on the other hand, is
a way of ensuring that a species will be able to reproduce and disperse itself
even under conditions that are unsatisfactory for the growth of certain of
its mutant strains. In sexual reproduction, spores form as the result of meiosis; in heterokaryosis, asexual conidiospores form and are much like the
conidiospores that would have been formed if the mycelia that fused to form
the heterokaryon had been able to grow separately.

In the following exercise, you will demonstrate heterokaryon complementation for vitamin deficiency, and will find how useful heterokaryons can
be as tools for genetic analysis.

A. Procedure   Wild type or "normal" pink *Neurospora* will grow and reproduce satisfac-
torily on a "minimal" medium*. The variant forms used in this experiment
("pantothenicless" and "nicotinicless" mutants) will grow on minimal me-

---

* One containing only the bare essentials for growth.

dium only if it is supplemented by the vitamins they cannot synthesize. All three forms will grow luxuriantly on a "complete" medium, containing all the required nutrients and an assortment of vitamin and other supplements.

You will attempt to make cultures on minimal medium, complete medium, and minimal medium supplemented by pantothenic or nicotinic acid. Next week, the results of these experiments will tell you which vitamin each mutant requires, and will show you that the two mutants can indeed complement one another and grow on minimal medium. This will enable you to consider some questions concerning the inheritance of vitamin deficiencies and of color.

Since every student cannot prepare all forty or more plates used in this experiment, your instructor will assign you or your group to the preparation of one or more of the plating "blocks" shown on the opposite page (FIGURE 18-8). Plates of fresh media are designated M (minimal), $M + N$ (minimal and nicotinic acid), $M + P$ (minimal and pantothenic acid), and C (complete); they will be found on your table or at the supply desk. Your instructor will tell you where to find conidiospore suspensions of the mutants (designated *1* and *2*) and the pink wild type (designated *3*).

Heterokaryosis (plating block at far left) requires contact between germinating spores of the two mutant strains and may fail in some cases because of poor plating technique; therefore, more than one student in each group, and perhaps every student in the class, will be asked to carry out this part of the experiment. Each of the other twelve plating blocks must be done at least in triplicate in order to simplify interpretation of results if contamination by airborne spores should occur; if only two plates are made, and if they produce different results, it is sometimes impossible to know which of the results is due to contamination.

Before you begin your work, know exactly which part of the experiment is your responsibility, and be sure you are familiar with sterile technique (Appendix 7). Make careful note of any modification your instructor may add to those directions. Label the top of each plate with your name or a code number for your group, as designated by your instructor and the bottom with the type of inoculation it has received. After each one has been properly labeled, return all plates to the instructor for incubation.

B. Results

1. Describe the results of each plating, making sure that you see every type of plating, not just those you made yourself. Use dissecting microscopes when necessary, and discuss your results with other students in the class.

2. Which vitamin does each mutant require?

3. Some of your plates will not turn out as expected. If you find such unexpected results, consider two factors in trying to explain them:

a. Consider possible sources of contamination. Make a comparison of the unexpected growth with growth of the same strain on other media.

b. Consider other practical reasons (error of technique, etc.) why the actual results differed from expected results. If growth did not occur where it was expected, examine other plates of the same kind.

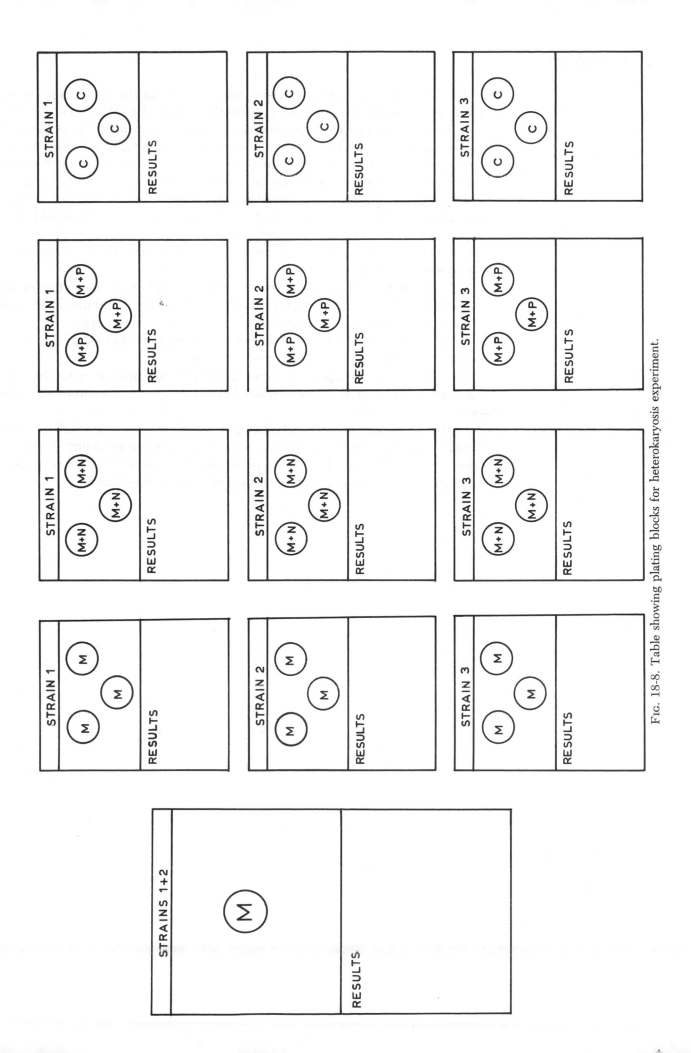

Fig. 18-8. Table showing plating blocks for heterokaryosis experiment.

4. Are the two vitamin-deficient strains mutant at the same point (gene locus) on the same chromosome (i.e., are they allelic)? Explain your answer.

5. Both mutants we have used are albinos. Albinism is the expression of one or the other or both of two mutant genes, called *al-1* and *al-2*. These two genes control the formation of different enzymes necessary in the synthetic steps leading to the production of normal pink *Neurospora* pigment. How can you tell, from your complementation experiment, that the strains are albino at the *same* gene locus? What difference in results would you expect if they were albino at different loci?

6. What is the purpose of growing the two mutants on "complete" *Neurospora* medium?

7. What reason(s) would you advance to explain a small amount of growth of strain *1* on minimal medium? A pink growth on minimal medium streaked with strain *1*? Pink growth on the heterokaryon plates?

8. Would you expect heterokaryosis to persist if strains *1* and *2* were streaked together on "complete" medium? Explain.

9. What is the purpose of suspending the conidiospores for the heterokaryosis experiment in water? (*Hint:* Why not take conidiospores directly from an agar plate?)

10. Heterokaryosis compatibility is controlled by a pair of alleles different from that which controls sexual (mating type) compatibility. Furthermore, strains capable of heterokaryosis are often not capable of sexuality. Can you suggest an adaptive advantage to this arrangement?

19 ANGIOSPERM
DEVELOPMENT

Biological Science, pp. 539-546.
Elements of Biological Science, pp. 336-341.

The flowering plant is far from a simple colony of cells; it is a complexly organized unit whose growth, development, and behavior are precisely regulated and synchronized. The plant's synchronous growth and development, and its ability to respond adaptively to environmental fluctuations, are achieved largely through the action of chemical agents called **plant hormones,** substances generally produced in the apical meristems of root and shoot and then transported to other parts of the plant by the vascular system or by diffusion.

In this unit, you will explore some of the normal events in plant growth and development and will initiate long-term observations of development in a flowering plant. In the following unit, we shall study experimentally some of the hormonal controls that natural selection has shaped to the plant's advantage as regulators of behavior and development.

PART I.
EMBRYONIC
DEVELOPMENT

Like most other multicellular organisms, a flowering plant begins life as a fertilized egg, or **zygote.** In the ovary of the parent plant, this cell develops into a small **embryo.** Together with food materials and adjacent ovarian tissue, the embryo is formed into a **seed.** The ovary, which may contain many seeds, develops (sometimes in association with other ovaries in the same flower) into a **fruit.** Both seed and fruit, as we shall later see in more detail, are important adaptations that help ensure the success of terrestrial reproduction.

The embryo often enters a period of dormancy before resuming growth (**germination**) under the proper conditions of light, moisture, temperature, and oxygen supply. When germination does occur, additional cells are produced in meristematic zones; the growth and differentiation of these cells, and their organization into plant organs — roots, stems, leaves, flowers, etc. — result in a structural pattern characteristic of the species.

1. Examine a soaked germinating bean seed. Locate and carefully remove the **seed coat;** this tissue protects the embryo and is often modified to absorb water or to facilitate dispersal of the seed by air, animals, or other agents. The fruit in which the seeds are borne may be similarly modified. On the bean seed, note the small, rounded **hilus** in the shallow concavity on one side of the seed; this structure serves to absorb water, while the rest of the seed coat is impermeable to water. Some seeds have a very hard water-impermeable seed coat that must be broken or abraded before water absorption can occur. What value might there be in such a seed coat?

2. Split the seed open and locate the two large **cotyledons,** leaves

modified for food storage; the cotyledons are part of the embryo, and the food they store is principally in the form of starch. The presence of two cotyledons is characteristic of dicots (Subclass Dicotyledoneae of the flowering plants; see Appendix 2). What are some of the other differences that distinguish monocots from dicots?

3. The embryonic plant axis is located between the cotyledons. The region sheathed at its tip in embryonic leaves is the **epicotyl,** and the opposite end of the axis, on the other side of the cotyledonary attachment, is the **hypocotyl.** Each terminates in an apical meristem. The epicotyl and part of the hypocotyl form the shoot (stem) of the plant; the remainder of the hypocotyl forms the root. Notice the position of the cotyledons on the maturing bean plants, on demonstration.

The original food in the seed was **endosperm,** a tissue derived by cell division and growth from the same cell that gave rise to the egg; during development of the bean embryo, its cotyledons absorbed all the endosperm. If a castor bean seed is available, dissect it carefully, and note the thin, veined cotyledon; the rest of the interior of the seed, apart from the embryo, is endosperm. *Do not eat* the castor bean seed; it is poisonous.

PART II. GROWTH AND DIFFERENTIATION OF THE ROOT AND SHOOT

Primary growth, as you recall from Topic 7, is the result of cell division in *apical* meristems, and of differentiation in the cells thus produced. *Secondary* growth is the result of cell division in *lateral* meristems (vascular cambium and/or cork cambium), and the differentiation of these cells into xylem, phloem, or cork. Primary growth increases the length of an organ, while secondary growth results in increase in girth. We shall investigate the zones of primary growth in a root and some of the patterns to which root growth gives rise; we shall then turn our attention to shoot development.

A. Root development and differentiation
THE LOCUS OF PRIMARY GROWTH IN A ROOT

1. The pattern of primary growth in a root may easily be studied by marking the root at known regular intervals and then comparing the position of the markings with the original positions after 24-48 hours. *Read the directions carefully* and obtain all the necessary equipment before proceeding.

a. Obtain a piece of thin glass or Plexiglas 3″ x 3″ and 4 germinated corn or pea seedlings, each with straight root about 2 cm long. Cover the glass plate with wet paper toweling cut to fit, and fasten the seedlings over the toweling with a rubber band, as shown in FIGURE 19-1. Being sure *not* to let the seedlings dry, mark them as directed in b.

b. Blot any excess moisture from the seedlings. With a marker like that shown in FIGURE 19-1 (made from a paper clip or soft wire and thread), mark the root with India ink nine times at 1-mm intervals, *starting at the root tip.* Paint the thread with India ink, or other black drawing ink, using the ink applicator or a dropper; wipe excess ink from the thread, and quickly make a thin mark on the root. You can make the marks easily if a millimeter ruler is slipped under the root during the marking operation. (*Note:* Be *sure* to wipe excess ink from the thread; if you do not, the marks will later smudge and be difficult to interpret.)

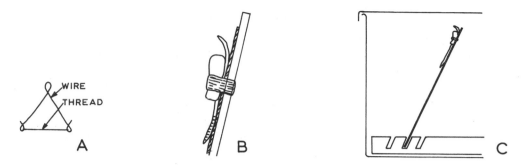

Fig. 19-1. Root-growth experiment. (A) Marker. (B) Diagram showing seedling attached to plate with rubber band. (C) Plate in holder, in moist chamber.

c. With a solvent marking pencil, mark the glass on the back with your name or seat number. Place the whole setup in a moist chamber at the supply desk, slipping it into one of the 60°-angled slots in the holder on the chamber bottom. In Table 1, make a scale sketch of the root, several times enlarged, showing the position of the markings and the length of the root relative to the size of the seed.

d. After 48 hours (or alternatively, using another student's specimen set up 48 hours ago), sketch the root again, to the same scale used for the previous sketch. Measure the distance between the markings; record in Table 1, and use the averages to make markings on the new sketch. Where did the most elongation occur? How far behind the root tip has elongation stopped taking place? If your initial lines were neat and clean, how do you account for smudging that has taken place a short distance behind the root tip?

Would a control have enhanced the value of this experiment? What would or could be used as a control?

PATTERN AND LATERAL ROOT FORMATION

2. As we shall see shortly, branches on a stem form from specialized meristems called lateral buds. Formation of lateral roots is much more random, since the laterals form through meristematic activity of the *pericycle* (see Topic 4), one of the cellular layers in the root adjacent to the vascular tissues. *Secondary* roots formed in this way may repeatedly rebranch, by the same method. Examine a demonstration slide showing formation of a lateral root; identify the pericycle and the lateral root.

What adaptive advantage might there be in greater flexibility in the pattern of branching in the root, as compared to the regular pattern of branching in a stem? What advantages would there be in regular branching of the stem?

B. *Stem development and differentiation*

(*Note:* Unless the weather is inclement, most of this study can be conducted out-of-doors; most of the following observations can be made on bare over-wintering tree twigs or on conifers.)

Stems, like roots, grow in length from apical meristems, and in girth either by cell enlargement (most monocots) or by growth from lateral meristems (like the vascular cambium and cork cambium in many dicots).

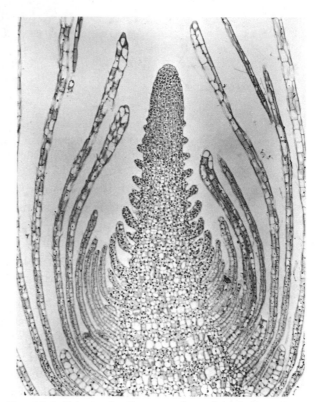

Fig. 19-2. Longitudinal section through apical meristem of *Elodea*. (*Courtesy Thomas Eisner, Cornell University.*)

Figure 19-2 shows the apical meristem of *Elodea,* a water plant, sheathed by developing young leaves and by **leaf primordia,** budlike structures that will give rise to leaves. Notice that the primordia become smaller as they near the stem-tip; differentiation in stems, like that in roots, proceeds as a region grows farther from the center of meristematic activity that produced it.

1. Apical meristems are of two kinds on most stems; one or more **terminal buds** may occur at the tip of a shoot, and **lateral** (axillary) **buds** occur in the upper angle between leaf petioles and stem axis. (If there is no foliage on the plant you are examining, locate buds adjacent to **leaf scars,** regions where leaves were previously attached.) In a temperate perennial plant, such as a tree, both types of bud enter a period of dormancy in the summer or fall, and renew their growth the following spring. Examine a twig several years old that shows both kinds of bud.

2. Each season's growth is delimited by a group of several **bud scale scars,** tissue that remains after the scales covering each bud fall off in the spring. Like annual rings, bud scale scars are indicators of age in stems. How many years growth are represented by the section of stem you are examining?

The presence of lateral buds is diagnostic of stems. Look, for example, at a compound leaf on demonstration; this leaf could be mistaken, on superficial examination, for a section of leafy stem, but notice the absence of

Fig. 19-3. Portion of a red maple twig. (*Courtesy Carolina Biological Supply Company.*)

buds. Growth of a lateral bud produces a *branch*, a new section of stem with an entirely new set of meristems. Whether a lateral bud develops depends upon its proximity to the apical meristems of other branches and of the main shoot; hormones transported from active apical meristems inhibit development of lateral buds, but this influence diminishes with distance from the active meristems.

Note that leaf scars are clearly evident only in the youngest sections of most twigs. Why? (*Hint:* for the same reason that bud scale scars are useful indicators of age only in young sections of a woody stem.) Label all parts of the maple twigs shown in FIGURE 19-3. How many years' growth is represented by these twigs?

3. Notice on a deciduous tree, or on a branch provided by the instructor, that growth of leaves is usually restricted to twigs formed in the current growing season; if you are examining leafless trees, all next year's leaves will be formed on the sections of stem axis that emerge from terminal and lateral buds on last year's twigs. Conifers and other evergreens are exceptions to this rule since although their leaves (needles) fall, they do not fall all at once, as is the case in deciduous trees; leaves of more than one growing season may be present simultaneously. Considering the geographical distribution of conifers and deciduous trees, what do you think might be the significance of the leaf-form and leaf-fall pattern differences between these two types of trees?

4. Examine demonstrations of modifications in stems and leaves; each one will be accompanied by an explanatory card. Of what advantage would each of these adaptations be?

PART III. GROWTH IN A FLOWERING PLANT

Over the next several weeks, you will be making daily observations of normal development in the garden bean *Phaseolus vulgaris*, or a related species. This plant grows rapidly, and in four to six weeks you will be able to gather enough data to make a substantial number of generalizations about the nature of plant growth. In this laboratory, you will be given a kit, with a container of soil, a watering vessel, two bean seeds, and a plastic bag. Take this unit home and follow the directions below for sowing and caring for the plants. Before proceeding, read over the section on data collection very carefully.

A. Procedure for setting up the plant

1. Soak the seeds for 2-4 hours in tap water in a paper cup. Do *not* soak seeds for an extended period.

2. Smooth over the soil in the pot and make two shallow depressions in the soil with your finger. Put the seeds into these depressions, and cover with soil to a depth of not more than a quarter inch.

3. Roll the plastic bag down so that it encloses only the bottom of the pot, to prevent damage to the sill or table on which the pot will be kept.

4. Find a place in your room *at* or *near* a window where the plant will receive sufficient light. *Do not* choose a high-traffic area, where the plant is likely to be knocked over, or an area directly atop or adjacent to a radiator or hot-air vent, where the plant is likely to be injured by the hot air. If necessary, put up a sign to discourage roommates and others from touching or damaging the plant.

5. Water the plant *daily* with the watering vessel or otherwise as directed by your instructor. Moisten the soil, preferably with small droplets, but do not soak it; remember that the roots require oxygen. Each day when you water, put a check mark and the date on both data sheets (Tables 2 and 3). Make measurements and take other data consistently, at the same time of day if possible.

6. If you *cannot* water the plant for a period of two or more days, arrange for someone else to water it. If you cannot locate another person to care for the plant, turn up the plastic bag and tape it shut loosely, to reduce evaporation from the soil while you are away. Water the plant liberally before you leave, but do not soak the soil.

B. Collection of data

You must collect data regularly and consistently. Help yourself by keeping a data sheet, ruler, and pencil near the plant. If there is more than one way of taking a measurement, decide on the way you will take it, make a record of your decision, and be prepared to defend it. If you work in pairs or groups on single plants or groups of plants, and you are cooperating with other students to gather data, be *certain* to keep the data sheet in a protected place (such as a plastic bag) near the plants, and make sure that each member of the group initials the proper place on the data sheet when he makes his measurements.

1. In Table 2, keep a daily record of the date, the height of the plant in centimeters, and any unusual events that occur (for example, if the plant is kept in good condition long enough, it will flower). All other records are to be kept in Table 3, and they will be taken at three-day intervals, beginning as soon as the plants have come up.

You will need to understand clearly the use of the following terms:

a. The *height* of the plant is that distance from the soil to the tip of the main shoot. When the plant outgrows your ruler, use the ruler in conjunction with string, or use a meter stick or a yardstick graduated in centimeters.

b. Leaf *blade length* is a measurement taken along the main longitudinal axis of the leaf blade. *Blade width* is a similar measurement taken at right angles to the length measurement, at the widest part of the leaf. *Petiole length* is a measurement from the base of the blade to the main stem axis. *Leaf pair 1* is the first pair of leaves to differentiate above the cotyledons; *pair 2* is the pair above *1*, etc.

c. A *node* is a region where a pair of leaves and lateral buds arise from the main stem or from a branch. *Node 1* is where the first leaf pair is attached.

d. An *internode* is the section of stem between an adjacent pair of nodes. Count the terminal stem section as a differentiated internode only when it is clearly bounded at bottom *and* top by pairs of unfurled, though not necessarily full-size, leaves.

2. On Graph A, plot the height of the plant vs time. When was the most rapid relative growth? In a separate curve on Graph A, plot the number of internodes vs time.

3. On Graph B, plot the sets of leaf blade length measurements vs time as three separate curves. On Graph C, do the same for the leaf width data; on Graph D, plot the petiole length data. How would you describe the pattern of leaf growth, as indicated by these data? How does the leaf growth pattern compare to the pattern of growth in height of the whole shoot? Are two or more leaves ever simultaneously in a rapid growth phase?

4. On Graph E, plot as separate curves the length of each measured internode vs time. How do these data compare to those for height, leaf blade length, leaf blade width, and petiole length?

5. Describe and explain any other events (e.g., flowering) that took place during your period of observation.

In Appendix 2, record the organisms with which you worked in this laboratory, if they are not already recorded.

DISTANCE BETWEEN MARKINGS = 1mm
INITIAL SKETCH (SCALE =)

ROOT NO.	TIP-1	1-2	2-3	3-4	4-5	5-6	6-7	7-8	8-9	
	DISTANCE BETWEEN MARKINGS									
1										
2										
3										
4										
AVG.										

FINAL SKETCH (SCALE =)

TABLE 1. THE LOCUS OF GROWTH IN A ROOT.

INITIAL	DATE	HEIGHT	COMMENTS	INITIAL	DATE	HEIGHT	COMMENTS

TABLE 2.

INITIAL	DATE	NUMBER OF DIFFERENT INTERNODES	LEAF PAIR 1 (mm)			LEAF PAIR 2 (mm)			LEAF PAIR 3 (mm)			LENGTH OF LATERAL (BUD OR BRANCH) IN mm AT NODE NUMBER							
			BLADE LENGTH	BLADE WIDTH	PETIOLE LENGTH	BLADE LENGTH	BLADE WIDTH	PETIOLE LENGTH	BLADE LENGTH	BLADE WIDTH	PETIOLE LENGTH	1	2	3	4	5	6	7	8

TABLE 3.

INITIAL	DATE	HEIGHT	COMMENTS

INITIAL	DATE	HEIGHT	COMMENTS

TABLE 2.

INITIAL	DATE	NUMBER OF DIFFERENT INTERNODES	LEAF PAIR 1 (mm)			LEAF PAIR 2 (mm)			LEAF PAIR 3 (mm)			LENGTH OF LATERAL (BUD OR BRANCH) IN mm AT NODE NUMBER							
			BLADE LENGTH	BLADE WIDTH	PETIOLE LENGTH	BLADE LENGTH	BLADE WIDTH	PETIOLE LENGTH	BLADE LENGTH	BLADE WIDTH	PETIOLE LENGTH	1	2	3	4	5	6	7	8

TABLE 3.

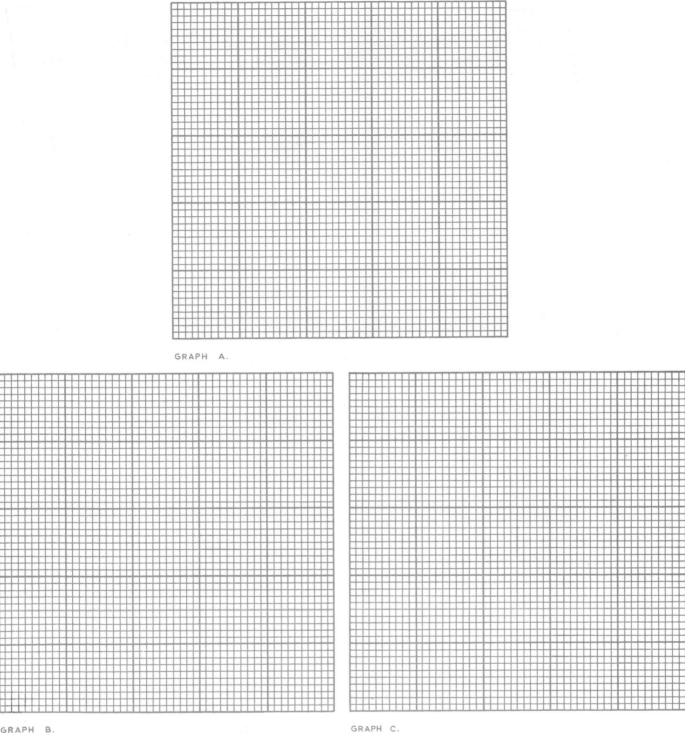

GRAPH A.

GRAPH B.

GRAPH C.

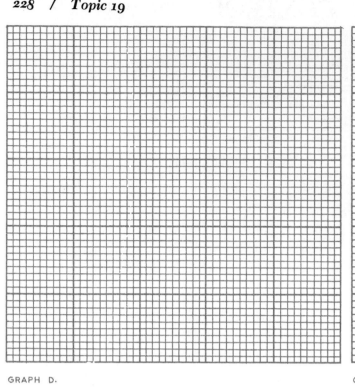

GRAPH D.

GRAPH E.

20 HORMONAL CONTROL OF PLANT DEVELOPMENT

Biological Science, pp. 260-274.
Elements of Biological Science, pp. 165-171.

Though we commonly associate the term *behavior* with the overt activities of animals, we are constantly reminded that plants can respond with remarkable versatility to environmental change. Plant stems, for example, turn actively toward light; roots turn toward gravity. Leaf petioles bend and turn, exposing the leaf blades to sunlight at the most effective angles. Most such adaptive movements are mediated by hormones. Plant hormones play an important governing role in a wide variety of other integrative and developmental processes, such as the origin, aging, and abscission (dropping) of leaves, flowers, and fruits; the development of species-specific patterns; and the appearance and course of cellular differentiations in growing regions of the plant.

Some of the primary effects of hormone action are well known. Hormones may stimulate or inhibit cell enlargement or cell division, with consequent changes in the shape, size, or orientation of the plant part in question; hormones may stimulate or inhibit cellular differentiations. Often several hormones interact complexly to produce effects different from those that would be produced by the activity of individual hormones. The precise action of a plant hormone depends, of course, on the hormone in question, its concentration, and the nature of the responding tissue. The actual mechanisms by which hormones act are poorly understood, although they are presently a topic of active research.

Apart from their inherent interest and importance as the principal regulators of plant growth, hormones find a wide variety of commercially valuable applications. By using the appropriate hormone, for example, one can induce or break dormancy in some seeds or tubers, increase the tenderness of celery stalks by accelerating their growth, or produce seedless fruits. A knowledge of how hormones act differentially on different species of plants makes possible increasingly effective weed control, and an understanding of how hormones function to influence plant pattern often makes possible the control of pattern to suit agricultural or horticultural requirements.

In this unit, you will work experimentally with the chemical control of plant growth, development, and pattern.

(**Note to the instructor:** Topic 20 is written for students who will be able to check regularly on experimental plants kept in the laboratory and in the greenhouse. If laboratory logistics prohibit students' regular attention to their experiments, or if greenhouse facilities are not available, please refer to the Instructor's Guide for suggested modifications.)

PART I. THE INFLUENCE OF AUXINS ON STEM AND ROOT GROWTH

The discovery of naturally occurring plant growth regulators dates largely from the experiments of Charles Darwin, who showed that the tip of the coleoptile (sheath) covering the young shoot of an oat plant was necessary for a phototropic response. It was eventually discovered that this response involved a chemical substance, manufactured in and transported from the plant tip to other parts of the plant. This substance was named **auxin**; later, it was isolated and found to be indoleacetic acid (**IAA**). For many years, auxin was synonymous with IAA; it has been found, however, that a variety of both naturally occurring and synthetic compounds have similar effects. Many such compounds have elements of chemical structure in common with IAA.

IAA and other auxins act by stimulating cell elongation; they affect the uptake of water and the elasticity of the cell wall. By some type of active transport that is not understood, they accumulate on the shaded side of a differentially-illuminated shoot, bringing about a more rapid elongation of cells on that side of the shoot than on the other, and hence a bending of the shoot toward a light source. If the auxin source (the shoot tip) is excised and then replaced off center, or if the tip is placed on a small gelatin block which is then placed off center on the shoot, this response will occur even in the dark (FIGURE 20-1). In each case, cell elongation occurs on the side of the shoot where the auxin source was replaced.

The placing of an auxin-containing gelatin block on the cut tip of the coleoptile is the basis for a test for auxin activity, called the *Went growth curvature test.* Any substance showing coleoptile-bending activity is classified as an auxin. Because the concentration-response relationship is well known and because the response is so sensitive to small differences in auxin concentration, the growth curvature test is commonly used as an indicator, or **bioassay,** of auxin concentration. This general principle — use of an organism to indicate the presence or concentration of a substance in amounts too small to be detected by conventional microanalyses — finds application in many fields of biology.

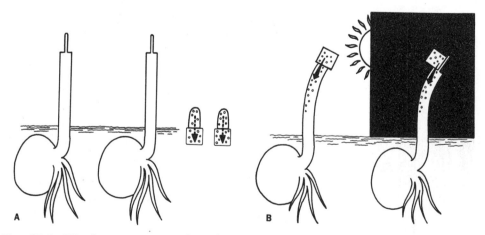

FIG. 20-1. Went's experiment. When the tips of coleoptiles are cut off and placed on blocks of agar for about an hour (A), and one of the blocks alone is then put on a stump (B, left), the stump will resume growing and will respond to light; if a block is placed off center on a stump in the dark (B, right), the stump will grow and will bend away from the side on which the block rests. Apparently a hormone has diffused from the tips into the blocks, and this hormone can then diffuse from the blocks into the stumps.

Fig. 20-2. Razor-blade cutter for making sections of uniform size.

Because success of the curvature test requires rigidly controlled environmental conditions, we shall illustrate bioassay by using a similar but not so sensitive test, the *straight-growth test*. In addition, we shall see that shoots and roots respond quite differently to IAA.

A. The straight-growth test

In this test, straight sections of coleoptile or stem, of known length, are placed in auxin solutions of unknown concentration. After a specified period of time, the sections are removed and their new lengths measured. To determine the unknown concentrations, these data are then compared with known information about the relationship between auxin concentration and growth for the plant in question.

1. Working individually, or in groups as directed by your instructor, obtain eight petri dishes. Label six dishes with the numbers 3, 4, 5, 6, 7, and 8, to correspond to 10^{-3}, 10^{-4}, 10^{-5}, 10^{-6}, 10^{-7}, 10^{-8}M concentrations of IAA. On the seventh dish, write "unknown" and on the eighth, write "control." Add your name to each dish-top. Obtain the appropriate dilutions of IAA, the unknown solution, and the control solution (one containing no IAA) from your instructor; half fill each petri dish with the solution indicated on its label.

2. Obtain several etiolated (dark-grown) shoots of Alaska pea on a wet paper towel. Carry out all subsequent handling on the towel, and enclose the shoots in it to prevent desiccation when they are not actually being handled. Using the apparatus illustrated in Figure 20-2, cut 24 5-mm sections from the *youngest internode* (that straight region just below the terminal curvature, as shown in Figure 20-3). As the sections are cut, put them quickly into the petri dishes until you have three such sections in each dish. Cover the dishes, put them in a dark place, and leave them for 12-24 hours. Record the initial section lengths, which should all have been the same, in Table 1.

3. At the end of the incubation period, record in mm the final length for each section and enter the data in Table 1. Compute the *average* elongation for each group, and plot these averages against IAA concentration on the graph provided below the table.

Which concentration of IAA was most effective in promoting elongation of the sections? Is there any evidence that higher concentrations are inhibitory?

Fig. 20-3. Dark-grown pea shoot.

What is the approximate concentration of the unknown IAA solution? Why is the "control" dish necessary? Did the stem sections in it show any growth? How did the control differ from the experimentals?

B. Root-growth test

1. Obtain eight petri dishes; label six dishes with the numbers 5, 6, 7, 8, 9, and 10, to correspond to 10^{-5}, 10^{-6}, 10^{-7}, 10^{-8}, 10^{-9}, 10^{-10}M concentrations of IAA. Label the seventh dish "unknown," the eighth "control," and put your name on all eight dishes. Place a disc of filter paper in each dish, and soak it with 5 ml of the appropriate IAA dilution, unknown, or control (no IAA) solution.

2. Obtain 24 healthy germinating seedlings of Alaska pea, with roots of approximately the same length (about 1 cm). Keep the plants wrapped in a wet paper towel except when actually manipulating them. Measure the root length of three of the seedlings, and record the data in Table 2, in the first block. Average these measurements. Place the three seedlings into the dish numbered 5.

3. Repeat this procedure for the other five concentrations and for the unknown and the control.

4. After 24-48 hours, remeasure and record the three primary root lengths (ignore any lateral roots) in each concentration set, and average them. Plot your data as for Experiment A, with root elongation on one axis and IAA concentration on the other.

In which of the solutions did the greatest increase in root length occur?

How would you describe the reaction of the roots to this range of IAA concentrations?

Which of the two tests (shoot elongation or root elongation) represents a more sensitive assay of IAA concentration? Be prepared to defend your answer.

**PART II.
AUXIN-INDUCED
TROPISMS**

The substantial difference in response of roots and shoots to IAA reflects a differential sensitivity to auxin that natural selection has shaped to the plant's advantage in an important behavioral response — **geotropism.** Roots exhibit *positive* geotropism, i.e., they grow toward gravity. Since seeds are often scattered randomly in the soil and therefore do not always germinate with the root end downward, positive geotropism is obviously an adaptive response. Stems exhibit a *negative* geotropic response, which is similarly adaptive, since it tends to push them into the light above the soil.

Both responses can be explained on the basis of auxin distribution in the organ concerned. In the absence of other stimuli, such as light, and if the organ is horizontal, auxin produced by either shoot or root meristem tends to accumulate in the lower portion of the organ near an active meristem.

In the stem, with its relatively high threshold of auxin sensitivity, this stimulates a greater elongation in the lower cells and thereby brings about an upward bending. In the root, with a much lower threshold of auxin sensitivity, a similar auxin concentration *inhibits* elongation on the lower side of the root, so that more elongation occurs on top; bending is therefore in a downward direction.

A. Root geotropism

1. Obtain a 250-ml beaker from the supply desk, and line it with several thicknesses of wet paper toweling. Fill the beaker loosely with moistened sphagnum moss or vermiculite, and obtain half a petri dish, or an appropriately trimmed piece of Parafilm, to serve as a cover. From the germination dishes at the supply desk, obtain four corn seedlings, each with a root about 1 cm long; wrap them in a wet paper towel, and keep them wrapped except when you are manipulating them for the experiment. The seedlings must not be allowed to dry.

2. Measure to the nearest mm and record in Table 3 the length of the root of one seedling. Place this seedling between the toweling and the glass wall of the beaker, root oriented directly downward. With a wax marking pencil label the outside of the container A, adjacent to the seedling.

3. Repeat this procedure with the three other seedlings, measuring each seedling and orienting it as indicated below. Space the seedlings equally around the beaker.

Seedling B: measure root; orient vertically upward.

Seedling C: measure root; orient horizontally to the right or left.

Seedling D: prior to measuring the root, excise 1 mm at the root tip; orient it horizontally to the right or left.

In Table 3, make a sketch of each seedling to show the appearance and orientation of its root; record the root lengths.

4. Make observations after 12, 24, and 48 hours. Indicate changes in the length and orientation of the roots with sketches.* At the end of the experiment, measure the main (primary) root accurately and record the measurement in Table 3.

Explain the changes in each seedling on the basis of auxin distribution in the root.

B. Demonstrations of other tropisms

Examine demonstrations of geotropism and phototropism in stems, and make notes on each demonstration.

PART III. THE CHEMICAL CONTROL OF PATTERN IN PLANTS

One of the most challenging problems facing students of both plant and animal development involves the explanation of how pattern (characteristic shape, size, contour, and appearance) comes into being. Leaves are a simple example: Why are they of different shapes in different plants? What selective advantage can there be in maintaining a particular leaf pattern? or a particular shape to the plant as a whole? What is the basis for branching pattern, and why does "tapering" tend to be more pronounced in some plants (e.g., conifers) than in others? What controls species-specific size? What causes dwarfness in plants?

We know that plant hormones play an important role in controlling pattern; the branching pattern of roots and shoots, for example, is in part the result of a phenomenon called apical dominance. Lateral buds are sen-

---

* Approximate for the first two sets of observations; do not remove the roots.

FIG. 20-4. Effect of gibberellic acid on cabbage. The plant at left is normal. The one at right was treated with gibberellic acid. (*Courtesy S. H. Wittwer, Michigan State University.*)

sitive to inhibition by auxin. Since the auxins are produced in active apical meristems, those lateral buds close to such meristems, and thus the branches to which those buds will give rise, are inhibited. The overall plant pattern, of course, depends upon the simultaneous activities of many apical meristems (hundreds or thousands of them in a large tree) and probably on many other factors.

The gibberellins are one class of hormones that seems in some plants to be closely linked to the control of size. First discovered in a fungus that infected abnormally long rice seedlings, gibberellins are now known to occur naturally in plants. They influence many aspects of plant development apart from size: dormancy in seeds and tubers, the initiation of flowering and fruiting, etc. Though they probably act in conjunction with auxins, they are not active in the growth curvature tests or any of the other standard assays for auxin activity. One type of gibberellin, gibberellic acid, shows a dramatic effect in its stimulation of rapid stem elongation in dwarf mutants and some other plants that normally undergo little stem elongation (FIGURE 20-4); such mutants will not respond to auxins. In the following experiment, you will explore the responses of dwarf and normal corn plants to gibberellic acid.

1. In the greenhouse, or the laboratory, if your instructor so directs, mix equal parts of vermiculite and garden soil sufficient for four 4-inch flower pots. Place this mixture in the four pots, to a level about one inch below the top. Moisten, but do not soak, the soil mixture in each pot.

2. Obtain ten germinated "normal" corn seedlings, and place five of them gently into depressions in the soil in each of two pots. Cover both sets of seedlings with about a quarter inch of soil-vermiculite mixture, orienting the green shoot upward through the covering soil layer. Label the pots with your name, date, class, and the designations "Normal-Experimental" or "Normal-Control."

3. Repeat the above procedure with ten dwarf corn seedlings, using pot labels marked "Dwarf-Experimental" or "Dwarf-Control." Be sure to cover the seedlings with soil-vermiculite mixture and to orient them as before.

4. Put the pots in two widely separated greenhouse areas designated "Control Plants" and "Experimental Plants." Following your instructor's directions, water the plants thoroughly. Make sure that all information is clearly marked on the pot labels, and that it will not run off when you water the plants.

5. In the following laboratory period, weed out two of the seedlings from each pot, leaving the three healthiest-looking specimens. Tag and number each seedling, and measure its height in centimeters from the soil to the top of the shoot. Record these data in the column marked "initial length" in Table 4.

6. Using a syringe with a fine needle or a perfume atomizer, spray both dwarf and normal *experimental* plants with a freshly made water solution of gibberellic acid (100 mg/liter). Spray the plants until the leaves have droplets that will almost run off, but avoid wetting the soil with the hormone mixture, and try to apply equal amounts of gibberellic acid solution to both experimental groups. With a *different* syringe or atomizer, spray the controls with an approximately equal volume of plain water.

7. Measure the height of each plant on each of the four days following treatment, and again seven days after treatment. Keep the plants watered, and stake them if necessary. Record all data in Table 4.

8. At the end of the experiment, analyze your results in whatever way you consider appropriate, comparing normal and dwarf plants in their response to gibberellic acid, and comparing the experimental and control plants in each group.

How soon after application did the response to gibberellic acid occur?

Advance an hypothesis to explain the action of gibberellic acid in correcting dwarfism, and suggest possible means of testing your hypothesis.

TABLE 1. EFFECT OF IAA CONCENTRATION ON SHOOTS.

A	INITIAL			
	1	2	3	AVG.
10^{-3} IAA				
10^{-4} IAA				
10^{-5} IAA				
10^{-6} IAA				
10^{-7} IAA				
10^{-8} IAA				
UNKNOWN				
CONTROL				

B	FINAL (AFTER HRS)			
	1	2	3	AVG.
10^{-3} IAA				
10^{-4} IAA				
10^{-5} IAA				
10^{-6} IAA				
10^{-7} IAA				
10^{-8} IAA				
UNKNOWN				
CONTROL				

AVERAGE ELONGATION

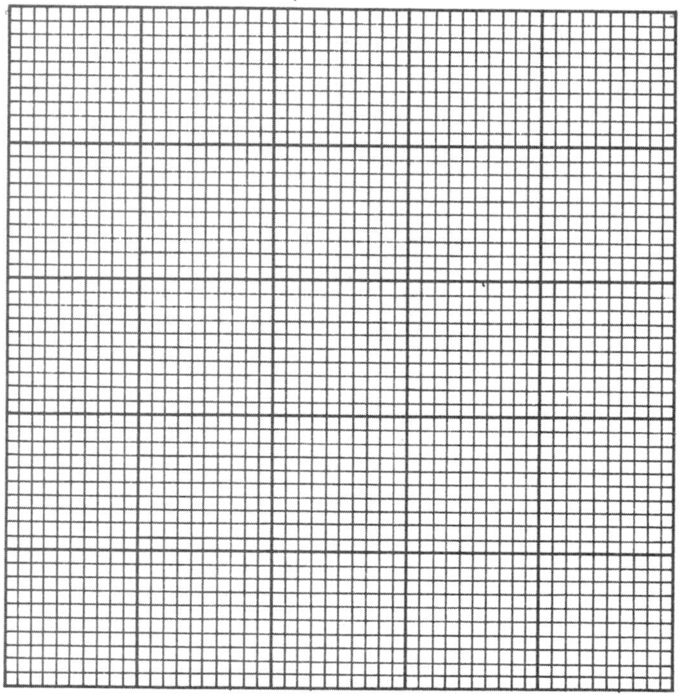

GRAPH 1.

TABLE 2. EFFECT OF IAA CONCENTRATION ON ROOTS.

A	INITIAL				B	FINAL (AFTER HRS.)				AVERAGE ELONGATION
	1	2	3	AVG.		1	2	3	AVG.	
10^{-5} IAA					10^{-5} IAA					
10^{-6} IAA					10^{-6} IAA					
10^{-7} IAA					10^{-7} IAA					
10^{-8} IAA					10^{-8} IAA					
10^{-9} IAA					10^{-9} IAA					
10^{-10} IAA					10^{-10} IAA					
UNKNOWN					UNKNOWN					
CONTROL					CONTROL					

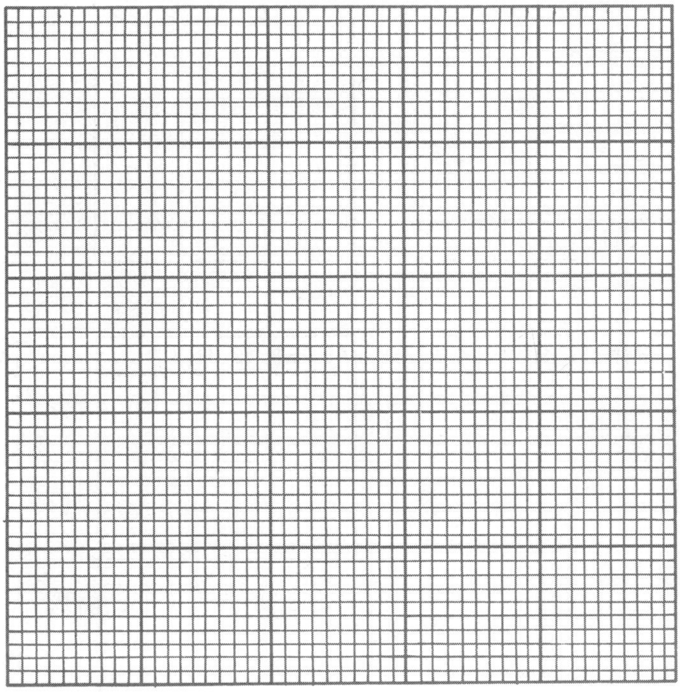

GRAPH 2.

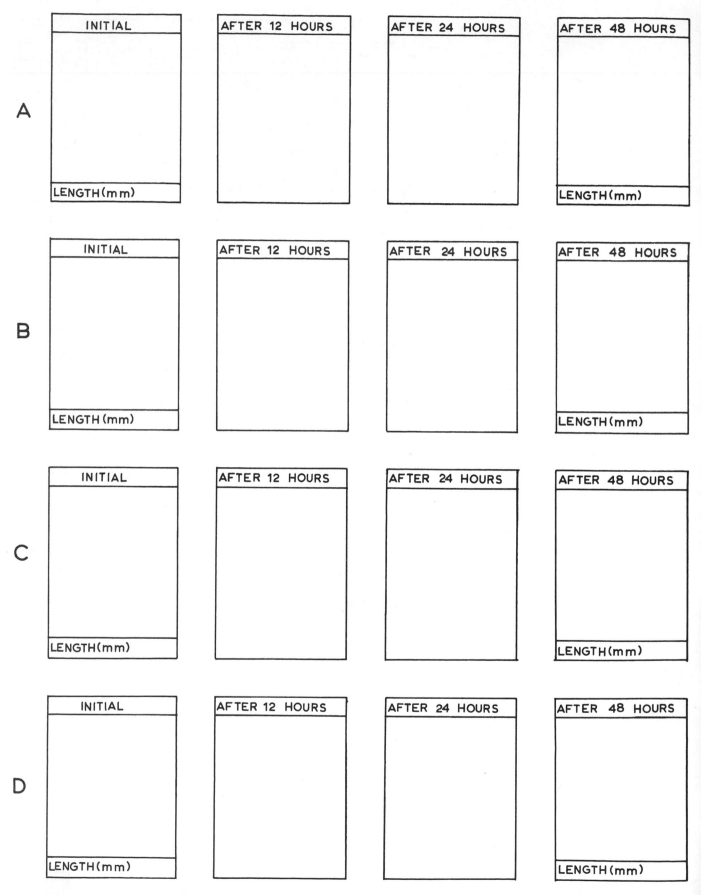

TABLE 3. GEOTROPISM EXPERIMENT

CORN VARIETY	TREATMENT	INDIVIDUAL	INITIAL LENGTH (cm)	LENGTH FOLLOWING TREATMENT (cm)				
				DAY 1	DAY 2	DAY 3	DAY 4	DAY 7
NORMAL	G.A. EXPERIMENTAL	1						
		2						
		3						
		AVERAGE						
	WATER CONTROL	1						
		2						
		3						
		AVERAGE						
DWARF	G.A. EXPERIMENTAL	1						
		2						
		3						
		AVERAGE						
	WATER CONTROL	1						
		2						
		3						
		AVERAGE						

TABLE 4. EFFECT OF GIBBERELLIC ACID ON NORMAL AND DWARF CORN SEEDLINGS.

TOPIC **21** ANIMAL DEVELOPMENT

Biological Science, pp. 546-567.
Elements of Biological Science, pp. 341-350.

Development is a continuous process; in sexual organisms, it starts when a sperm fertilizes an egg, and ends with death. Between these two extremes occur a multitude of progressive changes in form and function, ranging from early embryonic development (during which the pattern of the organism is elaborated from a mass of cells) to changes in sexual maturity and aging. Cellular growth and differentiation, metamorphosis (changes in appearance and way of life associated with growth in insects and many other animals), and regeneration (replacement or repair of a lost or injured part of an organism) are also developmental processes. All organisms develop, and developmental processes are as much the result of adaptation and natural selection as are anatomical, physiological, and behavioral characteristics.

In this laboratory, you will have an opportunity to compare early embryonic development in several living organisms and to observe metamorphosis and regeneration.

Early embryonic development is particularly significant because it is at this time that the pattern of the organism originates. The following brief summary of stages in embryonic development is provided to help you follow examples given later in the laboratory.

1. **Fertilization** involves the fusion of nuclei and other events associated with the union of egg and sperm cells (gametes). The product of fertilization is called a **zygote.**

2. **Cleavage** is the term applied to the series of mitotic divisions undergone by the zygote. No growth occurs during early cleavage. The pattern of cleavage — whether the whole zygote or only a part of it divides — varies in different species of organisms according to the amount and distribution of yolk (food material) contained in the egg.

3. **Blastulation** is the formation of a mass of cells that, in most organisms, become arranged into a thin-walled sphere. The central cavity of the blastula is called the **blastocoel.**

4. **Gastrulation** is characterized by morphogenetic movements that lead to the formation of the **archenteron,** or primitive gut, and to the formation of three distinct *germ layers* of cells, the **ectoderm, mesoderm,** and **endoderm,** from which the organs of the embryo are later elaborated.

5. **Morphogenesis** is the development of characteristic internal and external form; it comes about by the shaping of the germ layers through differential growth, movement, and association of cells in those layers.

We shall compare these processes in four animals: a sea star, a common echinoderm; a sea urchin, another echinoderm; a frog; and a chicken. Sea star and sea urchin embryos have simple and direct blastulation and gastrulation; in frog and chicken embryos these processes are complicated by the presence of a large amount of unevenly distributed yolk.

PART I. SEA URCHIN DEVELOPMENT (LIVE MATERIAL)

Sea urchins are scavengers of shallow coastal waters. Take this opportunity to observe adult urchins. Notice the tube feet, each tipped by a small sucker. These enable the urchin to move over the rocks. The spherical body is covered with long spines. Gently poke one of these spines; what happens? Of what advantage are the spines? On the lower surface examine the large mouth with its five teeth. These are used for rasping algae from rocks and tearing the urchin's other food. The sexes are separate, and gametes (eggs or sperms) are emitted from five pores on top of the animal directly into the ocean, where fertilization occurs. The pores are sometimes difficult to see on a living animal, but are readily observable on a cleaned skeleton.

The sexes are not externally distinguishable, but the animals can be sexed and gametes obtained by effecting gamete release in several ways. One easy method involves passing a ten-volt alternating current through the animal. This brings about immediate shedding of gametes, which stops when the flow of current ceases. Another method involves injecting potassium chloride solution into the body cavity of the animal.

A. Gametes

The instructor will demonstrate how gametes are procured. Make a wet mount of each; if necessary, use vaseline to prevent evaporation. Be *very careful* not to mix sperms and eggs or the droppers used to transfer them to slides. (*Note:* It will be necessary to adjust the light carefully in order to see the sperms.)

B. Fertilization

1. Obtain some fertilized eggs and make a wet mount. Support the the cover slip with vaseline or with fragments of cover slip.* Successful fertilizations are indicated by the presence of a membrane outside the fertilized egg, and a clear space between the membrane and the egg. Unfertilized eggs do not have the space or the membrane.

2. The actual process of insemination of the eggs occurs much too rapidly for you to observe, but you should see many unsuccessful sperms attached to the membrane of the fertilized eggs. If you wish to observe fertilization and the formation of the fertilization membrane, you may do so by mixing a drop of sperm suspension with a drop of egg suspension** on a slide just prior to examination.

C. Cleavage

The rate of development varies with temperature and with the urchin species; a variety of other environmental factors may also influence it. If your eggs were properly fertilized, you can expect the first cleavage to begin 50–100 minutes after fertilization, or later, if the temperature is lower than 20°C. Thirty minutes after fertilization, make fresh wet mounts of fertilized eggs; examine the cytoplasm for evidence of change. *Do not* leave the slide on the microscope stage, unless you turn the light off; heat from a microscope lamp will seriously interfere with development and may kill the em-

---

* It is essential to support the cover slip when you observe fertilized eggs, as pressure exerted by an unsupported cover slip will prevent normal development.

** Egg and sperm suspensions of proper concentration in natural sea water will be provided by your instructor.

bryos before they begin to cleave. Continue at intervals to make wet mounts, with the cover slips properly raised, until you begin to see cleavages.

The cleavages are very easily observed in sea urchin embryos, and with good lighting you may be able to observe some of the internal events that precede the actual cleavages.

The yolk of the echinoderm egg is almost equally distributed throughout; its small amount does not interfere with cytokinesis. It is therefore possible for the entire zygote to cleave.

Watch the process of several cleavages carefully and make sketches. Keep *careful* records of time.

The table* below shows approximate schedules of development for two species of sea urchins. Remember that these are approximations, and that the rate of development you observe may be influenced by differences in the room temperature, oxygen supply to the embryos, etc. You may be asked to use batches of previously fertilized eggs to work out a time schedule of development for the species you are using.

STAGE	*Strongylocentrotus* (WEST COAST)	*Arbacia* (EAST COAST)
Formation of fertilization membrane	2–5 min.	2–5 min.
First Cleavage	70–90 min.	50–70 min.
Second Cleavage	100–120 min.	78–107 min.
Third Cleavage	130–160 min.	103–145 min.

D. Blastulation

If developing blastulae are available, make a wet mount and observe them. Notice that they are hollow during this stage. Echinoderm embryos become ciliated, rotate, break out of the fertilization membrane, and begin to swim.

E. Gastrulation

Gastrulation in the sea urchin embryo comes about by an infolding or invagination of one side of the blastula, called the vegetal pole. The cells of this region gradually work inward and upward until they lie in contact with the inner side of the cells of the opposite side of the blastula. Thus, the cavity of the blastula, the blastocoel, is almost lost, and the new cavity formed becomes the gut. It is called the archenteron. **Mesoderm** in the sea urchin is formed by out-pocketings (evaginations) of the **endoderm** (lining of the archenteron), which eventually break away from the endoderm. The outer layer of the gastrula is called **ectoderm.** All the stages in echinoderm early development are summarized in FIGURE 21-1, a series of photographs of living sea star embryos.

---

* Table modified in part from Costello, D. P., et. al., *Methods for Obtaining and Handling Marine Eggs and Embryos*, Marine Biological Laboratory, Woods Hole, Massachusetts (1957).

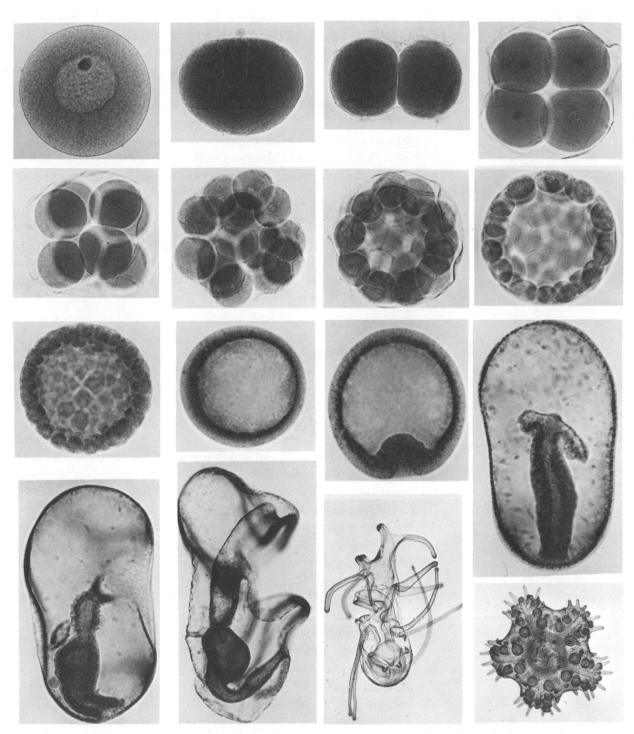

FIG. 21-1. Stages in sea star development. (*Courtesy Carolina Biological Supply Company.*)

PART II. SEA STAR DEVELOPMENT (PREPARED SLIDES)

A. Unfertilized egg

Fertilization through gastrulation in the sea star are practically identical to the same stages in the sea urchin. Make sketches as you examine each stage.

The undivided egg is a nearly spherical cell whose large nucleus and nucleolus are clearly visible. A small amount of **yolk,** or stored food, is present as many small particles distributed throughout the cytoplasm.

B. Early cleavage

The two-cell stage, produced by the first mitotic cell division, is easy to identify. The two cells remain attached to each other; their nuclei are usually very difficult to see, as is the case with most later stages.

Examine a four-cell embryo. Compare the size of each of its cells with that of the undivided ovum. Do the same for the eight-cell stage. Is there any growth in size? Where does the embryo obtain raw materials for respiration?

Many successive mitotic cleavages produce a many-celled, grapelike cluster called a **morula** (16 or more cells). Note that there has still been little increase in the total size of the embryo.

C. Blastula

As cleavage continues, the embryonic cells become arranged in a sphere, the **blastula.** In sea stars the blastula walls are usually one cell thick; its central cavity is called the **blastocoel.** (Remember that the specimens on your slides are whole embryos and that you must focus carefully to obtain optical sections showing the internal characteristics of this and later stages.)

D. Gastrula

Shortly after formation of the blastula, a small depression, or invagination, begins to form at one point on the surface of the embryo; this is the beginning of the critical process called **gastrulation.** If you examine **gastrulas** carefully, you can see that the cells that formed the blastular wall at the end where invagination occurred are larger than those at the other end of the embryo. The difference in size is not very great in sea star embryos but is more pronounced in many small animals. The smaller cells make up the animal pole of the embryo and the larger cells make up the vegetal pole; invagination typically occurs at the vegetal pole. As gastrulation proceeds, the invagination becomes larger and larger.

Locate embryos showing the initial steps of gastrulation and others showing many successive degrees of invagination. The latest stages on your slides are embryos in which the inner end of the invagination has begun to expand. (Remember that embryos are randomly oriented on your slides and therefore you will view some longitudinally, some diagonally, and some in cross section. Include among your drawings gastrulas shown in both longitudinal and cross sections.)

Gastrulation in the sea star produces an embryo with two primary cell layers, an outer **ectoderm** and an inner **endoderm.** The third primary layer, the **mesoderm,** soon begins to form. Gastrulation, then, eventually results in an embryo with three roughly delineated primary germ layers, an essentially obliterated blastocoel, and a new cavity that is continuous with the outside through the **blastopore** and will become the lumen of the digestive tract.

PART III. FROG DEVELOPMENT

Eggs are produced in the ovary of female frogs during the spring and summer months when food is available and the animals are active. Mature eggs are then held in the ovary during the winter months, and released naturally the following spring. Release occurs during a contact of the male and female

Fig. 21-2. Male frog clasping female frog (amplexus). (*Courtesy Carolina Biological Supply Company.*)

called amplexus (FIGURE 21-2), but fertilization and development are external. Frogs brought into the laboratory during the winter or early spring months can be artificially induced to release eggs by injecting them with frog pituitary extract. This procedure apparently simulates the natural hormonal changes that occur in spring. Eggs are released from the ovaries into the coelom, and are passed by ciliary action to the oviduct. The eggs may remain in the oviduct for several days, but they can be forced out through the cloaca by gentle pressure on the frog's abdomen.

Active sperms can be obtained by dissecting and macerating the testes of a wintering male frog, or by squashing the dissected testes between glass slides. The sperms are then diluted in pond water, checked for activity, and used to fertilize eggs. About an hour before the beginning of the laboratory period, your instructor fertilized some frog eggs. You may obtain them from dishes at the supply desk.

A. Gametes

1. Obtain a drop of diluted frog sperm from the dish at the supply table. The sperms are very small, and you may be able to see only their sickle-shaped heads. Many of the sperms will be non-motile, but others will display a spiral forward motion due to the activity of their flagella. (You may occasionally see flagella if the lighting is very carefully adjusted.)

2. Obtain some unfertilized eggs from the culture dish at the supply table, cutting them away from the rest of the egg mass with forceps and scissors. Put them into a small dish and cover with pond water. Observe with the dissecting microscope. Notice that the eggs are partially pigmented; the black side contains very little yolk, and is called the **animal hemisphere;** the unpigmented, yolky side is called the **vegetal hemisphere.** This unequal distribution of yolk causes some important differences in cleavage and gastrulation between the eggs of the sea urchin and the frog.

Unfertilized eggs are held in random orientation by their sticky jelly coats, and a mass of them will appear mottled because some have the animal hemisphere upward and some have the vegetal hemisphere upward. When it is fertilized, the egg shrinks slightly, and a fertilization membrane very much like a sea urchin's appears. The eggs are then free to rotate within their membranes, and they shift so that the animal pole is uppermost. If you turn your inseminated eggs over when you get them from the supply desk, those that are fertile will rotate in ten to fifteen minutes.

B. Cleavage

1. The first cleavage will be evidenced by the appearance of a furrow across the zygote's upper surface, the animal pole. Depending upon the water temperature, this will occur 150–180 minutes after fertilization. To be certain that you see the furrow, observe your eggs at about five-minute intervals, beginning two hours after fertilization. Once it has begun, the cleavage is rapid.

The crinkling of the egg surface along the cleavage furrow and the disappearance of the furrow after the division is completed are external evidences of mechanical forces involved in animal-cell cytokinesis. Are comparable forces involved in plant-cell cytokinesis?

2. The second cleavage occurs within thirty minutes after the beginning of the first, and will begin at the upper, or animal, pole of the egg. Has the first division been completed when the second begins?

C. Later embryonic development

FIGURES 21-3 and 21-4 and your text will help you to understand later development in the frog. Because of the large amount of yolk, gastrulation cannot occur as it does in the sea urchin embryo, but the end result is the same as in the sea urchin: the three germ layers and archenteron are formed.

The instructor may save your embryos for observation next week, or he may provide embryos from previous laboratory periods for your observation of later stages in development. If models of frog development are available, be sure to study them sometime during this period.

PART IV. CHICK DEVELOPMENT

In many animals, all embryonic cells are incorporated into the developing new individual. In the development of terrestrial vertebrates, however, a very substantial number of embryonic cells develop into *temporary* structures that make possible embryonic development on land. Reptiles, birds, and mammals have reproductive and developmental adaptations that provide for the metabolic needs, water balance, and general protection of the embryonic individual as it develops within some protective structure (either an egg, as in reptiles and birds, or the uterus of the female parent, as in mammals). You have already studied the mammalian placenta, an adap-

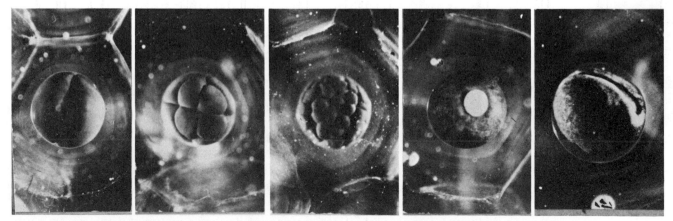

FIG. 21-3. Photographs of frog development; whole embryos. (*Courtesy Carolina Biological Supply Company.*)

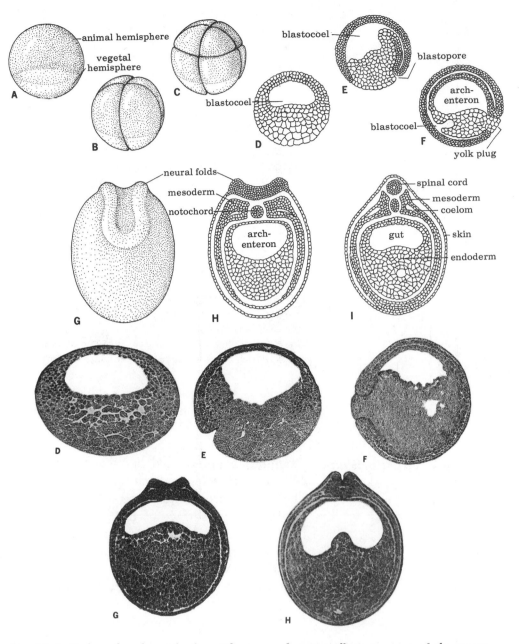

FIG. 21-4. Early embryology of a frog. The upper drawings illustrate some of the stages in early development, sketched whole and in section. The lower photographs show sections through some stages, and are lettered to correspond to stages in the drawings. (A) Zygote. (B-C) Early cleavage stages. (D) Longitudinal section of blastula. (E-F) Longitudinal section of two stages in gastrulation. (G) An early neurula, showing the neural folds and neural groove. (H) Cross section of a neurula after formation of mesoderm. (I) Cross section of a later embryo, showing spinal cord and notochord. (*Photographs D, E, F, G: Courtesy Carolina Biological Supply Company. Photograph H: Courtesy Ward's Natural Science Establishment, Inc., Rochester, New York.*)

tation that provides for the growth of the embryo within the uterus of the female parent. Part of the placenta and all of the extraembryonic membranes associated with the embryo, including the amnion, chorion, and allantois and the umbilical cord connecting the embryo to the placenta, are structures composed of cells derived from the fertilized egg. The chick

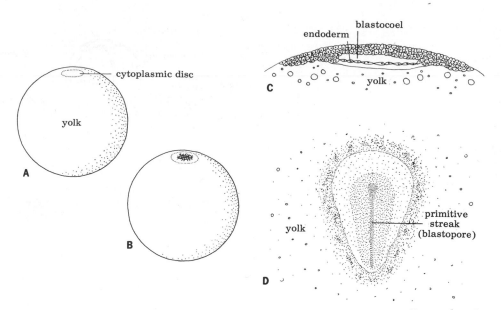

Fig. 21-5. Some stages in early development of a chick. (A) Zygote. A small cytoplasmic disc lies on the surface of a massive yolk. (B) Early cleavage. There is no cleavage of the yolk. (C) Section through a blastula. (D) Surface view of a gastrula. Involution of cells along the midline of the embryo during gastrulation produces a clearly visible primitive streak, which is essentially a very elongate blastopore.

embryo is a convenient one in which to study embryonic membranes and to see some of the important morphogenetic processes in more detail.

A. Organization of the egg

Examine an unfertilized hen's egg provided by your instructor. The egg cell, or **ovum**, released from the hen's ovary consists of the yellow yolk mass plus a small area of yolk-free cytoplasm called the **germinal disc**, which is located on the surface of the yolk. Because of the large amount of yolk, cleavage occurs only in this disc, and the embryo grows by absorbing raw materials for its metabolism from the yolk (FIGURE 21-5).

During its passage through the oviduct, several accessory features are added to the ovum. (*Note:* The term "egg" is best reserved for the shelled entity laid by the hen, while the egg cell, that structure released from the ovary, is called an "ovum.") In the first section of the oviduct, a viscous, stringy albumen is secreted and adheres to the ovum. As the ovum moves along, rotation twists the stringy albumen into a pair of strands, which surround and project beyond the ovum. In the next oviduct region, a more watery albumen is added. Finally, two shell membranes and a calcareous shell are applied to the outer albumen.

The egg is a remarkable piece of architecture. The hard, protective shell is porous, permitting gas exchange with the environment. Albumen serves as a water reservoir for the early embryo and a source of food for the later embryo; it also contains substances that retard the growth of bacteria. The yolk of the ovum contains proteins, carbohydrates, lipids, and vitamins that sustain the metabolic needs of the rapidly growing young chick during its normal 21-day incubation.

Since eggs are held in the terminal portion of the oviduct for variable periods before they are laid, the stage of development at the time of laying varies with each egg. However, cleavage, blastula formation, and gastrulation have usually been accomplished by the time the egg is laid.

Because of the high concentration of yolk in the ovum, cleavage is partial, occurring only in the germinal disc. Gastrulation differs considerably from that in the sea urchin or frog, because the physical forces to which the embryo is subject are very different, and because the blastula is a flat plate of cells rather than a hollow ball (FIGURE 21-5). However, the end result (formation of archenteron and three germ layers) is the same as in the other types of embryos. The yolk of the chick egg remains extraembryonic, but becomes enclosed within a membrane, the **yolk sac,** which grows out from the body of the embryo; this contrasts with the frog embryo, in which the yolk is included within certain cleavage cells and eventually becomes a part of the digestive tube.

B. Opening the egg Each group of students will be provided with a 33-hour, a 48-hour, a 72-hour, and a 96-hour embryo. Study each embryo with reference to the appropriate descriptive section below and to the appropriate figures; a dissecting microscope is most useful for this purpose. Compare your observations with those of others in your group, so that you see all four stages.

1. To remove your embryo from the shell, tap an egg on the edge of a culture dish about half full of warm Ringer's solution, and force open the halves of the shell under the solution. Be careful not to let the sharp edges of the shell tear or puncture any of the delicate membranes. The yolk mass should shift in the solution so that the embryo is uppermost. Why?

Add more Ringer's solution if necessary to support the embryo properly.

2. After study of the embryo on the yolk, it may be cut off the yolk and transferred to a watch glass for further observation. To remove the embryo from the yolk, grasp the ovum membrane outside the embryonic disc with your forceps. Disturbing the yolk as little as possible, cut com-

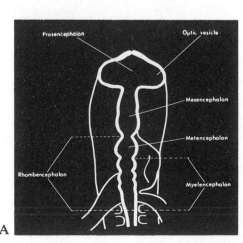

FIG. 21-6. 33-hour chick embryo. (A: *Courtesy Carolina Biological Supply Company. B: Courtesy Ward's Natural Science Establishment, Inc., Rochester, New York.*) **B**

pletely around the embryonic disc with your finest scissors. Place a watch glass in the culture dish, and gently float the embryo away from the yolk mass onto the watch glass. Remove the watch glass from the culture dish; arrange the embryo so that it is free of distortion and securely attached to the watch glass (strips of filter paper are very good for holding down the edges of the embryonic disc). Add a little warm saline to the embryo.

C. 33-hour embryo (FIGURE 21–6)

At this stage the embryo looks like a light streak in the center of the germinal disc, and you will not be able to see details without removing it from the yolk and examining it with a microscope. Its most outstanding feature is the folding of tissues at the anterior end to form the three major areas of the brain: forebrain, midbrain, and hindbrain. An outline of the foregut should be visible at this time. Anterior to the embryo proper lies a region called the proamnion, so named because it will later join with similar lateral and posterior folds to form the protective amniotic sac. The **neural fold** (forming the hindbrain and spinal cord) forms a groove posteriorly, continuing into the **primitive streak,** the region of the embryo through which cells move into the interior during gastrulation.

The **notochord,** forerunner of the vertebral column, is a supportive rod that can be seen running longitudinally through the center of the embryo. Lateral to the notochord is a series of 10–12 **somites,** which later develop into such mesodermal derivatives as the skeleton and striated muscles. Anterior to the somites, locate the **vitelline veins,** which will eventually carry food from the yolk to the developing embryo. Parts of these vessels develop into the heart. In order to bring out details, stain the embryo with neutral red found at the supply desk.

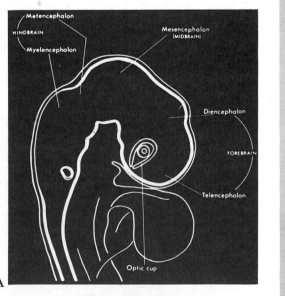

FIG. 21-7. 48-hour chick embryo. (A: *Courtesy Carolina Biological Supply Company.* B: *Courtesy Ward's Natural Science Establishment, Inc., Rochester, New York.*)

D. 48-hour embryo (FIGURE 21–7)

The anterior end of the 48-hour chick embryo has changed considerably from that of the 33-hour embryo. The head has turned so that the right side

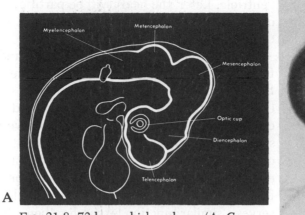

FIG. 21-8. 72-hour chick embryo. (A: Courtesy Carolina Biological Supply Company. B: Courtesy Ward's Natural Science Establishment, Inc., Rochester, New York.)

is up; this torsion of the body will continue until the whole embryo has come to lie on its left side. How far has the torsion proceeded at this stage? Count the number of somites, beginning with the most anterior one. Which is the first somite to show a full dorsal surface to the observer? Note that the anterior part of the head is bent toward the posterior part of the head. This bending, or cranial flexure, occurs in the region of the hindbrain. Identify the parts of the brain.

Notice the formation of the heart. Is there any beating or circulation? Have the vitelline veins grown out over the yolk? Has blood formed? Notice, in the pharyngeal area, the development of **gill slits.** What is their significance? Notice the increased folding of the amnion over the head region.

E. 72-hour embryo
(FIGURE 21–8)

Torsion and a distinctly C-shaped form are evident. Notice that the embryo is covered by a thin-walled sac, the **amnion.** How far posteriorly does the amnion go? Gently touch the embryo with a probe in order to observe the protection the amnion provides. How many somites have been developed? What changes have taken place in the brain?

Observe the size of the heart and its various chambers. The larger ante-

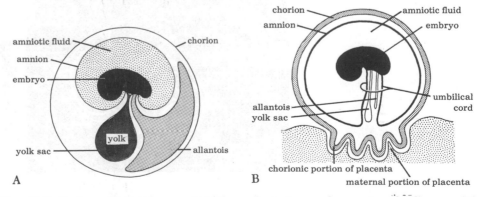

FIG. 21-9. Embryonic membranes. (A) The embryonic membranes in a bird's egg. (B) diagram of human embryonic membranes and placenta.

rior portion is the single atrium, and the ventral posterior portion is the ventricle. Later, the heart will be partitioned into four chambers. Notice the development of the vitelline arteries and veins. Trace the passage of blood through the beating heart.

Notice the development of anterior and posterior limb buds, which will form, respectively, wings and legs. Between the hind limb buds, the **allantois,** which is small at this stage, projects ventrally from the gut. It is one of the four extraembryonic membranes, and functions as a depository for excretory wastes; it also distributes blood vessels. The **chorion,** formed at the same time as the amnion, serves with the allantois in the transport and exchange of respiratory gases. Note in FIGURE 21-9 that the allantois of a bird embryo expands beneath the chorion; the allantois carries blood vessels, a function of which the chorion is incapable. In most mammals, the allantois serves a similar function, expanding beneath the chorion and thus helping to form a highly vascularized double membrane.

F. 96-hour embryo (FIGURE 21–10)

The increased size and vascularity at this stage are obvious. As you opened the egg, you saw blood vessels pressed close against the shell. What is their function? Toward the tail end of the embryo, you should be able to see the fluid-filled allantois. It eventually expands between the embryo and the chorion. As the embryo metabolizes the food material of the yolk, waste products accumulate in this sac. The embryo is now completely surrounded by the amnion.

The liver may be observable as a light yellow organ filling the anterior portion of the abdominal cavity. Most of the rest of the viscera will appear pinkish-white except for the spleen, a small purple organ.

Notice the increased brain size and differentiation, and the increase in the blood supply to the brain region. The eyes have now become pigmented. Also notice the gill slits and limb buds.

The instructor may be able to provide older embryos for interested students.

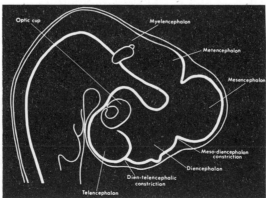
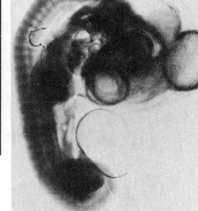

FIG. 21-10. 96-hour chick embryo. (*Courtesy Carolina Biological Supply Company.*)

**PART V.
EXTRAEMBRYONIC
MEMBRANES IN
MAMMALS**

Your instructor will demonstrate a uterus taken from a pregnant female pig and will discuss some aspects of its anatomy and of pregnancy and birth in mammals. The following outline is provided to help you follow his discussion.

Structure:
1. Uterine horns
2. Body of the uterus
3. Ovaries
4. Oviducts, with funnel and ostium at the distal end

Events in fertilization and early development:
1. Release of eggs from the ovary, and their passage down the oviduct
2. Release of sperm into the vagina at copulation
3. Course of sperm up through the uterus into the oviduct, where fertilization occurs
4. Course of the fertilized eggs down into the uterus, where they eventually implant

The fertilized egg gives rise not only to the embryo, but also to the three extraembryonic membranes, which may be seen by cutting open a portion of the uterine wall where an embryo is located. The innermost membrane, the amnion, is clear and non-vascularized, and serves as a protective fluid-filled cushion for the embryo. The other two membranes are fused into one, the chorio-allantoic membrane. Note the large number of blood vessels in this membrane, and its foldings which interdigitate with foldings in the uterine wall. The circulation in the embryo and in the chorio-allantoic membrane are connected by the umbilical cord.

The chorio-allantoic membrane and a portion of the uterine wall together form the **placenta**; materials are exchanged between the capillary system in the uterine wall and that in the chorio-allantoic membrane. Note the relatively loose connection between the parts of the placenta in the pig; this connection is much more intimate in the human being and causes some rupture of blood vessels at the time of birth, which would not occur in the pig.

**PART VI.
METAMORPHOSIS**

Many animals go through a larval stage that bears little resemblance to their adult form. The series of developmental changes that convert an immature animal into its adult form is called **metamorphosis.** The rather drastic metamorphosis of a larva into an adult frequently involves extensive cell division and cell differentiation, and often also morphogenetic movements.

On demonstration are the life stages of an insect that undergoes complete metamorphosis. Notice the changes from egg to larva, pupa, and adult. As the larva grows, it molts several times. This growth through a series of larval stages does not bring it closer to the adult appearance. After larval development is completed, it enters the pupal stage, during which it becomes enclosed in a case or cocoon. During the pupal stage, most of the larval tissues are destroyed, and new adult tissues and organs develop from small discs of tissue that were present but undeveloped in the larva.

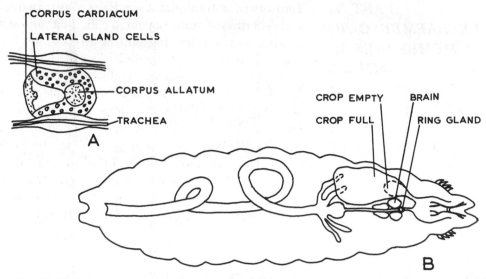

FIG. 21-11. Blowfly larva showing ring gland (A) and approximate size of empty and full crop (B).

What controls metamorphosis in an insect? To answer this question we must first examine the changes in structure that constitute development, and we must study critically the changes in physiology and behavior that accompany development. From such study, the following facts emerge: It is the endocrine organs — principally the brain and the ring gland — that control insect development. If we remove them or interfere with their secretions, normal developmental events do not take place. The ring gland, as you can see by examining FIGURE 21-11A, is a three-part structure, consisting of the corpus allatum, the corpus cardiacum, and the lateral gland cells*. The brain secretes a hormone that triggers secretion of a hormone (ecdysone) from the lateral gland cells. The corpus allatum secretes a third hormone (juvenile hormone).

Hormonal activity occurs in cycles, one in each larval instar; the time of hormone secretion and activity is known as the critical period. Once this period has begun, molting is bound to occur. In the third instar, or last larval stage, however, when the larva is about full size, it stops eating and tries to find a dry habitat. In this stage, the pattern of hormonal secretion changes; the corpus allatum no longer secretes its hormone, although the brain and the lateral-gland hormones are secreted on schedule. Instead of molting, the animal pupates, and metamorphosis begins.

If you have followed the discussion to this point, you realize that we have only begun to answer the question we set out to explore; further, we have raised or implied a great many more questions that we did not at first anticipate. For example: Is the brain hormone's *only* function to control the secretions of the ring gland? Why does the pattern of secretion change in the third larval instar? Does molting depend upon secretion of the corpus allatum hormone? Is it the absence of this hormone in the third instar that initiates pupation? If so, was the corpus allatum hormone inhibiting the lateral-gland hormone, or was there some other type of interaction between

---

* In some insects, these cells are called the *paratracheal* glands, and in others the *prothoracic* glands.

them? Is the lateral-gland hormone necessary for pupation, or is the removal of the corpus allatum hormone sufficient in itself?

It is now obvious that the first question we asked was really more complicated than we anticipated. Clearly, we cannot answer all of the new questions, but they serve to illustrate the kinds of queries that arise from the investigation of an apparently simple question. We might have found the information about secretions in many textbooks, but a further investigation is necessary in order to answer some of the other questions.

One method for determining which hormone is functional in *initiating* metamorphosis, and which in *inhibiting* metamorphosis is to isolate various parts of the body in such a way that the hormones will not be secreted into those parts, but will remain in others. Such isolations can be accomplished by ligaturing the body with fine thread at the proper stages in glandular secretion.

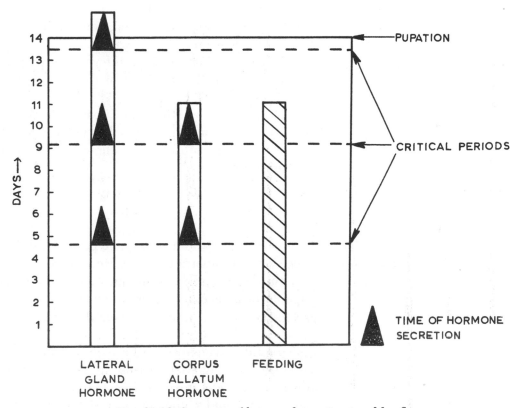

FIG. 21-12. Sequence of hormonal secretion in a blowfly.

A. Procedure

The blowfly is an example of an insect that undergoes complete metamorphosis. This animal is very useful as experimental material since it is easy to handle and maintain in the laboratory. You will be provided with ten blowfly larvae that are in the early third-instar stage. The larvae stop feeding, and stop producing the corpus allatum hormone three days before pupation, and go into the last critical period twelve hours before pupation (FIGURE 21-12). To determine whether the larvae are at the right stage, examine them under the dissecting microscope; the crop should be nearly empty (FIGURE 12-11B shows a full and empty crop). Using the fine thread provided, ligature various parts of the body you deem necessary. Place the larvae in a storage container. Label this container, and set it aside until the following week.

Consider the following in making decisions about your isolations:

1. The anterior end can be distinguished by the black mouth hooks.

2. A ligature tied below the fourth segment will isolate the brain and ring gland in the anterior end.

3. A ligature two segments from the anterior end will isolate both the brain and the ring gland in the posterior part of the body.

4. A ligature tied after the critical period will determine whether or not the ligature had any physical effect on pupation.

Be sure to run proper controls where you feel they are necessary.

B. Results

If pupation occurs, the larval skin changes into a hard, brown, barrel-shaped case (shrunken, black coloration is a degenerate condition). With fine forceps and some care in dissection, you may be able to remove the pupal case to find the developing adult within.

From your data, and the information previously given, what evidence can you give for the functions of the lateral-gland and the corpus allatum hormones?

How could you stimulate pupation in the part of the animal isolated from the endocrine organs if it has not yet pupated?

How could you determine the effect of corpus allatum hormone alone on metamorphosis?

PART VII.
REGENERATION

Regeneration plays an important part in the life of an organism. By this process many important tissues are replaced. The regeneration of lost blood and the healing of broken skin or bone are examples. Some species are capable of regenerating entire organs or regenerating an entire organism from a part.

Regeneration is very similar to embryonic development. Unspecialized cells rapidly divide, each eventually becoming some particular part of the organism. Regeneration entails the same problem as does embryonic morphogenesis, where unspecialized cells differentiate. During both embryonic development and regeneration, a strong polarity exists where the anterior region develops at a greater rate than the posterior. In the following study you will have an opportunity to compare rates of development in different regenerating regions.

You will work with fresh-water planarian *Dugesia*. We have selected these animals for the study because they have an extremely high potential for regeneration. Planarians have been known to regenerate a whole new organism from a middle section. As a rule, the more anterior the origin of the section the more it is likely to regenerate entirely. Why?

You will be provided with a planarian, vial(s), and a separate container of pond water. During the next two weeks, you will watch this planarian regenerate at cuts you design.

A. Cutting the worm

1. Obtain a planarian from the stock container at the supply desk.

2. Carefully place the planarian on a moistened cork or on a piece of glass. When the worm extends itself on the flat surface make the necessary cuts with a razor blade. It is best to perform all operations while observing the worm with a dissecting microscope. Suggested cuts are outlined in FIGURE 21-13.

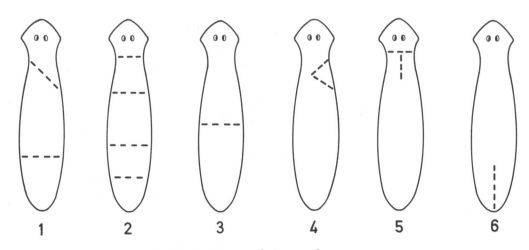

FIG. 21-13. Suggested cuts on planarian.

3. Transfer the cut sections into the vial(s) provided. Fill the vial(s) with pond water, place the cork or cap on tightly, and take the vial(s) along with you for close daily observation.

B. Care of the worm

1. During the next two weeks of close observation, do not feed the planarian. Planarians that have been operated on are hardly in a position to begin feeding right away; giving them food will only foul the water, leading to death and disintegration of the worms or fragments.

2. Change the water daily. Use only clean spring or pond water provided by your instructor. Keep in your room, in a cool place, protected from strong light.

3. Any cuts intended to separate two or more portions without detaching them completely from the rest of the worm will have to be renewed every few hours for at least half a day, to prevent their re-fusing.

4. Keep a record of your observations. Unless the cuts made in the operations are recorded in sketches, you are likely to lose track of the origin of some pieces by the time they have undergone at least some regeneration.

5. Compare rates of regeneration between different sections, studying other students' data where necessary. Is there a gradient between the anterior and posterior regions?

Look up the phylum and class of each of the animals you have studied in this period, and if they are not already included, enter each of them in the appropriate space in Appendix 2.

QUESTIONS FOR
FURTHER
THOUGHT

1. Compare early cleavage in a sea urchin, a frog, and a chick.

2. Since the parents of sea urchins, sea stars, and many other marine animals never come into contact, gamete release in adjacent animals must be reasonably simultaneous. What factors do you suppose trigger the simultaneous release of gametes under natural conditions?

3. Relative to their adult size, birds' eggs are very large when compared to those of echinoderms. What problems of survival necessitate the large size of a bird's egg, and how are these problems solved by echinoderms?

4. Why is rate of development a function of temperature?

5. From what two structures is the mammalian placenta derived?

6. What is a germ layer? What structures are formed from each germ layer?

7. In some fly larvae, growth occurs by cell enlargement rather than by cell division. What adaptive value would there be to this arrangement?

8. What animals other than arthropods undergo metamorphosis? What features do their life histories have in common?

Biological Science, Chapter 17.
Elements of Biological Science, Chapter 17.

Throughout this course we have attempted to interpret all topics in the light of modern evolutionary theory. This has given us a framework within which we can better understand factual detail and perceive its significance. We have emphasized that the theory of evolution, one of the unifying principles of biology, involves two basic ideas: (1) that the characteristics of populations of organisms *change* over periods of time, and (2) that the direction and rate of evolutionary change is determined by natural selection. You have studied the genetic basis for evolutionary change and should now understand how natural selection brings about shifts in the composition of the gene pool of the population. This laboratory should clarify further some of the concepts in evolutionary theory.

In this laboratory you will examine a series of demonstrations illustrating evolutionary principles. The exhibits are arranged in the order of their treatment in this topic; you should study each carefully and answer the questions asked. Sometimes there is no absolute answer for a question, and you are then expected to be able to defend whatever answer you select. The purpose of this exercise is not to have you memorize specific facts concerning the particular animals and plants exhibited; rather it is to help clarify in your mind some basic concepts of evolutionary theory by giving you personal contact with actual examples of evolution. The organisms in these exhibits were chosen simply because they provided good illustrations of important ideas and because they were easily obtained; many other organisms might just as well have been used. You should, therefore, concentrate more on the general principles illustrated than on the names and facts involved.

You may work together in groups of two or three if you wish, but you are asked to converse as quietly as possible and to move in an orderly fashion from exhibit to exhibit in sequence; in this way confusion can be held to a minimum.

Some of the exhibits listed below may not be available to you, and others not listed here may have been added. Follow your instructor's directions concerning these.

**PART I.
PALEONTO-
LOGICAL
EVIDENCE FOR
EVOLUTION**

Fossils provide scientists with some of the best evidence regarding the past history of life on earth, and they sometimes make it possible to trace in detail the evolutionary changes that have taken place in various lines of organisms. Some of the major trends in organic evolution should have become apparent to you as you used Appendix 2, and will be further clarified when we take up the different groups of organisms later in the course. These major trends did not become apparent to biologists all at once, but have slowly emerged from the painstaking integration of hundreds of different bits of evidence accumulated through the years by a multitude of individual research efforts, each dealing with the evolution of a specific group of organisms.

The evolutionary history of the horse family is particularly well known because its fossil record is more complete than most.

1. Examine the large mount illustrating horse evolution. This display shows the characteristics of the skull, forelegs, hind legs, and teeth of four horses; each is shown in its proper geological period (i.e., Eocene, Oligocene, Miocene, etc.). Between the lower Eocene *Eohippus* and the modern *Equus* you can recognize many fossil stages. The chief changes you should note are:

   a. increase in overall body size

   b. lengthening of the distal portion of the legs

   c. reduction of lateral digits

   d. increase in length of the anterior part of the skull

   e. increase in depth of the molars, and increase of their grinding lophs (ridges)

   f. increase in resemblance of the premolars to the molars

   g. completion of the post-orbital bar of skull (i.e., bar behind eye socket)

Doubtless, change has also taken place in many other characteristics; we know, for instance, that there has been a progressive increase in the size of the neopallium. This was certainly not determined directly, since soft tissues do not become fossilized readily; but cranial casts can be used to calculate such a character.

2. Examine the series of fossils of other organisms on display and read the explanatory cards beside them.

PART II. DIVERGENT EVOLUTION

In studying the evolution of horses, you focused attention on a few representatives, each from a different geological epoch. Stacked one next to the other, they may give the misleading impression that the evolution of the horse family progressed in a single direct line, starting with *Eohippus* and terminating with *Equus*. This oversimplification, if not properly qualified, amounts to a falsification. In reality, things happened in a far more complex, yet more meaningful way. At any one time there lived not just one type of horse, but a diversity of them. Some types were destined to become extinct because they failed to evolve satisfactory adaptations, whereas others, which did evolve them, were able to survive by successfully meeting new environmental requirements. The particular representatives we chose for study belonged to specific strains that are known to have survived through change. In other words, they did not belong to evolutionary "blind alleys." That is why when we consider them alone, disregarding all others, we are left with the impression of a straight, direct ascendancy. It is important to note that only a fraction of the total number of species of any one group of organisms living at a certain time is destined for survival. We know this has been true in the past, because we have ample substantiating evidence from fossils; and there is certainly every reason to believe that it will also be true for species living today.

At any one time, then, most groups show considerable diversity. A series of demonstrations have been set up to help you understand such adaptive variation, which results from divergent evolution, i.e., increasing dissimilarity between two or more populations that were initially one. (What is the difference between divergent evolution and parallel evolution?)

We shall consider diversity at many levels of the taxonomic hierarchy; you should, therefore, be familiar with the chief categories of this hierarchy. These are:

Kingdom
>   Phylum or Division
>   >   Class
>   >   >   Order
>   >   >   >   Family
>   >   >   >   >   Genus
>   >   >   >   >   >   Species

Such a classification scheme is a method whereby biologists pigeonhole and extract meaningful pattern from the innumerable diverse organisms with which they must work. In essence, it is simply an outline, much like one you might use to help you study a history or a chemistry course. Such a classification could be based on almost any characteristic of the organisms being studied, but scientists have chosen to use common ancestry as their classification principle. Can you name four or five different major sources of information a biologist might use in determining ancestry?

A. Divergence between phyla of a single kingdom

3. Examine the demonstration showing members of five different animal phyla. Presumably all animals share a common ancestry at some remote time in geological history, but they have diverged greatly and can now be divided into a number of large categories (phyla) based on degree of relationship. The rather gross characters used for separation at this high level of the taxonomic hierarchy should be apparent to you from even a superficial examination of the five specimens shown here.

> a. Would you expect that an earthworm belongs to the same phylum as specimen A, B, C, D, or E? On what characters do you base your decision?

> b. Would you expect that a crayfish belongs to the same phylum as specimen A, B, C, D, or E? Again, give reasons for your decision.

B. Divergence within a phylum

4. Examine the demonstration of the classes within the phylum Mollusca. Most members of this phylum have shells, though the forms of the shells vary greatly. Most cephalopod molluscs, however, have no external shell, though their ancestors, known now only from fossils, were shelled.

a. Contrast the shells of class A (Amphineura) with those of the other classes.

b. Contrast the shells of class B (Gastropoda) with those of the other classes.

c. Contrast the shells of class C (Pelecypoda) with those of the other classes.

d. Contrast the shells of class D (Scaphopoda) with those of the other classes.

e. What adaptive advantages might the members of class E (Cephalopoda) have gained by losing their shells?

Within some of these classes, more diversity can be seen on demonstration. This is divergent evolution within classes instead of between classes.

C. Divergence within a class

Before you proceed any farther, perhaps it would be well to clarify the meanings of a few terms commonly used in evolutionary discussions. Two structures (or two characteristics of any type, for that matter) are said to be **homologous** if they evolved from a *common* ancestral structure. If two characteristics are functionally very similar but evolved from *different* ancestral characteristics, they are said to be **analogous.** It should be clear that it is always necessary to specify in what sense either of these two terms is used in any given instance. The flippers of whales and seals, for example, evolved from the front legs of land-mammal ancestors. Thus they are homologous in the sense that both are forelimbs, with the same basic bone structure as other vertebrate forelimbs, but the modifications that make them flippers are not homologous, and they must be said to be analogous with regard to their flipper characteristics.

Two other terms, **specialized** and **generalized,** also need clarification. These terms are used in indicating the relationships of organisms, or particular characteristics of organisms, to their environment. Specialized means adapted to a special, usually rather narrow, way of life; generalized means broadly adapted to a greater variety of different types of environments and ways of life. For example, the rear legs of a frog are specialized for a particular type of locomotion on land; the rear legs of a dog, while specialized in many ways themselves, can be regarded as more generalized than those of a frog.

The terms **primitive** and **advanced** are used in comparing organisms, or characteristics of organisms, to the ancestral type. Primitive means first, original, like the earliest member; advanced means the opposite. A word of caution is in order here — the terms "primitive" and "advanced" do *not* imply any value judgment that one is "better" than the other, nor do they imply that one is more complex than the other; they refer *only* to relative sequence in time.

5. When studying the molluscs in Exhibit 4, you saw that main evolutionary lines within a phylum can diverge so much that they come to be considered as separate classes; and you saw also that there is divergence within each class. Now examine representatives of the vertebrate class

Reptilia. All reptiles are, of course, related at the class level, yet they show striking differences that provide a good example of strongly divergent evolution. The example would be even more impressive if we could include the many diverse dinosaur types that once roamed the earth, all of which were reptiles.

a. The fossil record indicates that the ancestors of the reptiles were four-legged amphibians without shells (their appearance resembled that of modern salamanders). Would you say that the legs of turtles and lizards are homologous structures? Explain your answer.

b. Which term, homologous or analogous, best describes the relationship of the shell of a turtle and that of a snail?

c. Which of the reptiles on demonstration would you consider the most "primitive" in overall body plan? Why?

d. Which would you consider the most "advanced" in overall body plan? Defend your answer.

e. Remembering that reptiles evolved from the amphibians, which would you say is more primitive, for a reptile to have legs or to lack them?

f. Which is more specialized, to have legs or to lack them?

6. Another striking example of divergence within a class is seen in fish. On display are representatives of some rather unusual types of fish, which should give you some idea of the great diversity within this class.

Examine each of the fish and briefly mention ways in which it differs from the perch (on demonstration), an example of the more usual fish form.

7. Examine the demonstration of the forelimbs of various members of the class Mammalia, noting both similarities and differences. (Three forelimbs, bird, frog, and turtle, included in the display are not mammalian.) Identify each bone in the human arm, remembering your study of the skeleton in Topic 13, and then try to locate the same bones in the limbs on display. You will see that bones differ greatly in length, thickness, degree of fusion with each other, etc. in the different animals. Having compared the bones of the different species (you may find it will help you to make a table showing your findings), try to explain them on the basis of the way of life of the animals involved. The object here is not to memorize the bones of the specimens examined, but to get an understanding of the way the same basic structures can be molded by evolution into a great variety of diverse types.

D. Divergence within an order

8. Examine the display skulls of various members of the mammalian order Carnivora. All members of this order possess certain skull characteristics in common, particularly the development of the canine teeth. This display includes representatives of four carnivore families: dog, cat, bear, and mink. Some of the differences between these should be apparent to you.

a. Compare the four families with regard to general shape of skull, tooth characteristics, etc.

b. What animals (not shown here) would you think are in the "dog" family?

c. What animals (not shown here) would you think are in the "cat" family?

9. The rodents comprise another order of mammals. Examine several rodent skulls.

a. Compare these with the skulls of carnivores, paying particular attention to the teeth.

b. What do these differences indicate about rodents' diets?

10. A third mammalian order, Artiodactyla, includes all hoofed animals with an even number of toes: cow, deer, sheep, goat, giraffe, pig. Examine the skulls on display and note similarities and differences.

a. Contrast the teeth of members of this order with those of carnivores.

b. What can you tell about the diet of members of this order by examining their teeth?

11. A fourth mammalian order, Primates, includes lemurs, monkeys, apes, and men. Examine skull casts of a gibbon, orangutan, gorilla, and chimpanzee, and compare these to the human skull.

a. Characterize the most striking differences between skulls of this order and those of carnivores and rodents (note the teeth particularly).

b. Which of the other primate skulls on display do you think most closely resembles the human skull? Why?

c. What major changes would be necessary to convert one of these skulls into one like that of man?

12. The insect order of beetles, Coleoptera, provides another good example of adaptive radiation within a single order. Examine the case containing many different beetles. The block in the lower right hand corner contains other insects, not members of the Coleoptera, which you may use for comparison in answering some of the questions below.

a. What characters do all the beetles have in common that distinguish them from the other insects shown?

b. Each block of beetles contains members of a single family. How do the first and second families differ from each other?

c. How do the third and fourth families differ from each other?

d. Examine the six specimens on the pinning block outside the case. All of these are members of the class Insecta, but not all are beetles. Judging from your generalizations in (a), which are Coleoptera?

E. Divergence within a family

13. You should already have some idea of families as a result of your examination of carnivores. Members of the cactus family provide a good example of divergence within a family. All clearly resemble each other more

than they resemble dandelions, oak trees, or grasses; but in spite of their basic similarities, the differences between the species on display are obvious.

14. Several members of a family of butterflies are on display. Both similarities and differences between them should be obvious to you.

15. An example of a family with diverse members is the one that includes pineapple, Spanish moss (not really a moss), and many epiphytic plants (aerial plants growing on other plants but not parasitic on them; very common in the tropics). Examine members of this family.

F. Divergence within a species

16. Examine the display showing 24 different shells of the bivalve mollusc *Donax variabilis*, the cochina clam. Note the striking variations in both color and pattern. Not only are all these shells of one species, but all were collected at one time and at a single locality. This, then, is an example of intraspecific variation that is not correlated with geography.

G. Sexual dimorphism

17. One type of intraspecific variation that should be distinguished from the types already seen is sexual dimorphism, i.e., the male and female of a single species look different. Examine the two sexes of the birds or insects on display. What might be the adaptive significance of the differences you see?

PART III. CONVERGENT EVOLUTION

A constant problem facing biologists is determining when similarity of appearances indicates actual phylogenetic relationships, and when it is simply a reflection of similar selection pressures acting on unrelated organisms. When the latter is true, we term it convergent evolution. A series of demonstrations has been set up to help you understand convergence.

18. Examine wings of bat and wings of bird. The two vertebrates evolved wings independently, though in each case the forelimb was the chief structure modified. Compare these wings with those of an insect, an animal in a different phylum from the other two. The insects' wings did not evolve from legs, and are very unlike the wings of birds and bats in basic structure.

 a. Should you use the term "homologous" or "analogous" when comparing the wings of birds and insects?

 b. In what sense are the wings of birds and bats homologous?

 c. In what sense are they not homologous?

19. Jointed legs have arisen convergently twice in the animal kingdom, in the phyla Arthropoda and Chordata. Note that such legs would be functionally efficient only in animals with hard skeletal systems.

 a. Are the legs of grasshoppers and squirrels homologous or analogous?

 b. Use the appropriate one of these terms with reference to the legs of lobster and man.

c. Use the appropriate one of these terms with reference to the legs of raccoon and man.

20. Many unrelated animals have convergently evolved "wormlike" bodies. Examine a series of such animals including earthworm, various insect larvae, millipeds, centipeds, snake, salamander, etc.

For what sorts of habitats is this body shape particularly well adapted?

21. Mimicry is an especially interesting type of convergent evolution. Here selection causes unrelated species to evolve similar appearances involving protection from predators. The species that are naturally protected by virtue of some "unpleasant" characteristics such as taste, smell, sting, etc. are called the model species; they often are brightly colored or conspicuous in some other way, thereby making it easier for predators who have experienced their unpleasant features to recognize them and avoid them in the future. Species that are not naturally protected by some unpleasant character of their own often mimic the appearance of model species and consequently avoid predation because the predators can't distinguish them from the unpleasant models. This phenomenon is fairly common among insects, as the display mounts indicate. The displays show the model species on the left with the mimic species to their right. Often only one sex of a given species is a mimic, the other sex appearing entirely different (providing more examples of sexual dimorphism).

There are two chief types of mimicry, Batesian and Mullerian. In the first type, only the model species is distasteful (i.e., tastes bad, or smells bad, or stings, etc.), and the mimic species, which has no such defensive mechanism, gains protection from predators by being mistaken for the model.

In Mullerian mimicry, all of the species involved are, essentially, both models and mimics. That is, all have some defensive mechanism. But if each had its own characteristic appearance, the predators would have to learn to avoid each species separately, thereby making the learning process a demanding one and also necessitating the sacrifice of some individuals of each species to the learning process. If, however, species convergently evolve more and more toward one appearance type, they come to constitute a single prey group from the standpoint of the predators, and avoidance is more easily learned.

22. A well-known example of convergent evolution between two different plant families is that involving cacti and euphorbias. The two types of plants look so much alike that the average person normally calls both "cacti," yet they are really not closely related. Can you think of reasons why these two plants, which live in the same sort of habitat, might have evolved the particular characters that both possess in common?

TOPIC **23** NONVASCULAR PLANTS

Biological Science, pp. 718-736, 742-749.
Elements of Biological Science, pp. 457-469,
473-476.

Earlier in this course, as we reviewed the general problems facing all living organisms, our attention was often directed to the remarkable adaptations with which the vascular plants have met the challenges of terrestrial existence. While we marvel at these adaptations and while we recognize the importance of vascular plants in terrestrial communities as shelter and food for numerous other organisms, we should not allow ourselves to lose sight of the fact that vascular plants represent only a small segment of the plant kingdom. The much more numerous **nonvascular plants** are the topic of today's laboratory.

Though often structurally simpler than their vascular relatives, the nonvascular plants, which often are confined to aquatic or very moist habitats, have equally complex and intricate adaptations. In Topic 18, you saw some of these adaptations when you studied the fungi, a group of nonvascular plants of major importance in the degradation and cycling of raw materials between living and nonliving systems. The photosynthetic nonvascular plants — principally the algae — rank far above terrestrial plants as the earth's major synthesizers of carbohydrate; they therefore occupy an important position as the primary producers in aquatic food chains. Finally, the nonvascular plants illustrate important structural and reproductive trends that give us some idea of the "raw materials" from which the adaptations of living terrestrial plants eventually were fashioned by natural selection. In the next unit, we shall examine some of these adaptations that made successful life on land a reality for the vascular plants.

PART I. DIVISION CYANOPHYTA (Blue-green algae)

Although some classification schemes do not formally consider them plants (in this book, for example, we shall classify them as Monera, a separate kingdom), the blue-green algae are of interest as examples of rather primitive structure and reproduction, and are probably illustrative of a type or organization through which more complex cells passed in the course of their evolution.

A. Oscillatoria

This alga is commonly found on stones or on the surface of damp earth.

1. Obtain some *Oscillatoria* and make a wet mount for microscopic examination. (If the culture is on agar, be sure *not* to take any agar when you take the specimens.) New cells are added to the filament through binary fission. Vegetative reproduction occurs when filaments break apart, often following the formation of a gelatinous region between a pair of cells. Such

regions may often be found under high power. Do cells of *Oscillatoria* divide in more than one plane?

Sexual reproduction is unrecorded in the blue-green algae, but it probably will be discovered when they are thoroughly investigated.

2. Look closely at a single cell, and notice the absence of nucleus and chloroplasts. Structurally much like bacteria, these organisms are called procaryotic because neither the genetic materials nor the photosynthetic pigments are localized in discrete bodies in the cytoplasm; much of the supplementary membrane structure (endoplasmic reticulum, mitochondria, etc.) typical of the more familiar eucaryotic cells is absent from these organisms.

The cells in an *Oscillatoria* filament are all alike, and the filament might indeed be regarded as a colony of individual plants. What criteria would you use to distinguish colonies from true multicellular organisms? Many blue-green algae are filamentous or are bound together in prominent gelatinous sheaths. What advantages might there be in this type of organization (in contrast to individual free-living cells)?

3. Find an area where several of the filaments cross, and observe the peculiar gliding locomotion. The mechanism of this movement is unknown. Many other algae show movements of the individual cells, colonies, or cell parts (such as chloroplasts, which, in some algae, orient to the changing angle of the incident sunlight). How might such movements be adaptive?

B. Gloeocapsa

This alga is normally found in thick mats on damp rocks. Examine living specimens. What is the appearance of the cells? How does the cellular structure compare to that of *Oscillatoria*? Are there isolated single cells, or only colonies? How many cells occur in the colonies? What holds them together? Is the binding material apparent around single cells or only around the colony? Sketch several specimens of *Gloeocapsa*.

PART II. DIVISION CHLOROPHYTA (Green algae)

The green algae are of special interest because they exhibit such a wide variety of forms and because comparisons of their biochemistry with that of the vascular plants strongly implies that these algae were ancestral to terrestrial plants.

A. Chlamydomonas

One of the simplest green algae — probably much like that from which more complex, multicellular forms arose — is the unicellular *Chlamydomonas* (FIGURE 23-1). Its internal structure is usually visible only under very high magnification (1000X and up) but swimming individuals can be seen clearly under low and high powers. Make a wet mount for observation of their locomotion.

LIFE CYCLE

1. *Chlamydomonas* is haploid. It reproduces asexually by the mitotic production of **zoospores**, and sexually by the mitotic production of **gametes.** Vegetative cells, zoospores, and gametes are practically indistinguishable; note that *both* types of reproductive cells are produced by mitosis.

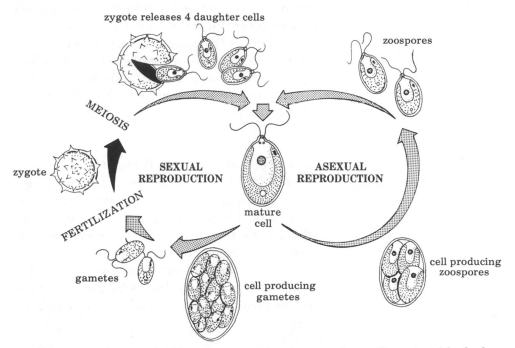

zygote releases 4 daughter cells

zoospores

MEIOSIS

zygote

SEXUAL REPRODUCTION

ASEXUAL REPRODUCTION

FERTILIZATION

mature cell

gametes

cell producing gametes

cell producing zoospores

FIG. 23-1. Life history of *Chlamydomonas*. This diagram shows all stages of both the sexual and the asexual cycle. Note the large chloroplast in the central vegetative cell.

Gametes fuse to form a **zygote,** the only diploid stage in the life cycle. The zygote immediately undergoes meiosis; the cells that result are freed and eventually develop into new vegetative plants. FIGURE 23-2 shows a schematic diagram of the life cycle, for comparison with life cycles of other plants.

SEXUAL REPRODUCTION: ISOGAMY

2. Sexual reproduction probably originated through the union of haploid cells of the same size and structure, a condition called **isogamy.**

Isogamy can be demonstrated in the laboratory by using specially prepared cultures of opposite mating strains of *Chlamydomonas. Being careful not to contaminate either dropper with the opposite culture,* obtain single

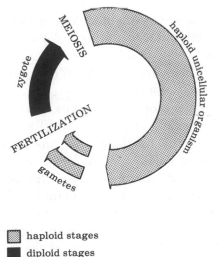

MEIOSIS

zygote

haploid unicellular organism

FERTILIZATION

gametes

haploid stages

diploid stages

FIG. 23-2. Schematic diagram of *Chlamydomonas* life cycle for comparison with that of other organisms. Note that the zygote is the only diploid stage. This type of life cycle was probably characteristic of the first sexually reproducing unicellular organisms, and it may be the type from which all other types arose.

drops of each culture (marked + and −) and allow them to mingle on a slide. Add a cover slip and examine under low and high power.

Many of the gametes clump together; shortly thereafter, they leave the clumps in pairs, fused together at the flagellar ends. Very close observation will reveal numerous stages in fusion after the pairing-off.

Your instructor may provide a film of sexual reproduction in *Chlamydomonas.*

B. Heterogamy and the division of labor

The green algae we shall examine in this section display two important trends. One is a tendency toward **heterogamy,** sexual reproduction that involves interaction between gametes that are clearly different. The difference can be of size or motility; the most extreme form of heterogamy, in which the female gametes are large, nonmotile eggs and the male gametes are small, motile sperms, is called **oögamy.**

The other tendency involves the organization of cells into colonies or multicellular plant bodies, with associations between the cells and consequent division of labor; cells become specialized for particular tasks.

THE VOLVOCINE SERIES

1. There are several green algal genera which, taken together, form a series of colonies of increasing size, internal differentiation, and tendency toward heterogamy. Individual cells of these colonies are much like *Chlamydomonas.* We shall examine *Pandorina* and *Volvox,* two members of what is called the volvocine series (because of its terminal member, *Volvox*). Your instructor may supply you with other members of the series.

a. Make a wet mount or examine a prepared slide of *Pandorina.* Colonies vary in size from 4 to 32 cells; in the larger colonies there should be a clear difference in cell size at the "anterior" and "posterior" ends. If you have living colonies, watch the movement. Does one end move consistently forward? Which end has the larger cells?

Asexual reproduction in *Pandorina* is much like that in *Chlamydomonas;* sexual reproduction, however, is heterogamous: the gametes are of different size, though both are flagellated.

b. Make a wet mount of *Volvox* or examine a prepared slide of sexual reproductive stages; do both, if possible. *Volvox* colonies (or should they be called multicellular organisms?) vary in size from several hundred to as many as 40,000 cells arranged in the wall of a hollow sphere and interconnected by cytoplasmic strands.

Asexual reproduction occurs through the formation of **daughter colonies,** which develop mitotically from single vegetative cells and are eventually freed into the interior of the parent sphere. If you have living colonies, look for some with daughter colonies in the interior of the parent sphere. What do the parental cells look like? When the parent colony dies and breaks apart, the daughter colonies are freed. You may be able to find parent colonies in various stages of degeneration, with daughter colonies inside. What do you suppose initiates the breakdown of the parent colony?

Sexual reproduction is oögamous; eggs and sperms are formed

(FIGURE 23-3). *Volvox* shows a marked division of labor: only a few of the cells are active in gamete production, while the rest are permanently vegetative.

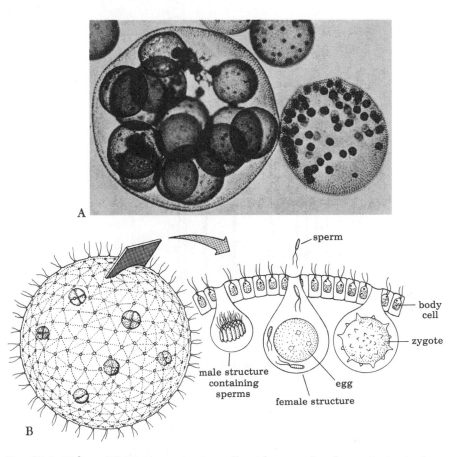

FIG. 23-3. *Volvox.* (A) Living vegetative cells with many daughter colonies in the interior of the parent colony. (B) Section through the surface of a colony, showing male and female reproductive structures. Sperms released by the male structures enter the female structure and fertilize the egg. After a period of inactivity, the zygote divides meiotically and the haploid cells thus formed then divide mitotically, producing a new daughter colony, which is eventually released. (A: *Courtesy Carolina Biological Supply Company.* B: *Modified from H. J. Fuller and O. Tippo,* College Botany, *Holt, 1949.*)

A FILAMENTOUS ALGA:
ULOTHRIX

2. Make a wet mount or examine a prepared slide of *Ulothrix*. Though relatively simple in appearance, this organism is clearly multicellular and not colonial; adjacent cells share common end walls, and there is some cellular differentiation. Locate the nucleus and single large chloroplast in vegetative cells; on a whole filament, find the **holdfast** cell, which attaches the organism to some substrate.

Any cell of the filament may reproduce asexually by becoming a **sporangium** (a spore-producing structure) and producing zoospores, each of which has four flagella. The zoospores may swim about for some time, then settle, attach, and grow into a new filamentous **thallus** (plant body). You have seen that zoospores can be produced by *Chlamydomonas* and by *Ulothrix*; they are a very common method of asexual reproduction among the

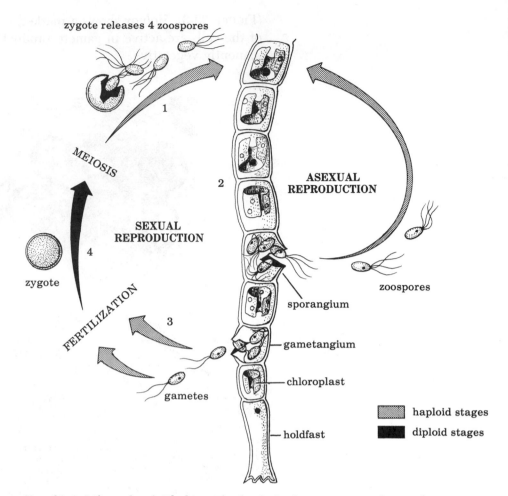

Fig. 23-4. Life cycle of *Ulothrix*. The haploid plant may reproduce either asexually or sexually, though a single filament would never reproduce both ways simultaneously, as shown here. Asexual reproduction is more common; certain cells of the filament develop into sporangia (spore-producing structures) and produce zoospores, which settle down and develop into new filaments. Under certain environmental conditions, the filaments may cease reproducing asexually and begin reproducing sexually; a cell becomes specialized as a gametangium (gamete-producing structure) and produces isogametes. Two such gametes may fuse in fertilization, producing a zygote, which divides meiotically and releases zoospores. What difference would there be between asexually and sexually-produced zoospores?

algae. What value would a *motile* asexual phase have for an attached algal species like *Ulothrix*?

Sexual reproduction is isogamous. The life cycle is shown in FIGURE 23-4; it is typical of the green algae. As you can see, it differs from that of *Chlamydomonas* only in having a multicellular haploid phase. The life cycle is schematically diagrammed in FIGURE 23-5.

SPIROGYRA

3. *Spirogyra*, an aberrant but fascinating form often studied by beginning students, is a multicellular green alga with a filamentous thallus. Examine a wet mount or prepared slide of this alga; in vegetative cells, locate the **spiral chloroplast** (studded with numerous **pyrenoids,** starch-production sites), and the large **nucleus,** situated in the center of the cell. Cytoplasm runs in strands from around the nucleus to the walls of the cell.

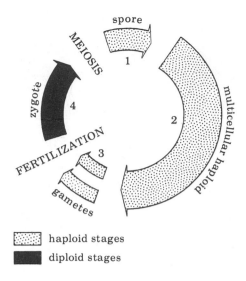

FIG. 23-5. Life cycle of *Ulothrix*. This cycle is characteristic of most multicellular green algae; note that multicellularity is present only in the haploid phase.

Spirogyra exhibits a peculiar form of heterogamy, in which the cells of two filaments of opposite mating strains **conjugate,** or become joined, by outgrowths of adjacent cells (FIGURE 23-6). The entire cell body of each partner cell is then transformed into an amoeboid gamete. Although the gametes look alike, only one of them moves through the tube formed at conjugation; after the migration, the nuclei fuse, a zygote is formed, and later, meiosis takes place. Three of the four nuclei produced by meiosis degenerate, and the fourth forms a new vegetative filament. How would this degeneration of nuclei be adaptive to the plant?

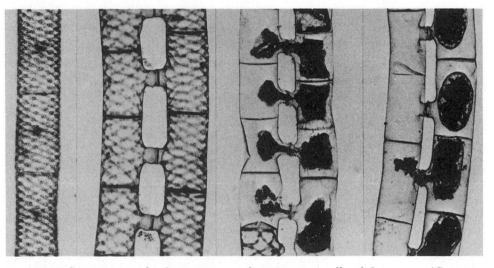

FIG. 23-6. Photomicrograph of vegetative and conjugating cells of *Spirogyra*. (*Courtesy Carolina Biological Supply Company.*)

If your slide shows conjugation, locate the stages in formation of the conjugation tube, formation of gametes, migration of gametes, and formation of zygote. Label the photomicrograph; sketch, adjacent to it, intermediate stages you can locate on your slide.

ULVA

4. You have seen two routes taken by the green algae in the development of multicellularity: formation of motile spheres in the volvocine series, and formation of filaments in *Spirogyra* and *Ulothrix*. A third route resulted

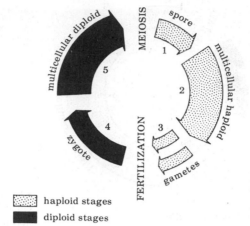

Fig. 23-7. Life cycle of *Ulva*. The gametophyte (multicellular haploid stage) and sporophyte (multicellular diploid stage) are equally prominent.

in the formation of nonmotile associations of cells with platelike thalluses. Since these algae usually live and are attached in areas of relatively rough water, there is often differentiation into blade, stalk, and holdfast.

Examine live *Ulva* (sea lettuce) or *Enteromorpha*, a close relative of *Ulva*. Obtain a small piece of the blade, make a wet mount, and observe the arrangement of cells.

The *Ulva* life cycle (Figure 23-7) differs somewhat from that of the other genera so far considered in that it has a multicellular *diploid* phase. While all the other sexually reproducing organisms you have seen so far exhibit alternation between haploid and diploid phases, only one of the phases ever becomes multicellular. In many of the more advanced algae, however, *both* phases are multicellular. The multicellular *haploid* phase is called a **gametophyte**, because it produces gametes; the multicellular diploid phase is called a **sporophyte**, because it produces spores. Such a life cycle is said to exhibit **alternation of generations**. (*Note:* The terms "gametophyte" and "sporophyte," though widely applied otherwise, are properly used only when there is alternation of *multicellular* generations.)

**PART III.
DIVISION
PHAEOPHYTA
(Brown algae)**

A. Laminaria

Examine a demonstration specimen of *Laminaria* or one of the other kelps; the thallus is a sporophyte. The gametophyte is much smaller; sexual reproduction is oögamous. In the sporophyte, notice the rootlike holdfast, stemlike **stipe**, and expanded leaflike blade. The cellular construction of these parts is much more complex than in an alga like *Ulva*; the stipe of some kelps has an outer surface tissue (epidermis), a middle tissue (cortex) containing many plastids, and a central core tissue. Some kelps have a meristematic layer similar to the cambium of the higher vascular plants, and, in a few species, there is a phloemlike conductive tissue in the central core. In short, these brown algae are complex plants that have convergently evolved many striking similarities to the vascular plants.

B. Fucus

Fucus, a common brown alga in northern coastal waters, is unusual among plants in having a life cycle that omits the multicellular haploid phase; gametes are therefore formed, by meiosis, directly from the diploid phase. In

FIG. 23-8. Thallus of *Fucus*, bearing receptacles. (*Courtesy Carolina Biological Supply Company.*)

this respect, *Fucus* and its close relatives have a life cycle closely comparable to that of animals.

1. Examine *Fucus*, alive if possible. Notice the holdfast, if it is still present, and its firm attachment to rock. The thallus bears **bladders** (what is their function?) and, at the tips of fertile plants, also bears **receptacles** (FIGURE 23-8). Examine the receptacles under a dissecting microscope, and note the numerous tiny openings. Each opening leads into a cavity, called a **conceptacle,** where the sperms or eggs are located.

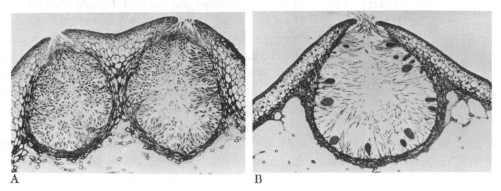

A   B

FIG. 23-9. *Fucus* conceptacles. (A) Male. (B) Female. (*Courtesy Carolina Biological Supply Company.*)

2. Make hand sections of male and female receptacles and locate thin sections of conceptacles. (Your instructor will distinguish the male and the female plants for you if the species you are using is dioecious; if it is monoecious, section receptacles until you find conceptacles comparable to FIGURE 23-9.) Notice that most of the cells in the conceptacle are sterile, colorless filaments. Terminal cells on some of the filaments become fertile; in male conceptacles they are called **antheridia,** and in females, **oögonia.** Many sperms develop from each antheridium and several eggs from each oögonium. Gametes are extruded from the receptacles of fertile plants, and fertilization occurs externally. The zygotes settle, attach, and develop into new diploid plants. The life cycle is diagrammed schematically in FIGURE 23-10.

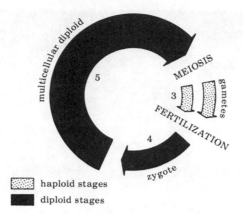

haploid stages

diploid stages

FIG. 23-10. Schematic life cycle of *Fucus*.

3. If the material available to you is not fertile, examine prepared slides of sections through male and female conceptacles; the instructor will prepare a demonstration of fertilization in *Fucus* if he is able to obtain live fertile material.

4. If you have time, make hand sections across various parts of the blade to study its internal structure. The plant body is composed of masses of branching filaments, most of which are colorless; only the outermost filaments are photosynthetic.

PART IV. DIVISION BRYOPHYTA (*Mosses and liverworts*)

It is little wonder that so many organisms live in the water for most or all of their lives, for water is more secure and stable than any other habitat. The terrestrial environment, in contrast, is harsh and insecure, constantly changing, and therefore extremely hazardous and hostile to most living organisms. We are not surprised to find that its successful inhabitants have diverse and often complex adaptations that enable them to cope with the trials of terrestrial existence. We shall find, in the remainder of this unit and in the next laboratory, that many of the attributes of terrestrial organisms cannot be understood properly without some knowledge of the problems and challenges they must face and meet in order to survive.

The difficulties facing a terrestrial organism are numerous. It must be able to obtain an adequate supply of water, both when water is abundant in the environment and when it is not. It must have a means of physical support, unnecessary in buoyant water, and if it is an animal it must be able to move about effectively. It must somehow reproduce in an environment that seldom provides sufficient water for flagellated sperms to swim and offers little security for the developing zygote. Finally, a terrestrial organism must be able to withstand many changes in temperature, humidity, wind, and light, and must be able to adapt to the other environmental fluctuations to which it is frequently subject.

All the modern multicellular photosynthetic plants that spend any part of their life cycle on land — including plants as diverse as mosses and trees — enclose their zygotes and early embryos in a protective jacket of cells, and are therefore grouped together in the artificial category "Embryophyta." Algae, in contrast, are characterized by exposed embryos or by embryonic development entirely apart from the parent plants.

We have very little information about the ancestors of terrestrial plants. Largely on the basis of biochemical similarities, the algal division Chlorophyta is supposed to have been ancestral to land plants, but we have only the scantiest fossil evidence to suggest the manner in which a transition might have taken place. The least complex living Embryophyta, and the only nonvascular members of that group, are the mosses, hornworts, and liverworts, of the Division Bryophyta.

A. Vegetative structure of moss

Examine an individual moss plant under the dissecting microscope. The green, "leafy" plant, familiar among the flora of water's edge, moist rock, and woodland, is the multicellular haploid phase, the gametophyte. Note the stemlike axis of the plant, and its leaflike appendages. The aerial parts of the plant are cuticle-covered, an important adaptation to terrestrial life. Remove a "leaf" and examine in a wet mount. Are there guard cells and stomata?

The gametophyte is anchored in the soil and obtains water by means of **rhizoids,** outgrowths at the base of the axis. Moss gametophytes lack vascular tissues, which sharply limits their size and the sort of habitat they can occupy.

B. Sexual reproduction in moss

1. The distribution of mosses is limited to moist places because they require water for sexual reproduction. Male organs (**antheridia**) and female organs (**archegonia**) occur, in season, at the tips of the gametophyte plants. Examine slides of longitudinal sections through these regions, and locate the sexual organs. Antheridia are club-shaped, and are filled with developing sperms; archegonia are vase-shaped, and in perfect longitudinal sections, you will be able to find an egg at the base of the archegonium. Many sterile filaments occur among the sexual organs in both sections. Label FIGURE 23-11.

Flagellated sperms, released from the antheridia, swim through the dew that frequently covers the gametophytes, or splash in drops of water from the male to the female plant. Once on the female plants, sperms are attracted

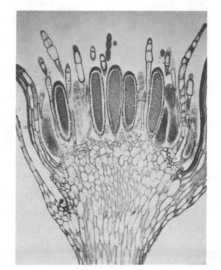

FIG. 23-11. Longitudinal section through antheridium of *Mnium* moss. *(Courtesy Ward's Natural Science Establishment, Inc., Rochester, New York.)*

to the eggs by chemical exudates from the archegonia. Fertilization occurs in the base of the archegonium. The zygote, deriving nutrient from the female plant, gives rise mitotically to a capsule-tipped stalk. Meiosis occurs in the capsule, forming thousands of spores. Stalk and capsule together constitute the diploid sporophyte.

2. Examine the sporophyte on one of the specimens. If the capsule has not already released its spores, carefully remove the top of the capsule and examine the interior with a dissecting microscope. Just below its apex, the capsule bears a ring of tiny teeth which, with periodic changes in humidity, bend inward and outward, dispersing the spores.

How would a high-growing sporophyte benefit the moss plant?

3. Spores, when they land in the proper place, germinate into filamentous young gametophytes, structurally reminiscent of the filamentous green algae. Branches of the filaments growing down into the soil become the rhizoids, and branches growing upward develop into the familiar "leafy" axes. Under the dissecting microscope, examine an agar plate upon which moss spores have been germinated. Find and sketch as many developmental stages as you can. A mature moss gametophyte (with attached sporophyte), and a schematized bryophyte life cycle, are shown in FIGURE 23-12.

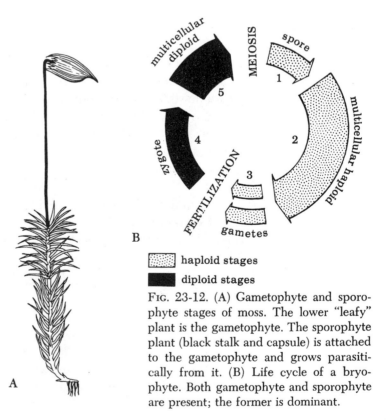

FIG. 23-12. (A) Gametophyte and sporophyte stages of moss. The lower "leafy" plant is the gametophyte. The sporophyte plant (black stalk and capsule) is attached to the gametophyte and grows parasitically from it. (B) Life cycle of a bryophyte. Both gametophyte and sporophyte are present; the former is dominant.

C. Vegetative structure of a liverwort

Liverworts occur in habitats much like those of moss, except that the former require more water. Examine vegetative plants of *Marchantia*, using a dissecting microscope and making hand sections. Examine prepared slides, if available. Is the whole *Marchantia* thallus photosynthetic? What function is served by pores in the top of the thallus? Compare the adaptations to terrestrial life of *Marchantia* with those of the moss.

Fig. 23-13. Thallus of *Marchantia*, with antheridium-bearing stalks. (*Courtesy Carolina Biological Supply Company.*)

D. Sexual reproduction in liverworts

The sexual cycle in liverworts is almost exactly like that of moss, except that antheridia and archegonia are borne on special aerial portions of the gametophyte plants. The sexual organs themselves are much like those of moss.

1. Examine the antheridium-bearing part of a gametophyte, comparing it to Figure 23-13. Examine the archegonium-bearing part of a gametophyte, comparing it to Figure 23-14. As in moss, the zygote formed at fertilization (where?) develops into the sporophyte; the liverwort sporophyte hangs downward from the archegonium-bearing part of the gametophyte. How are the spores dispersed?

2. If available, be sure to see a film of sperm transfer and fertilization in *Marchantia*.

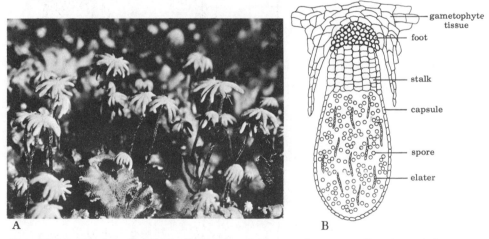

A   B

Fig. 23-14. *Marchantia*. (A) Thallus with archegonium-bearing stalks. One of the mature stalks, at arrow, is bearing young sporophytes on its undersurface. (B) *Marchantia* sporophyte with foot embedded in gametophytic tissue, short stalk, and capsule. The capsule contains spores and elaters; the latter are elongate structures that aid in dispersal of the spores. (*A: Courtesy Carolina Biological Supply Company.*)

QUESTIONS FOR FURTHER THOUGHT

1. What advantages are there to sexual reproduction by isogamy? By oögamy?

2. Define alternation of generations, and explain its significance. Which of the following organisms exhibits alternation of generations: *Fucus, Ulva, Spirogyra, Chlamydomonas, Laminaria*, moss, *Homo sapiens*?

3. What evidence can you suggest to defend the notion that blue-green algae are plants? What evidence suggests that they are not plants?

4. Algae occur in many different colors, some so striking as to be quite unlike the usual green plants. What do you suppose is the adaptive significance of difference in coloration, particularly in the case of algae that live attached to rocks relatively close to shore?

5. The majority of shore-dwelling algae have holdfasts, differentiations of the plant base that cement the organism to rock or some other substrate. What benefit might the plants gain from such structures? What special adaptations might you expect to find in algae that grew only in mid-ocean?

6. Do holdfasts, or any other structures in algae, function in transport within the plant? Are they involved in nutrient procurement? How does an alga solve the problems of nutrient procurement that are met by the roots of a terrestrial plant?

7. What unique features of bryophytes enable them to survive on land? What features make them poorly adapted to land?

8. How would you describe the association between the moss gametophyte and the moss sporophyte?

9. What asexual reproductive features characterize the Bryophyta? Of what advantage is it to terrestrial plants to have both asexual and sexual reproduction?

24 THE MOVE TO LAND— VASCULAR PLANTS

Biological Science, pp. 752-762.
Elements of Biological Science,
pp. 480-488.

In the last unit, you studied the Bryophyta, which despite their relative simplicity and essential dependence on water have nevertheless taken important steps toward successful terrestrial life. The aerial parts of bryophytes are cuticle-covered, and they retain their zygotes, embryolike, in the female reproductive structure, or archegonium. Furthermore, the bryophyte zygote gives rise to an aerial spore-bearing structure; this **sporophyte** effects dispersal.

While bryophytes thus have some features important to terrestrial existence, a full measure of success on land calls for other adaptations. Both vegetative and reproductive functions must be completely freed of their dependence on water, yet water must be efficiently procured and transported to all parts of the plant body; fertilization must somehow take place despite the absence of an aquatic medium for swimming sperms; and developing embryos must be protected from desiccation during their development and dispersal. Effective competition with other plants for dispersal and for sunlight often gives the larger plant a selective advantage, yet large stature demands major supportive and vascular adaptations.

In this unit, we shall consider some of the adaptations with which several important subgroups of the Division Tracheophyta (vascular plants) have met the challenges of terrestrial life.

PART I. FERNS (Subdivision Pteropsida)

Like most vascular plants, ferns exhibit an alternation of multicellular generations in which the diploid multicellular sporophyte is the dominant phase.

A. Structure of the fern sporophyte

1. Examine a mature fern sporophyte, the familiar leafy woodland plant. Locate the **rhizome** (an underground stem) and the **fronds**, each of which is a complexly subdivided leaf. Are the fronds cuticle-covered? Do they have stomata and guard cells? If sufficient material is available, make hand sections of frond axis and of rhizome, and stain with phloroglucinol-HCl. (If these materials are unavailable, examine prepared slides.) Examine microscopically. Phloroglucinol-HCl stains lignin a bright red. (Lignin is a cellulose derivative that occurs in the walls of xylem vessels, sclerenchyma, and other strengthening cells.) Notice the marked difference between the arrangement of vascular tissues in the fern and in flowering plants, examined in Topic 7. Is secondary growth possible in a fern? How does the sporophyte survive the changing seasons?

Using your general knowledge of stem structure and your observations of rhizome sections, label FIGURE 24-1.

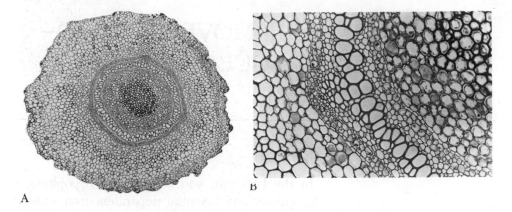

A   B

FIG. 24-1. Cross sections through fern rhizome. (A) Low power, showing whole rhizome. (B) High power, showing a portion of the vascular region. (*Courtesy George H. Conant, Triarch Incorporated, Ripon, Wisconsin.*)

2. On the underside of some or all of the fronds, notice brownish **sori** (singular, sorus); each one is a cluster of many **sporangia.** The structure of the sorus varies from species to species, but it often includes an umbrella or shelflike structure, the **indusium,** that protects the sporangia until they are mature. In FIGURE 24-2, locate and label sporangia and indusium. Remove a small sorus-bearing region from a frond; examine with a dissecting microscope, and then remove some of the sporangia for closer microscopic observation.

Each sporangium is crested by a special ring of cells (the **annulus**) with unevenly thickened walls; when a mature sporangium dries, the cells of the annulus lose water and their walls shrink; this results in the bending of the annulus as shown in FIGURE 24-3. Eventually the tension of the small amount of liquid water in the annulus cells becomes greater than the cohesiveness of water; the water turns instantly from liquid to vapor, restoring the annulus cells to their former size and shape and causing a violent snap of the annulus back to its former position. This movement is generally effective in dispersing the spores to great distances. If you have mature but unopened sporangia, with spores clearly visible inside, you may be able to see some of the dramatic annular movements by allowing the sporangia to dry under the heat from an incandescent lamp.

3. Spore-bearing fronds are called **sporophylls** (sporo-, *spore*; -phyll, *leaf*). In many ferns, all the fronds become sporophylls; in others, only a few

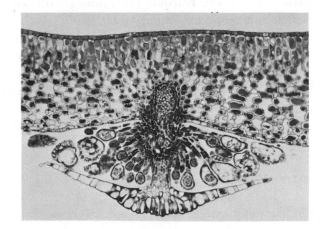

FIG. 24-2. Fern leaf and sorus, median longitudinal section. (*Courtesy Ward's Natural Science Establishment, Inc., Rochester, New York.*)

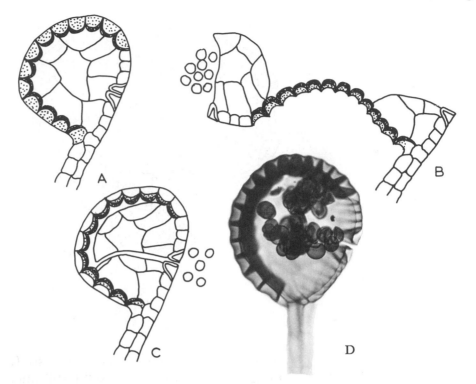

Fig. 24-3. Ejection of spores by a fern sporangium. (A) Intact sporangium. (B) Top of sporangium bent back as a result of loss of water from cells of the annulus. (C) Top of sporangium snaps shut, releasing spores. (D) Photograph of sporangium, showing annulus (*A-C: Modified from F. C. Steward, About Plants: Topics in Plant Biology. Addison-Wesley, 1966. Used by permission. D: Courtesy Carolina Biological Supply Company.*)

fronds become sporophylls. In some species the fronds that become sporophylls are so highly modified at maturity that they do not even resemble the vegetative fronds at all. Into which of these categories does your specimen fit? Examine other types on demonstration. We shall see that in angiosperms and conifers, the most fully successful terrestrial plants, sporebearing is always restricted to groups of a few highly modified leaves — flowers or cones — that bear little resemblance to foliage leaves.

B. The fern gametophyte

1. Fern spores are produced by meiosis in the sporangium, and are dispersed by the wind; spores that land in favorably moist environments germinate and undergo mitosis to form a heart-shaped multicellular gametophyte called a **prothallus.**

With a dissecting microscope, examine specimens of immature and sexually mature fern prothalli. Note their size and general shape; in the older, sexually mature specimens, notice prominent **rhizoids.** If enough living material is available, make wet mounts of the mature prothalli and locate the **antheridia** and **archegonia** (much like the sexual organs of moss and liverwort), borne on the underside of the prothalli (FIGURE 24-4).

2. Ferns are **homosporous,** i.e., they produce only one morphological kind of spore. Though the spores may be genetically variable (why?), each one gives rise to a haploid gametophyte that bears both antheridia and archegonia, and the gametes these organs produce are genotypically iden-

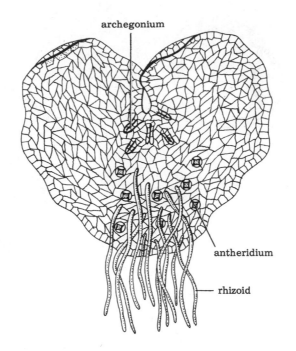

archegonium

antheridium

rhizoid

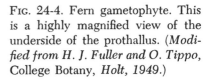

Fig. 24-4. Fern gametophyte. This is a highly magnified view of the underside of the prothallus. (*Modified from H. J. Fuller and O. Tippo,* College Botany, *Holt, 1949.*)

tical, since they are produced by mitosis. Clearly, it would be of advantage to the plants to discourage self-fertilization and encourage union between sperms from one gametophyte and eggs from another. By what possible mechanisms could the plant accomplish this? Considering the fact that self-fertilization defeats part of the selective value of sexual reproduction, can you suggest any *advantages* to homospory and the consequent production of both sexual organs on the same gametophyte?

In our examination of the reproduction of the higher vascular plants, we shall see that any advantages of homospory have been sharply offset by the much greater advantages of **heterospory** — production of different "male" and "female" spores, and hence different gametophytes.

3. The fern zygote (where is it formed?) develops into a sporophyte that quickly becomes independent of the gametophyte; the prothallus soon disintegrates. Examine, on demonstration, an old prothallus with a young sporophyte attached.

From one point of view, ferns are no better adapted to terrestrial life than are the bryophytes; even though the vascularized sporophyte can attain a larger stature, the fragile gametophyte can survive only in moist places. Remember, though, that sexual organs are formed on the *underside* of the *prothallus*, and that the gametophyte, though fragile, is also quite small and can grow close to the ground where it is not so susceptible to drying. Do you think the size and structure of the fern gametophyte would encourage the dispersal of ferns to and their survival in habitats not open to mosses and liverworts? Would a large mosslike gametophyte have the same advantages as the fern gametophyte?

C. The fern life cycle   The fern life cycle is diagrammed schematically in Figure 24-5, for comparison with other plant life cycles. We have noticed in ferns several trends that have been sharply accentuated in higher vascular plants:

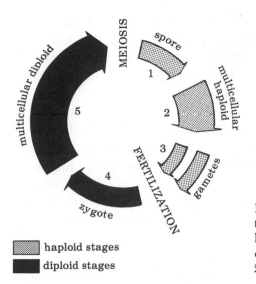

FIG. 24-5. Life cycle of fern. Both gametophyte and sporophyte are present; the latter is dominant. Compare this life cycle with that of a bryophyte (FIGURE 23-12B).

haploid stages
diploid stages

1. The diploid sporophyte is the dominant phase in the life cycle, and the haploid gametophyte is very much reduced in size. Though still multicellular in all terrestrial plants, the gametophyte in flowering and cone-bearing plants is an inconspicuous phase in the life cycle; it is not free-living and photosynthetic, as it is in ferns and mosses and liverworts.

2. Certain leaves of the sporophyte may become specialized for spore-bearing, while other leaves remain vegetative. This trend begins in the ferns and is carried to extremes in the flowering and cone-bearing plants, where a few groups of leaves are highly modified to bear spores. While ferns are homosporous, we have noted that heterospory is the rule in the more fully terrestrial seed plants.

PART II. SEED PLANTS (Subdivision Spermopsida)

In Topics 4 and 7, we considered the structural basis of water and mineral procurement, gas exchange, and transport in seed plants. In doing so, we reviewed many of the adaptations that have enabled these plants to attain their present dominant and successful position among the land flora. We saw, for example, that a complex vascular system carries materials throughout the plant; water and minerals are absorbed by the roots and carried upward to the stem and leaves in the xylem, while the products of photosynthesis are carried to other parts of the plant by the phloem. We saw that guard cells, interrupting the cuticle-covered leaf epidermis, serve to control the diffusional loss of water from stomata, and that natural selection has turned this loss (transpiration) to advantage in moving materials in the xylem. We saw that both stem and root have been strengthened by the presence of cells with secondarily thickened walls (such as sclerenchyma) and that some of the secondarily thickened cells (such as xylem vessels and tracheids) serve the double function of support and transport, permitting a much greater stature than is possible in nonvascular plants. Finally, we saw that secondary growth, by adding to the girth of stem and root, makes possible a larger plant body and provides the strength that often carries such plants through severe climatic stress.

In this section, we shall study the *reproductive* adaptations that have contributed importantly to the dispersal of seed plants and to their survival in a wide variety of different terrestrial habitats. We shall confine our study to two groups of seed plants, the conifers (Class Coniferae) and the flowering plants (Class Angiospermae).

A. Conifers

Hemlock, pine, and spruce are conifers with which you are probably already familiar. In each of these, the large green tree is the *sporophyte*, which is clearly the dominant phase in the life cycle.

To put reproductive adaptations of seed plants in their proper perspective, let us briefly summarize some of the salient differences between the reproduction of ferns and that of the seed plants.

▶ As in ferns, the seed-plant sporophyte is the dominant phase in the life cycle; the gametophyte, while present, is much reduced.

▶ In ferns, we noticed a tendency to restrict spore-bearing to special leaves (sporophylls). In seed plants, this tendency is much more evident; the spore-bearing parts of the plant are cones (in conifers) or flowers (in flowering plants). These structures represent groups of highly modified sporophylls.

▶ Unlike ferns, which exhibit homospory, the seed plants produce different "male" and "female" spores, often on separate parts of a plant or on separate plants. The "male" spores are often called **microspores** and the "female" spores **megaspores.**

▶ While fern spores develop into the haploid multicellular gametophyte, a green free-living structure, seed-plant gametophytes are not green and never become free-living. Instead, gametophyte development occurs *on the parent sporophyte*. The male gametophytes (pollen grains) are generally carried to the vicinity of the female gametophyte by wind or by animals. Fertilization and the development of seeds

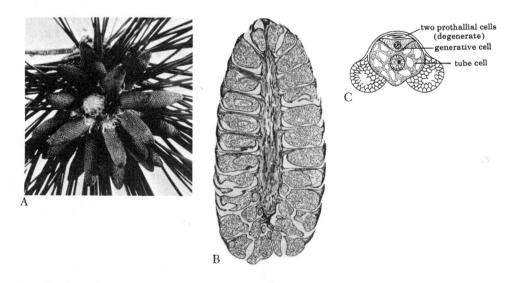

Fig. 24-6. Male pine cones. (A) Mature male cones. (B) Longitudinal section through a male cone. Each sporophyll (cone scale) bears a large sporangium that becomes a pollen sac. (C) Pollen grain (male gametophyte) composed of four cells, two of which are degenerate. (A, B: Courtesy Carolina Biological Supply Company.)

occurs on the parent sporophyte plant; the seeds are then released to "fend for themselves."

1. Examine a pine or other conifer branch with **cones** attached. Locate the very compact green female cones, and the smaller, lighter male cones. (Large green or brown cones are the older female cones; ignore them for the moment.) Both male and female cones are composed of **cone scales** (each one a sporophyll) arranged on a short section of stem. Each scale bears one or more sporangia, within which haploid spores are produced by meiosis.

In the male cone, each haploid spore develops into a four-celled pollen grain, the male gametophyte, Figure 24-6.

In the female cone, each scale bears two sporangia, each enclosed in a protective integument except for a small opening, the **micropyle** (Figure 24-7). Within the sporangium, meiosis occurs, forming four haploid megaspores; three of these disintegrate, and the fourth divides mitotically to form a multicellular mass, the female gametophyte. When mature, the female gametophyte produces 2 to 5 archegonia at its micropylar end; egg cells develop in each archegonium. Each sporangium, together with the gametophyte inside it and the integument surrounding it, is called an **ovule.**

2. When the male, or pollen, cones ripen, the sporangia open and the pollen is released and carried away by the wind. If available, examine microscopically some pine or spruce pollen; notice that each grain has winglike processes that probably assist in its dispersal. Each pollen grain is covered with a protective cuticle, and is thus protected against water loss.

Most of the pollen never reaches the female gametophyte, but some of it sifts down between the female cone scales and comes into contact with a sticky secretion at the micropyle of an ovule. As the secretion dries, the pollen grain is drawn into contact with the sporangium; it promptly begins to germinate, developing a tubular outgrowth, the **pollen tube,** which grows through the tissue of the sporangium and penetrates the female gametophyte.

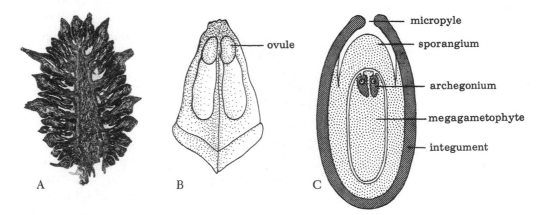

Fig. 24-7. Female pine cones. (A) Longitudinal section through a female cone. Ovules can be seen on the surface of the sporophylls, near their base. (B) Scale from female cone. The two ovules lie on the surface of the scale, near its point of attachment to the cone axis. (C) Section through an ovule, showing female gametophyte. (A: *Courtesy Carolina Biological Supply Company. B, C: Modified from H. J. Fuller and O. Tippo, College Botany, Holt, 1949.*)

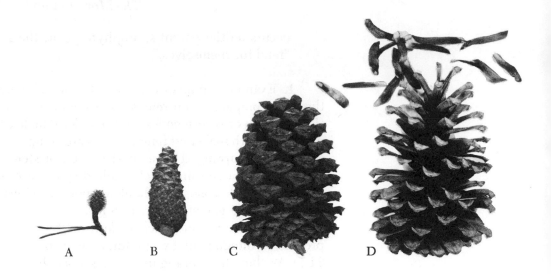

Fig. 24-8. Maturation of female pine cones. (A) Female cone in the first year of growth; (B) In the second year; (C) In the third year. (D) Female cone with seeds. (*Courtesy Carolina Biological Supply Company.*)

One haploid nucleus, derived from one of the pollen-tube nuclei, fertilizes the egg when the pollen tube reaches the archegonium. The resulting zygote develops into an embryo, which, along with food material derived from the female gametophyte, is enclosed in a protective coating derived from the old integument of the ovule. The entire structure is shed from the surface of the cone scale as a **seed.**

3. Development from pollination to shedding of the seed may take several years in some conifers; during this period, the female cone increases greatly in size. On a tree, you should be able to locate several different stages in female-cone development, with reference to FIGURE 24-8. In maturity, the female cone dries, the scales separate from one another, and the seeds fall out. This mature cone is the one often used in Christmas decorations.

If sufficient material is available, or if you are examining cones on a tree in the field, remove cones in several stages of development, and carefully dissect the scales. In the youngest cones, you will be able to find the ovules in pairs at the base of each scale; in developing cones, you may find ripening seeds; and in mature cones, you will probably find a substantial number of scales from which seeds have already fallen. Under the dissecting microscope, dissect a seed; locate the embryo and stored food material.

B. Flowering plants

While flowers and the plants that produce them are familiar to us all, it is not readily apparent that the flower, like the cone, is actually a group of spore-bearing leaves (sporophylls) so highly modified for the production and protection of male and female gametophytes that they no longer bear any resemblance to foliage leaves. Flowers thus carry to extremes a tendency to restrict spore-bearing to specialized leaves; we saw a few ferns that displayed this tendency, and we encountered it again, in much more pronounced form, in conifers. The manner in which flowering plants bear their sporangia is fundamental to their present highly successful and dominant position on land.

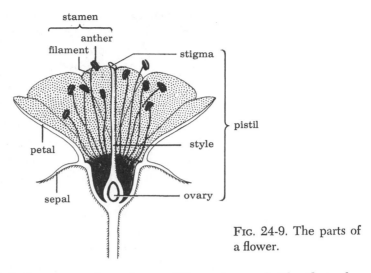

Fig. 24-9. The parts of a flower.

While flowering plants have a life cycle much like that of conifers, we may briefly review some of the differences between these groups before examining in detail the reproduction of flowering plants.

▶ While conifer ovules are borne on the *surfaces* of highly modified leaves (cone scales), ovules of flowering plants are borne *within* a group of even more highly modified leaves, in a section of the flower called the ovary.

▶ While conifer pollen reaches conifer ovules principally in wind currents, the pollen of flowering plants is often carried by some animal agent (frequently an insect); both the animal vectors and the flowers have developed striking adaptations that reflect their intimate relationships. Many flowering plants could not reproduce sexually without their animal pollinators.

▶ Finally, the ovary, or ovaries, of the flower develops into a fruit, or fruits, protecting the seeds and often facilitating the species' dispersal.

1. Examine a simple typical flower (e.g., gladiolus), and locate the four sets of modified leaves (Figure 24-9):

a. **Sepals,** which enclose and protect all the other floral parts during the bud stage. They are usually small, green, and leaflike, but in some species they are large and brightly colored. All the sepals together form the **calyx.**

b. **Petals,** internal to the sepals and often showy in animal-pollinated flowers. Together, the petals form the **corolla.** In some flowers, petals are reduced or absent altogether; how do you think these flowers would be pollinated?

c. The **stamens,** the male sporophylls, located just inside the circle of the corolla. Each stamen consists of a **filament** and a terminal ovoid **anther,** in which pollen is formed.

d. The **pistil,** representing several fused female sporophylls. Each pistil consists of an **ovary** at its base, a slender **style** (sometimes more than one) that rises from the ovary, and an enlarged apex called the

stigma. That the pistil represents several fused sporophylls in many species is usually evident from the multiple structure of the stigma and from the microscopic structure of the ovary, which is similarly divided. Make cross sections of the ovary of your flower; notice its subdivided interior and locate ovules (sporangia) attached to the wall of the ovary by short stalks.

Make similar sections of an anther, and examine microscopically the anther section and some of the pollen.

Flowers having all four sets of modified leaves are called *complete* flowers; flowers lacking one or more of them are called *incomplete*. Incomplete flowers lacking one of the sets of reproductive organs are called *imperfect* flowers, while flowers with both sets of reproductive organs are called *perfect* flowers. Examine flowers on demonstration; into which of the above categories does each fit? Identify as many of the sets of modified leaves as are present on each, and determine if possible what kind of agent and (if an animal) how large an agent effects pollination.

2. Angiosperm gametophytes are formed in the ovules and anthers in much the same way as in conifers. The female gametophyte, commonly a seven-celled, eight-nucleate structure, is formed from one of four haploid megaspores that result from meiosis in the ovule; the gametophyte is called an **embryo sac.** The male gametophytes, or pollen grains, are produced by meiosis much as in conifers.

Pollen deposited on the stigma develops a tube that grows down into the style, contacts the embryo sac (FIGURE 24-10), and releases two nuclei. One of these nuclei fuses with an egg cell (or nucleus); the resulting zygote develops into the embryo. The other pollen nucleus fuses with two or more other nuclei in the embryo sac and by mitotic divisions gives rise to the en-

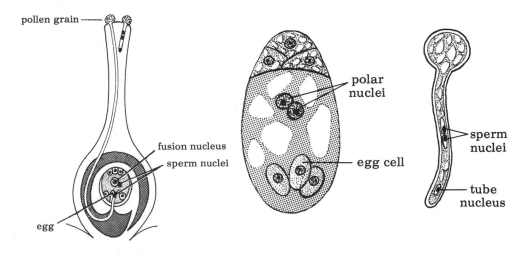

FIG. 24-10. Fertilization in an angiosperm. Pollen grains land on the stigma and give rise to pollen tubes that grow downward through the style. One of the pollen tubes shown here has reached the ovule in the ovary and has discharged sperm nuclei into it. One sperm nucleus will fertilize the egg cell, and the other will unite with the diploid fusion nucleus (derived from the two polar nuclei) to form a triploid nucleus, which will give rise to endosperm. Details of the female gametophyte (embryo sac) and of the pollen grain are shown in the two drawings on the right. Instead of being exposed, as it is in conifers, the angiosperm female gametophyte is completely enclosed within the ovule inside the ovary of the flower.

dosperm, a tissue that nourishes the developing embryo when the seed breaks dormancy and the young plant resumes growth.

3. The walls of the ovaries of flowering plants develop into the **fruit,** which encloses the **seeds.** The many types of fruits reflect the many different kinds of ovaries and a wide variety of adaptations for dispersal. *Simple* fruits are derived from a single ovary; *aggregate* fruits are derived from many separate ovaries of a single flower which later fused to form a single fruit. Examine the seeds and fruits on demonstration. Discuss with your classmates the mechanisms by which these seeds and fruits are dispersed, and be prepared to defend your suggestions in a class discussion.

Open a simple fruit, such as an apple, by cutting through the center, separating the top from the bottom half. By examining this cross section, decide how many sporophylls fused to form the ovary.

QUESTIONS FOR FURTHER THOUGHT

1. Compare the structure of the moss with that of a fern.

2. What is heterospory, and what is its significance?

3. Compare the gametophyte and sporophyte of a moss with those of a conifer, with respect to size and independence.

4. What are the major features that distinguish conifers from flowering plants?

5. What is a sporophyll? In what groups of plants are sporophylls most strikingly modified? What purpose do the modifications serve? Be able to defend your answers.

6. Where on an angiosperm plant is the gametophyte located?

7. Discuss some of the adaptations by means of which cross-pollination is encouraged in the flowering plants.

8. Mendel, in doing his experiments with garden peas, found that peas normally are self-fertilized, since maturation of the floral organs occurs before the flower has opened. What procedure would you follow in order to bring about cross-fertilization in such a plant?

TOPIC 25 PROTOZOA, COELENTERATA, AND PLATYHELMINTHES

Biological Science, pp. 765-781.
Elements of Biological Science, pp. 490-503.

The business of procuring appropriate food in sufficient quantity is a matter of the most compelling urgency for most animals, and many of them spend their entire lives in continuous feeding, with little or no time left for other activities. *Paramecium*, for example, is a continuous feeder, as are most of the protozoans. With nutrient procurement so high in priority for most animals, it is no surprise to find that natural selection has endowed animals with remarkable adaptations of structure, function, and behavior directly related to feeding; you examined several such adaptations in Topic 5. It is the purpose of this laboratory to introduce you to these and other adaptations in representative animals of several invertebrate phyla and, at the same time, to give you an opportunity to learn the general characteristics of these animal groups at first hand.

One major aim of this laboratory, and indeed of your entire laboratory experience, is to help you understand the lives of other organisms. Learning the names and functions of parts of animals may have some inherent interest, but it cannot become really meaningful until you understand the significance of each anatomical, functional, and behavioral adaptation. Such understanding and heightened appreciation of each species' role in the overall economy of nature can come only from careful thought about how particular adaptations help fit an animal to life in its particular habitat. Sometimes we shall pose questions to guide your thought, but they should be considered only as guides; most of your questions should arise from your own study of the materials provided. Sometimes answers will come from common-sense thinking, and sometimes from your own experiments; in either case, training yourself as a questioner will be rewarded by a much deeper and more meaningful comprehension of organisms than could ever be the case through rote memorization.

PART I. PROTOZOA

Paramecium and *Amoeba*, which you have examined previously, are classified as members of the Phylum Protozoa because each consists of a highly differentiated single cell.

Stentor, a relative of *Paramecium*, has organelles for feeding, digestion, locomotion, water balance, oxidative metabolism, and integration; in addition, it exhibits a remarkably complex behavior. These same activities, in a human being, would require the cooperative action of organ systems composed of many millions of specialized cells. In this unit, you will review some of the complex internal structure that characterizes members of the Phylum Protozoa, and you will explore several aspects of *Stentor's* behavior toward food.

FIG. 25-1. Living *Stentor*. This photomicrograph was made with oblique light. (*Courtesy Carolina Biological Supply Company.*)

A. Structure of Stentor

Obtain several animals in a depression slide, without a cover slip, or on a plain slide with a cover slip raised by vaseline or bits of glass. Study the structure of *Stentor*, and note the following with the aid of FIGURES 25-1 and 25-2.

1. trumpet-shaped body
2. frontal field (the region where food is brought in to be ingested)
3. band of membranelles (fused cilia)
4. oral pouch
5. gullet
6. food vacuoles
7. contractile vacuoles
8. chainlike macronucleus

Use the dissecting microscope and the low power of your compound

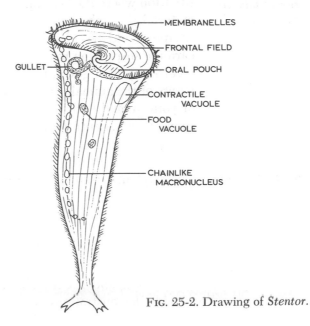

FIG. 25-2. Drawing of *Stentor*.

microscope; experiment with different types of lighting until you find a set of conditions that gives you clear visibility. Later, you will need to be able to observe the feeding current created by the beating of membranelles and cilia around the frontal field. Find animals that are attached; if the animal you are observing swims away, locate another one that is attached.

B. Feeding current

Lift one edge of the cover slip with a dissecting needle and, on the end of a toothpick, introduce a very small amount of yeast suspension stained with carmine or with Congo red; carefully watch the current created by the beating of the membranellar band. As you can see, this current serves to bring suspended particles into the gullet, and may be reversed or stopped altogether under certain conditions. Watch the current for some time, until you feel able to describe its shape and direction. Using simple arrows and lines, add this information to FIGURE 25-2.

C. Food selection

By carefully observing animals on the slide used in B, you may be able to detect evidence of discrimination between dye particles and yeasts. If you can see discrimination, where does the sorting occur? Is the selection complete, or are some indigestible dye particles taken into the animal?

A better demonstration of food selection will occur if unfed animals, on a new slide, are offered a combination of *Euglena* (a green flagellate) and indigestible carmine dye particles. By observing the formation of food vacuoles, it should be simple to determine whether selection is occurring, whether any indigestible particles are taken in, and if so, the approximate ratio between digestible and indigestible materials.

Why would food selection be adaptive for the animal? What "criteria" might it use in exercising selection?

D. Behavioral responses to non-nutritive material

It is possible that an animal like *Stentor* might, on occasion, find itself in a situation where the beating of the membranelles brings in only non-nutritive material. In such circumstances, a behavior pattern enabling detection and escape would be advantageous. With a clean toothpick, add some of the carmine suspension to a fresh mount of unfed *Stentor*. Carefully watch and record any changes in the feeding current, and in the posture and movements of the animal; describe the reaction sequence. Compare the reactions of your animal with reactions observed by others in the class. Are the responses consistent? How would each one be adaptive?

E. Other Protozoa

The Phylum Protozoa is customarily divided into four classes:

1. Flagellata
2. Sarcodina
3. Ciliata
4. Sporozoa

Review and list the characteristics of each class. Your instructor may provide demonstrations of protozoans in addition to those you have already

FIG. 25-3. *Hydra,* with bud. (*Courtesy Ward's Natural Science Establishment, Inc., Rochester, New York.*)

seen. Determine the class of each of these animals, and enter it in the appropriate space in Appendix 2.

PART II. ***COELENTERATA***	Coelenterates include forms such as hydroids, jellyfishes, sea anemones, corals, and relatives of these; almost all of them are marine. We shall examine one of the few fresh-water species, *Hydra,* to illustrate the general body plan, cellular construction, and behavior of members of this phylum.

A. Structure of Hydra

    1. *Hydra* is a small coelenterate that lives attached to submerged rocks, leaves, and twigs, its **tentacles** outstretched for passing prey. Obtain a specimen in a small petri dish, or place one on a slide with a well-raised cover slip. Allow the animal to relax, and examine it with the dissecting microscope and with a low power of the compound microscope. Note the **radial symmetry.** Locate on your specimen and label in FIGURE 25-3:

        a. basal disc (the animal's point of attachment)
        b. body column
        c. tentacles
        d. mouth region (at the base of the tentacles)
        e. nematocysts (small swellings localized largely in the tentacles)

    2. Food is taken in and waste voided through the same opening. The body wall consists of two prominent cellular layers, an outer **epidermis** and an inner **gastrodermis.** Locate and label these layers in FIGURE 25-4. If your specimen is mounted on a slide, you may be able to see the epidermis and gastrodermis with the compound microscope. In *Hydra* and most other members of its class (the Hydrozoa), a very small amount of jellylike material, the mesoglea (mesoderm), occurs between the two primary cell layers; in many other coelenterates (e.g., sea anemones) the mesodermal layer forms most of the animal's bulk.

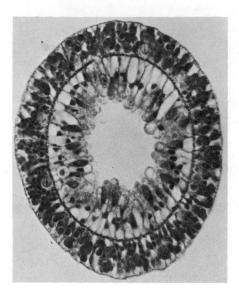

FIG. 25-4. *Hydra*, cross section. (*Courtesy George H. Conant, Triarch Incorporated, Ripon, Wisconsin.*)

The gastrodermis bounds a *gastrovascular* cavity, so called because both partial digestion and circulation occur there. No special circulatory system is present, but cilia on the gastrodermal cells cause food in the cavity to move about, thereby reaching all parts of the organism, including the interior of the tentacles. Digestion is partly extracellular; the gastrodermal cells phagocytize and digest intracellularly particles resulting from extracellular digestion.

The nervous system is in the form of a net, which is unpolarized; impulses can travel in either direction along the nerve cells. In *Hydra*, there is no marked nervous centralization, although there is some tendency for nerve cells to congregate about the mouth region; you can determine that the animal is more sensitive there by touching it gently with a teasing needle in a variety of regions including the mouth. The nerve net is modified slightly in some other coelenterates, such as jellyfishes and sea anemones, which are capable of more complex coordinated movement than is *Hydra*.

The epidermis contains **gland cells, nematocysts,** and **epitheliomuscular cells.** Fibers of the epitheliomuscular cells form the basis for movements in coelenterates, since there are no specialized muscle cells.

B. *Feeding behavior in* Hydra

PREY CAPTURE

1. All coelenterates are carnivorous predators, and capture live prey by means of nematocysts, specialized epidermal cells. When properly stimulated, nematocysts discharge threads that entangle or pierce and poison the prey. Following capture, the victim is ingested and digested; any indigestible remains are eliminated through the mouth.

a. Transfer your *Hydra* to a small petri dish or Syracuse dish and observe it under the dissecting microscope, preferably with a dark background and side lighting. Introduce some small brine shrimps (*Artemia*) that have been carefully rinsed free of sea water; watch the *Hydra* carefully for the next few minutes. Record your observations in as much detail as possible. (*Note:* Should the *Hydra* not capture and ingest one of the brine shrimp within a few minutes, return your animal to the instructor and obtain a fresh specimen.)

b. If additional specimens are available, you may wish to compare your results with the feeding responses of *Hydra* that have already been fed.

FEEDING RESPONSE

2. a. As you have probably already observed, the capture of prey is normally followed by a relatively complex series of movements in which the prey is transported to the widening mouth and then pushed into the gastrovascular cavity. If you did not notice these movements or did not try to differentiate them from other movements made by your specimen, watch another unfed animal capture and ingest prey. These movements constitute the "feeding response," and are apparently chemically controlled.

At present, it is unclear whether the chemical message arrives from the prey, or is a result of nematocyst discharge, or both. The feeding response can be induced by a variety of chemical agents, some of which are present in the body fluids of prey animals. A coelenterate would be exposed to these substances when its nematocysts pierced the prey. On the other hand, experiments have shown that an animal which has adapted to one of these chemical agents, and is no longer showing the feeding response, can be induced to respond when it is allowed to capture normal prey.

b. Your instructor will provide a solution of reduced glutathione, one of the agents known to induce the feeding response in *Hydra*, and one also present in the body fluids of prey animals. Set up a *properly controlled* experiment to test the *Hydra's* reaction to the presence of reduced glutathione. How you add the solution is for you to decide, but be sure not to add more than a few drops, and be careful to include appropriate controls. Record the results.

c. With a new sterile lancet, obtain a large drop of your own blood and let it dry on a slide. Discard the lancet immediately after use; do *not* leave it lying on the laboratory desk. After the clot has dried, break it up with forceps and offer one of the larger pieces to the *Hydra*. When the animal has "captured" the blood clot, let the clot go and observe the reaction.

C. *Feeding behavior in a sea anemone*

Obtain a starved specimen of a sea anemone (such as *Metridium* or *Anthopleura*) in a large culture dish of sea water.

Your instructor will provide clam juice and bits of clam meat, sand, glass rods and medicine droppers, and filter paper (which may be soaked in meat juice, sea water, saliva, or other substances). Determine the reaction of the tentacles to various forms of mechanical and chemical stimulation, and determine what stimuli are necessary for a feeding response. How does the anemone compare in this respect to *Hydra?* What events constitute the feeding response in the anemone?

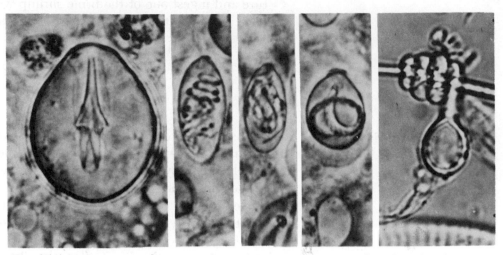

FIG. 25-5. Nematocysts. The one on the far left penetrates prey, while the others entangle prey with sticky threads. The one on the far right has been ejected and is shown attached to the bristle of a prey animal. (*Courtesy Carolina Biological Supply Company.*)

D. Nematocyst discharge

HYDRA

1. Transfer a *Hydra* to a slide with a small amount of water; add a cover slip and put the slide aside to dry slightly. Observe at intervals under the low and high power of the compound microscope. The appearance of several discharged and undischarged nematocysts is shown in FIGURE 25-5. Find as many types as you can on your slide, and make sketches.

SEA ANEMONE

2. The nature of the stimuli responsible for nematocyst discharge can be demonstrated by several simple experiments, best carried out with a sea anemone.

a. *Mechanical stimulation.* Isolate a tentacle from an anemone by using forceps to grasp the basal portion and scissors to remove it completely. Carry out this procedure as quickly and gently as possible in order to avoid nematocyst discharge due to excessive mechanical stimulation. Place the tentacle in sea water in a depression slide, and examine under low power. Stimulate the tentacle by gently rubbing it with a fine glass rod. Examine for nematocyst discharge.

b. *Chemical stimulation.* With as little disturbance as possible, introduce to the slide a drop of fresh clam extract or a bit of saliva dissolved in sea water; again examine the tentacle for evidence of discharge.

Under normal circumstances, neither chemical nor mechanical stimulation alone is sufficient to provoke nematocyst discharge. Combined, however, they will cause nematocysts to fire; you can demonstrate this by mechanically stimulating the tentacle in the presence of a chemical stimulant. Other factors, such as the nutritional state of the animal, also influence nematocyst discharge.

How would use of both chemical and mechanical sensory modalities, rather than just one or the other, be of advantage to the animal?

Fig. 25-6. Jellyfish. (*Courtesy Carolina Biological Supply Company.*)

E. Coelenterate diversity

Although the cellular structure and feeding behavior of *Hydra* and sea anemone are quite typical of coelenterates, the life cycle and many other features are not. Both animals are **polyps,** fleshy columns with attachment at one end and a mouth and tentacles at the other. Many coelenterate polyps live in colonies; each colony contains individuals specialized for several different functions, a condition known as **polymorphism.** The life cycle of some coelenterates includes a **medusa** stage in addition to the polyp. The medusa has the form of a jellyfish (FIGURE 25-6), is free-floating in the water, and usually produces gametes. Medusae are often produced by a polyp specialized for this purpose. *Hydra* and sea anemones are both specialized in their lack of the medusa stage; both produce gametes in the polyp stage.

1. Both polymorphism and the medusa stage are illustrated by *Obelia,* a common marine form. Examine slides of *Obelia,* and locate **feeding** and **reproductive** polyps. The feeding polyps are much like small *Hydra,* with mouth, tentacles, and gastrovascular cavity; the reproductive polyps are covered with buds of medusae, which they produce asexually. The various polyps in the colony are interconnected by a hollow stalk; all are therefore in communication with the gastrovascular cavity of the feeding polyps. Both stalk and individuals are largely covered by a hard material, the **perisarc;** in some species, the soft parts of the body can be withdrawn completely into the perisarc, and thus protected from predators. Label FIGURE 25-7.

Gametes are usually produced by the medusa, except in forms like *Hydra* or sea anemones which reproduce sexually in the polyp stage. In either case, gametes are usually shed directly into the ocean. Early development produces a small **planula** larva, which eventually settles and develops into a new polyp.

2. Examine demonstrations of other members of the three coelenterate classes:

    a. Hydrozoa
    b. Anthozoa
    c. Scyphozoa

Review and list the characteristics of each class, and enter each of the animals observed in this laboratory in the appropriate section of Appendix 2.

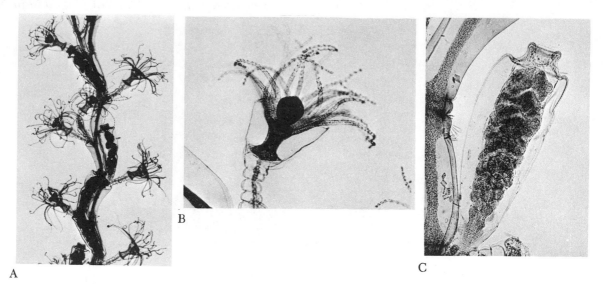

Fig. 25-7. *Obelia.* (A) Part of mature colony. (B) Feeding polyp. (C) Reproductive polyp. (A: *Courtesy George H. Conant, Triarch Incorporated, Ripon, Wisconsin. B, C: Courtesy Ward's Natural Science Establishment, Inc., Rochester, New York.*)

PART III. PLATYHELMINTHES (FLATWORMS)

Members of the Platyhelminthes, a phylum of dorsoventrally flattened worms, occur in both marine and fresh-water habitats; several have adaptations for living in very moist places on land. Some of the marine species are conspicuous and colorful; most of the fresh-water forms are small and retiring, and are generally not seen except by a close observer.

While their bodies are constructed on a relatively simple plan, flatworms differ from coelenterates in that their mesoderm is a much more prominent tissue layer; it forms muscle cells and associates with the other tissue layers to form distinct organs.

There are three flatworm classes. The free-living species comprise the Class Turbellaria; their feeding habits range from predation through scavenging of detritus. The other two classes, the flukes (Class Trematoda) and the tapeworms (Class Cestoda), are serious parasites of man and a wide variety of other animals. In this unit we shall examine a fresh-water planarian worm, to get a general idea of the structure and behavior of members of this phylum; then we shall turn to specimens from the other two classes, which nicely illustrate some of the special adaptations required by parasitic organisms.

A. Class Turbellaria (Free-living flatworms)
EXTERNAL STRUCTURE OF A PLANARIAN

1. The common planarian *Dugesia* inhabits fresh-water streams and ponds, scavenging upon small live or freshly dead animals. Obtain a specimen in a small petri or Syracuse dish of spring water. In subsequent handling of these animals, use *only* spring water; tap water contains chlorine and other chemicals that are toxic to these delicate animals. Observe:

> a. Gliding locomotion, powered by cilia that cover the animal's ventral surface and directed by muscular movements of the body. If you use a dissecting microscope, you will probably notice a slimy mucus trail; mucus is secreted by glands around the margin of the ventral surface of the animal, and it provides traction for the cilia.

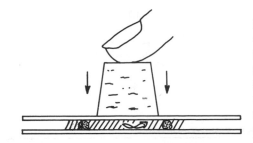

FIG. 25-8. Method of flattening a planarian worm. The worm is placed in water between two slides that are held apart by pieces of modeling clay. The slides are pushed together slowly by pressure on one of them, until the worm is barely flattened and immobilized; too much pressure will kill the animal.

b. Distinct bilateral symmetry, in contrast to the radial symmetry of coelenterates. What is the advantage of bilateral symmetry?

c. Dorsoventral flattening, with the dorsal surface arched slightly (best seen in prepared cross sections).

d. The head, somewhat pointed in *Dugesia,* and bearing a pair of pigmented eyes and a pair of lateral mechanosensory and chemosensory projections, the auricles. Notice how the moving animal "tests" the environment ahead of it by oscillating the head region.

INTERNAL STRUCTURE OF A PLANARIAN

2. To see internal structures, mount a living animal in spring water between two glass slides, held apart by small pieces of modeling clay (FIGURE 25-8). A little pressure on the top slide will immobilize the animal by compressing it; either dorsal or ventral side may then be studied under the compound microscope. *Too much pressure will kill the animal.*

a. The mouth is a small hole near the center of the ventral surface; it is sometimes difficult to see. The mouth leads into a short cavity containing the protrusible pharynx, which in turn communicates with the large gastrovascular cavity. Look for the pharynx by focusing just anterior to the mouth region.

The gastrovascular cavity branches into three main divisions; one runs anteriorly and terminates between the eyes, while the other two approximately parallel branches run posteriorly. All three branches subdivide repeatedly. To see the full extent of the gastrovascular cavity, examine a recently fed specimen or a slide made from such a specimen.

b. You may see a pair of ventral nerve cords by mounting and examining a living specimen of the white planarian *Phagocata,* or by examining a demonstration slide of *Bdelloura,* a marine flatworm. In the latter specimen the nerve cords show up clearly as a pair of longitudinal, often orange-colored, structures terminating anteriorly as the brain. Flatworms show a distinct tendency toward cephalization or concentration of nervous tissue at the anterior end of the animal; this trend is carried further by the more complex phyla.

c. Flatworm gas exchange and excretion occur largely by simple diffusion. Water-balance problems are solved by a flame-cell system. If you did *not* see flame cells in Topic 10, follow the directions on page 117 for making a slide.

FIG. 25-9. Light response chamber for testing response of planarians.

d. Reproduction is both sexual and asexual in most flatworms; sexual reproduction is seasonal, and the reproductive organs develop in each individual only as the season approaches. Male and female organs are present in each worm, but the animals mate with one another and cross-fertilization, through exchange of sperms, is thus assured.

BEHAVIOR
OF A PLANARIAN

3. a. *Feeding.* Flatworms are notoriously sensitive to disturbances in the water about them, but with care you may be able to induce your animal to feed. When food is encountered, the pharynx is protruded through the mouth; slight sucking action draws food particles into the gastrovascular cavity.

Place a small piece of hard-boiled egg yolk into a dish containing one or two specimens. Position the dish under a dissecting microscope, and shield it from light and from other disturbances. Then go on to another part of this unit, inspecting the dish at intervals to see if the worms are feeding. When you see the worms feeding, look for the protruded pharynx; if you cannot find it, lift one end of the worm gently with forceps, and try to observe the pharynx before it is withdrawn.

b. *Response to light.* You can test the flatworm's response to light by placing several specimens in a petri dish of pond water and offering them a choice between light and dark regions. Cover one half of the dish top with black construction paper, new carbon paper, or a book; arrange a light to brightly illuminate, *without overheating*, the other half of the dish (FIGURE 25-9). Start the test with more of the animals in the illuminated half of the dish; note their response after a few minutes, and repeat the test several times for consistency (i.e., reintroduce animals to the light, by turning the dish cover so that they are again exposed). Of what value would these responses be in nature?

You can test whether the eyes are the only means of light perception by carrying out the same experiment after cutting the animal into two pieces*, one with and one without eyes. Test both regions for a response to light. Be sure to try several different light intensities and allow ample time for a response to occur. Record and explain all responses to each different condition.

---

* Place the animal on a stiff white card, and cut it with a razor blade.

c. *Response to touch.* Touch the animal on the auricles and on other parts of the body; record the response in each case. Vary the nature of the stimulus (e.g., glass rod, teasing needle, water stream from medicine dropper) and the intensity of the stimulus. Design and carry out an experiment to discover whether integrating activity of the brain is necessary to elicit a response.

d. *Righting reaction.* Turn an animal over with a stiff paper card, and carefully observe its response. How do you think it senses orientation with respect to gravity? If you cannot answer this question, find another student whose worm is swimming upside down along the water surface; if you have such a worm, point it out to your instructor and to your neighbors. The worm is able to swim upside down on the water surface because of surface tension, a phenomenon previously discussed in Topic 2, page 23. By adding a small amount of detergent like Photoflo or Tween 20 to the water, you can reduce the surface tension and cause the worm to drop to the bottom of the dish. Does the upside-down swimming give you any hint about how the planarian senses orientation with respect to gravity?

B. Class Trematoda (flukes)
STRUCTURE OF AN ADULT FLUKE

1. *Prosthogonimus,* a common parasite of the oviducts and intestines of domestic and wild waterfowl, is typical of its class. Study a slide of *Prosthogonimus* with reference to FIGURE 25-10, and examine live specimens if available.

The animal is covered with a **cuticle,** which protects it from enzymes and toxic substances often present in the peculiar areas that many flukes inhabit (e.g., urinary bladder, intestines, etc.). Two prominent muscular **suckers** attach the animal firmly to its host. Why would strong attachment

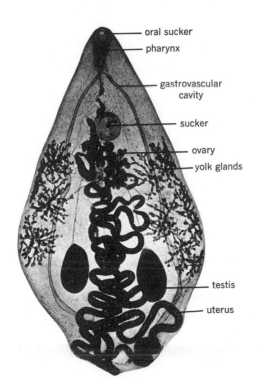

oral sucker
pharynx
gastrovascular cavity
sucker
ovary
yolk glands
testis
uterus

FIG. 25-10. *Prosthogonimus macroorchis,* a fluke that parasitizes the oviducts of the domestic hen and other waterfowl. (*Courtesy General Biological Supply House, Inc., Chicago.*)

be necessary? The digestive system is Y-shaped, as in planarians, but the single branch of the Y is merely a short esophagus, opening within the oral sucker; the other branches of the digestive tract are simple, long, unbranched sacs that run along most of the animal's length. The nature of the digestive tract should make it easy for you to guess the animal's diet. Essentially all the other prominent organs are reproductive structures, which you can identify in FIGURE 25-10. Other endoparasitic flukes are structurally much like *Prosthogonimus*.

FLUKE REPRODUCTION

2. Attached within its host, the adult worm produces hundreds of thousands of eggs, each enclosed in a resistant capsule. Within the capsule, a tiny **miracidium** larva develops. When the ciliated miracidium hatches, it swims about until it locates an appropriate **intermediate host,** often a mollusc; there it undergoes a series of asexual multiplications, each one producing more infective units, which are eventually released from the mollusc and seek new intermediate hosts. The precise number of asexual stages and the precise hosts involved depend on the species of fluke.

LIVING FLUKES

3. If possible, you will be provided with a living frog, in which adult flukes may be found at several locations, especially in the lungs and the bladder.

a. To find lung flukes, open the frog's body cavity, cut a lung at its base, and cut or tear it open. Pin the tissue to a wax-lined dissecting pan, and examine with a dissecting microscope. To locate bladder flukes, excise, open carefully, and pin the bladder (in a stretched condition) to a dissecting pan. Study its interior surface with a dissecting microscope.

b. When you locate a fluke, study its movements, and note how tenaciously it clings to its host. Examine the lining of the infected organ; note the areas of hemorrhage and scarred tissue where flukes have previously been attached. What damage is done to the host?

c. Carefully remove an adult fluke, put it on a slide in a drop or two of pond water, and add a cover slip. Place the slide under a dissecting microscope; then, with a piece of absorbent paper, touched to the edge of the cover slip, remove excess water until the worm is immobile and well flattened, but not crushed. During the study, make sure that you add enough water to replace that lost through evaporation. Locate and identify as many of the internal structures as possible.

d. Miracidia can be observed by teasing apart a mature bladder fluke in a small dish of pond water; the ripest eggs hatch almost at once upon contacting a hypoosmotic environment. Remove some of the miracidia to a slide for closer microscopic observation.

e. If materials are available, your instructor may provide other stages in the life cycle of the fluke.

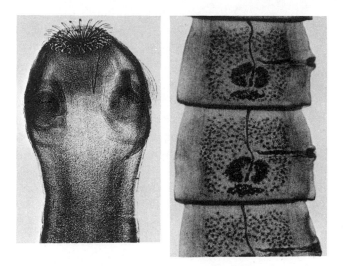

FIG. 25-11. *Taenia pisiformis,* the dog tapeworm. Left: Scolex. Note the hooks and suckers. Right: Mature proglottids. (*Courtesy Carolina Biological Supply Company.*)

C. Class Cestoda (tapeworms)

Tapeworms are usually intestinal parasites of vertebrates. Their life cycle involves one or two intermediate hosts, and they are highly modified for reproduction. Examine a living or preserved tapeworm; note the attachment point or **scolex,** the short neck, and the long, ribbonlike body, consisting of increasingly mature segments called **proglottids.** New proglottids are produced at the base of the neck; as they grow older a..d farther from the scolex, they become sexually mature. The sexual organs can be identified in FIGURE 25-11.

Each sexually mature proglottid is inseminated by a more anterior one; it then begins to develop eggs. Ripe proglottids, which may contain more than 100,000 eggs, leave the host in feces. They hatch if eaten by an appropriate intermediate host. The larvae encyst in the host's tissues; the adult form develops when one of these cysts, present in raw or poorly cooked meat, is ingested by the appropriate final host.

How does a tapeworm feed? How do its feeding habits differ from those of flukes? How does damage to the host differ from that done by flukes? Would the feeding habits of the tapeworm render it more vulnerable to elimination from the host?

Enter the animals observed in this part of the laboratory in the appropriate spaces in Appendix 2.

QUESTIONS FOR FURTHER THOUGHT

1. How does the requirement for locomotion influence the general physical shape of an animal? Compare the basic design of fast-moving and nonmotile animals. Does *Stentor* change shape when it stops swimming?

2. Distinguish between polyps and medusae. What is the function of the medusa in the coelenterate life cycle?

3. What role do you think diffusion would play in coelenterate metabolism?

4. What is polymorphism? How might it benefit a coelenterate?

5. What is mesoglea?

6. Summarize your findings in the experiments on feeding behavior in *Hydra* and sea anemone and suggest experiments to answer some of the questions raised by your studies.

7. Marine organisms that live near the shore are often less sensitive to small environmental changes than are fresh-water organisms or marine organisms that live far from shore. What adaptive value would a high tolerance to environmental change have for such animals?

26

ARTHROPODA, ANNELIDA, AND MOLLUSCA

Biological Science, pp. 789-803.
Elements of Biological Science, pp. 508-517.

We have used insects and earthworms many times in previous laboratories to illustrate special types of adaptations, and we have looked briefly at molluscan classes in the laboratory on evolution. In this laboratory, we shall look more carefully at the three major phyla to which these animals belong. Our purpose is to help you understand the distinctive characteristics of these phyla and to give you some idea of the diversity of animals they embrace.

PART I. ARTHROPODA (CLASS CRUSTACEA)

As you doubtless already know, the arthropod class Insecta is by far the largest class in the entire animal kingdom. Its species inhabit almost every sort of terrestrial habitat and many fresh-water habitats as well. They are the arthropods most familiar to the average person. But since we have already examined many aspects of the biology of insects in previous laboratories, we shall devote most of our attention here to the arthropod class Crustacea, another enormous group whose species also live in a great variety of habitats. Because most of those habitats are in the sea, crustaceans are not as well known as insects to most people, but they are very important in marine communities. Some Crustacea (e.g., crayfish) inhabit fresh water, and a very few (e.g., sow bugs) are terrestrial.

We shall examine in some detail a crayfish or a lobster. We must emphasize, however, that the characteristics of these animals are quite different from those of many other crustaceans, and you should not think of crayfish and lobster as "typical" members of their class. Indeed, the great majority of crustaceans are planktonic organisms that bear little resemblance to a lobster.

A. External anatomy

Obtain a crayfish or lobster in a dissecting pan. Note that the entire body is covered with a hard **exoskeleton.** The exoskeleton, which is composed of chitin (a polysaccharide) and calcium carbonate, is secreted by the underlying epidermis. What are some functions of the exoskeleton?

1. The body is divided into an anterior **cephalothorax** and a posterior **abdomen;** the exoskeletal shield covering the cephalothorax is called the **carapace.** In the adult animal, segmentation is evident only in the abdomen, but some other adult crustaceans and the early developmental stages of all arthropods show clearly that segmentation is part of the basic body plan. What other phyla show marked segmentation? Is it always evident in the adult stages? What function does segmentation serve in arthropods? In other segmented phyla?

2. Each segment, including those fused to form the cephalothorax, bears an appendage. Though very different in function, the appendages on successive segments develop embryonically in the same manner, and are therefore spoken of as **serially homologous.** As each appendage develops, it becomes structurally different from the others; in the adult, it is modified to carry out some particular function.

a. Notice the five pairs of large appendages attached to the cephalothorax. The most anterior pair are **chelipeds** (pincers), which function in the capture of prey, in defense, or (in the male) for holding a female during mating. The posterior four pairs are called **walking legs.**

b. Behind the walking legs, on the abdomen, are five pairs of **swimmerets.** In the female, the swimmerets are much alike; in the male, the anterior two pairs are stiffened and folded forward, and aid in the transfer of sperms during mating. Pick up an animal of each sex to see these appendages clearly. (Press your index finger down on the carapace and grasp the animal around the cephalothorax with the rest of your hand.)

c. The last abdominal segment bears a pair of broad, fan-shaped **uropods;** a terminal extension of the body, the telson, together with the uropods, forms a tail fan employed in rapid backward locomotion.

d. Anterior to the cheliped are six pairs of serially homologous appendages, concerned largely with the handling of food. With forceps gently locate each appendage and separate it from the others. The table below lists these appendages in order (from the cheliped forward):

APPENDAGE	FUNCTION
3rd maxilliped	
2nd maxilliped	
1st maxilliped	
2nd maxilla	
1st maxilla	
mandible	

Study each appendage carefully and, in the right hand column of the table, suggest its specific function. Later, you will have an opportunity to test these ideas; if you have time, you may later wish to remove these appendages and study their structure more carefully with a dissecting microscope.

e. Ahead of the mandibles is a final pair of homologous appendages, the antennae. On the basal segment of each antenna, locate a small raised nipple, the **excretory pore.**

The smaller **antennules** and the pair of stalked **compound eyes** are additional sensory organs that develop on the first segment. Early development shows that they are not homologous to the other appendages.

3. Between the bases of the third and fourth pairs of walking legs, the female has an exoskeletal swelling, the **seminal receptacle,** where sperms from the male are deposited during mating. At the bases of the second and fourth walking legs, respectively, locate openings of the female and male reproductive organs. The organs themselves may be seen in the internal dissection.

B. Behavior

1. Put a screen around a pan containing a living crayfish; arrange the screen so that it will hide most of your movements. Dangle an earthworm near the animal's mouthparts, and describe the crayfish's reactions in detail, noting particularly the activity of appendages around the mouth.

2. Observe the locomotion of a crayfish, paying particular attention to the activity of the walking legs. Are all four legs identically constructed? Are all four actually used in walking? Do the swimmerets function in locomotion?

3. By making threatening movements or by pushing the animal gently from the front, you may induce it to leap backward with a sharp flexure of the abdomen. Of what value is this reaction?

4. Make threatening movements in front of several living crayfish, until you see a "rearing up" response. Try several other specimens of the same and different sex, to test their response to the same stimulation. Is there any correlation between "aggressiveness" and the sex of the animal? If so, which sex is more aggressive and what function would this behavior serve in nature? What role does aggressive behavior play in societies where the male is the principal aggressor?

5. Turn your animal over in its pan; note carefully its responses and the appendages it uses as it rights itself. Do any of the appendages seem to be consistently more active than others when the animal is turned upside down? Can you change this pattern of activity by removing or immobilizing the active appendages? What organ is responsible for sensing changes in the animal's orientation with respect to gravity?

C. Internal anatomy

If you did not examine the gas-exchange system in Topic 6, page 70, do so before proceeding.

Anesthetize your specimen by placing it in a fresh- or sea-water solution of chloroform for ten minutes, or until it ceases to thrash about. Pin it securely, dorsal side up, in a wax-lined dissecting pan. Arrange the specimen so that it may later be viewed under the dissecting microscope without removing any pins.

With a teasing needle, scrape a small hole near the center of the dorsal side of the carapace. Alternately using needle, forceps, and fine scissors, carefully remove most of the dorsal carapace, as indicated by the broken line in FIGURE 26-1. Work forward and backward from your original hole, removing small pieces of exoskeleton as you go. Keep the instrument points *up;* tissues beneath the carapace are soft and easily injured. Once the animal has been opened, carry out as much of the dissection as practicable under saline solution, and make liberal use of your dissecting microscope.

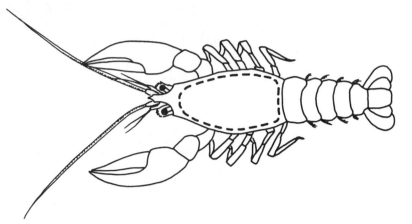

Fig. 26-1. Crayfish, dorsal view showing cut in carapace to expose heart prior to internal dissection.

CIRCULATORY SYSTEM

1. a. You will now be observing the dorsal wall of the **pericardial sac,** which encloses the **heart.** If the animal was not over-anesthetized, the heart will still be beating; if so, use forceps and scissors to carefully remove the pericardial sac. If not, proceed directly to inject the circulatory system, as described below.

Inject the heart with a concentrated dye solution, using a 1 cc dye-filled syringe and a fine-gauge needle. Insert the needle into the heart, but *not* all the way through it; hold the syringe in place, and rapidly depress the plunger. If the needle is properly in place, some dye will leave the heart through its six surface openings (**ostia**), but most of the dye should fill the arteries. Remove the top of the pericardial sac (if it is not yet removed) and pour some water over the preparation to remove excess dye.

Remember that the crustacean circulatory system is an open system. Blood from the body drains into the pericardial sac and enters the boxlike heart through the three pairs of ostia, during the heart's relaxation phase. The ligaments suspending the heart in the pericardial sac aid in expanding the heart's volume so that blood is pulled in during the relaxation phase. When the heart contracts, valves bounding the ostia prevent blood from leaving the heart, and force it through several arteries. The ostial valves thus operate much like valves in the vertebrate heart to maintain a one-way circulation.

b. If your injection was successful, the ophthalmic and antennary arteries were almost certainly injected. The former artery runs mid-dorsally from the anteriodorsal surface of the heart; the latter pair of arteries runs forward to serve antennae, anterior muscles, and other organs. Other arteries, which will be transparent unless injected, may be found posterior and ventral to the heart.

The arteries branch and rebranch, terminating in sinuses that surround the organs, and in tissue spaces within the organs. These sinuses communicate with one another, and the blood eventually makes its way back to the pericardial sac.

REPRODUCTIVE SYSTEM   2. Remove the heart by cutting the arteries and ligaments that suspend it. Also remove the thin floor of the pericardial sac, just below the heart.

> a. The reproductive organs of both sexes are approximately Y-shaped, with the upper branches of the Y extending forward and the lower branch running back some distance toward the abdomen or into it. The Y shape will be clearest in the whitish testes; near the center of the Y, a pair of sperm ducts arises, one going to the base of each of the last walking legs, where it opens to the exterior.

> b. The ovaries are likely to be orange in color and filled with eggs. The size of the eggs and of the ovaries is determined by the season. Trace one of the two fragile oviducts to its opening on the base of the second walking leg.

> c. If you remove the oviducts or sperm ducts and macerate them on a slide, you may be able to see living, mature sperms or eggs. If you find them, call them to the attention of other students.

DIGESTIVE SYSTEM   3. Food enters the digestive tract through the mouth, an opening between the mandibles; it then passes through a short **esophagus** into a **cardiac stomach,** a **pyloric stomach,** and an **intestine,** which opens in the last segment of the abdomen as the **anus.**

> a. The cardiac and pyloric stomachs are thin-walled, white structures near the dorsal midline of the cephalothorax. The former is the more anterior; by lifting up its anterior end, the esophagus is easily seen. The cardiac stomach bears in its interior a large number of chitinous plates, bristles, and teeth (together called the **gastric mill**) that serve to grind food. The grinding action is accomplished by the activity of many small muscles inserting on the wall of the stomach and originating on the exoskeleton. Locate some of these muscles and also the large muscles that operate the mandibles; the latter will have been cut during the removal of the carapace earlier in the dissection.

> b. Properly ground food is strained through a system of bristles into the pyloric stomach, and travels thence into the intestine or into one of the yellowish, or yellow-green, lobes of the digestive gland.
>
>   The digestive gland secretes enzymes into the cardiac and pyloric stomachs; some digestion probably also occurs in the intestine and in the tubules of the digestive gland. In addition to providing enzymes, the digestive gland stores food. To what vertebrate organ would it be comparable? To what structure(s) in plants?
>
>   Trace the intestine posteriorly through the abdomen to the anus; as you do so, notice blood vessels and muscles in the abdomen.

> c. After the digestive system has been studied in place, remove it and cut it open midventrally under water; study the arrangement of filtering and grinding plates and bristles. *Postpone this removal, however, until after the nervous system has been studied.*

4. a. Carefully remove any remaining carapace, including the projection between the eyes; avoid injury to underlying tissues. Lift up the cardiac stomach; then, just ventral to a point between the eyestalks, locate a pair of large ganglia. These ganglia constitute the **brain;** if the circulatory injection was successful, the brain will be pinkish. From the brain, short nerve tracts run to the eyes, antennae, antennules, and dorsal surface of the head; these will show up clearly if a small amount of 95% alcohol is placed on them.

b. The largest pair of nerves leaving the brain runs around the esophagus to join the **subesophageal ganglion;** the rest of the central nervous system consists of a **ventral nerve cord,** a double structure that runs from the subesophageal ganglion back into the abdomen. To see the cord, remove all the overlying organs posterior to the esophagus; also remove the calcified plates of the ventral cephalothoracic exoskeleton. These plates cover the nerve cord in the thorax, and they must be removed to expose it.

c. Antennae, antennules, and tactile bristles all over the body contain chemo- and mechanoreceptors. The **statocysts,** a single one of which opens on the basal segment of each antennule, serve to sense equilibrium; each statocyst contains hair cells that are stimulated by the movement of tiny sand grains when the animal changes its position in space. How do these organs compare to your own equilibrium receptors?

d. Examine one of the compound eyes under the dissecting microscope, noting the square facets that make up the eye's surface. The separate units that make up a compound eye are called **ommatidia;** you can see individual ommatidia if the eye is immersed in boiling water for a few seconds and then teased apart on a slide and examined microscopically. Compare your observations to the diagram in FIGURE 26-2. How does the arrangement and operation of receptor cells in a compound eye differ from that in a vertebrate eye?

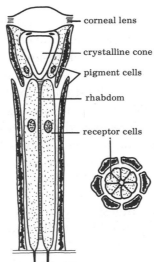

corneal lens

crystalline cone

pigment cells

rhabdom

receptor cells

FIG. 26-2. Longitudinal section and cross section of an ommatidium from an insect compound eye. The lens and crystalline cone focus incoming light rays into the rhabdom, which distributes the light into the eight receptor cells surrounding it. The pigment cells prevent passage of light from one ommatidium to another (*Redrawn from R. E. Snodgrass,* Principles of Insect Morphology, *McGraw-Hill Book Co., 1935. Used by permission.*)

EXCRETORY SYSTEM   5. The excretory (and in the crayfish, water regulatory) organs are a pair of large **green glands**, located just beneath the origin of the eyes. The soft, greenish portion is glandular, and is overlaid by a transparent, membranous bladder; the bladder opens to the exterior through the excretory pore. With a fine-gauge needle and syringe, inject a little dye or water into the bladder through the excretory pore to distend it so that it will be more easily seen.

D. Other Crustacea   Examine other crustaceans on demonstration, and compare them with the crayfish or lobster.

PART II. OTHER ARTHROPOD CLASSES   On demonstration are a variety of representatives of other arthropod classes. As you look at them, ask yourself in what ways they resemble crustaceans and in what ways they differ. Also compare each class with the other classes on demonstration. Pay particular attention to body regions, antennae, mouthparts, other appendages, and special features such as wings, spinnerets, etc. There is explanatory material beside each demonstration.

PART III. ANNELIDA   Inasmuch as we have already examined in some detail several aspects of annelid physiology (see pp. 104, 118, 143) we shall here simply look briefly at some of the characteristics that distinguish the classes within this phylum.

A. Class Polychaeta   Examine the polychaetes on demonstration. Can you see a well-defined head? What sorts of sensory structures does it bear? The body segments of these marine worms usually bear lateral appendages called parapodia. What is their function? Some biologists have suggested that arthropod legs evolved from annelid parapodia.

B. Class Oligochaeta   Examine the oligochaetes on demonstration. This is the class to which earthworms belong. The class also contains many fresh-water species. What are the principal differences between oligochaetes and polychaetes?

C. Class Hirudinea   Most leeches are specialized as external parasites. What are their most obvious special adaptations for parasitism? How do these animals move? Can you see any evidence of external segmentation?

PART IV. MOLLUSCA   The phylum Mollusca is the second largest in the animal kingdom. Though the members of the various molluscan classes differ considerably in outward appearance, they all have fundamentally similar body plans. The soft body consists of three principal parts: (1) a large ventral muscular **foot**, which functions in locomotion; (2) a **visceral mass** above the foot, which contains the various internal organs; and (3) a heavy fold of tissue called the **mantle,** which covers the visceral mass and which in most species contains glands that secrete a shell.

Examine members of the molluscan classes Amphineura, Gastropoda, Scaphopoda, Pelecypoda, and Cephalopoda on demonstration, and read the explanatory material beside each display. Compare and contrast the classes with regard to shell characteristics, method of feeding, and method of locomotion.

Compare molluscans with both annelids and arthropods with regard to gas exchange, circulation, excretion, nervous system, skeleton, and development.

QUESTIONS FOR FURTHER THOUGHT

1. What is the importance of plankton Crustacea in marine communities?

2. Discuss the assertion that the appendages of an arthropod tell you almost all you need to know about its habitat, what it eats, how it carries out gas exchange, how it moves, etc.

3. Describe the fundamental differences between a tapeworm, flatworm, roundworm, and earthworm.

4. What is the adaptive value of a segmented body?

5. The larvae of certain clams attach themselves to the gills of fish. What adaptive advantage would there be to this arrangement?

6. In a sense, arthropods and vertebrates may both be regarded as the most highly evolved representatives of their respective evolutionary lines. They have independently evolved many similar "solutions" to problems of survival. Discuss the ways in which these two groups resemble each other and ways in which they differ.

7. What do you think would have been the characteristics of the first arthropods?

Members of the phylum Echinodermata are exclusively marine. Among them are the sea stars, brittle stars, sea urchins, sea cucumbers, sea lilies, and their relatives. Some echinoderms are used as food in tropical and oriental cultures. Others (the sea stars) are important predators of commercially valuable molluscs, such as clams and oysters.

Members of this phylum are unusual among the invertebrates in possessing an internal skeleton. It is composed of numerous calcareous plates embedded in the body wall. The skeleton often bears bumps or spines that project from the surface of the animal. This characteristic has earned for the group the name "spiny-skinned" animals. A feature unique in the echinoderms is their water-vascular system, which you will examine in this laboratory.

Many aspects of the biochemistry, early embryology, and larval structure of echinoderms suggest that these animals are more closely related to chordates than to the other major invertebrate phyla, despite the fact that the adult structure of echinoderms shows little resemblance to chordates.

PART I. SEA STARS (CLASS ASTEROIDEA)

As a representative echinoderm, you will study a living sea star in some detail. Obtain a specimen in a deep dissecting pan. Immerse it in sea water for as much as possible of the examination of external anatomy, and in cold tap water for the internal dissection. Label FIGURES 27-1 and 27-2 as you proceed.

A. External anatomy

Larval echinoderms, which you examined in Topic 21, show bilateral symmetry; the adults often show a five-part radial symmetry, after a complex metamorphosis. Your sea star will probably have five arms or **rays**, attached to a **central disc**. This central region contains the main digestive tract and openings for the reproductive and water-vascular systems; the rays contain the bulky digestive glands, reproductive organs, and apparatus concerned with operation of the tube feet.

1. In the **oral** side of the central disc, find the mouth, protected by several pairs of spines; on the **aboral** side, opposite the mouth, find the prominent **madreporite,** a small, stony plate (often brightly colored) that forms the entrance to the water-vascular system. When the madreporite points away from you, the ray to its right is designated 1 and that to the left is designated 2; rays 3, 4, 5, and others if present, follow in counterclockwise order. On the aboral side of the disc, between rays 4 and 5 of a five-rayed sea star, you may be able to locate the tiny anus.

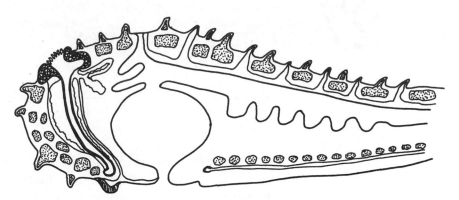

FIG. 27-1. Schematic longitudinal section through central disc and ray of sea star. Label all parts as you locate them in examination of external and internal anatomy.

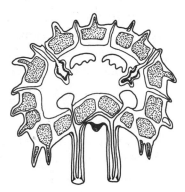

FIG. 27-2. Schematic cross section through a sea star ray. Label all parts as you locate them.

2. Examine the aboral surface of a ray and of the disc with a dissecting microscope. The animal is studded with calcareous spines; in FIGURE 27-1 note that these spines are projections from deeper-lying plates. The plates together form a protective endoskeleton, which originated from the embryonic mesoderm.

Close microscopic examination of the area around each spine will reveal a ring of **pedicellariae,** scissorslike devices. Stimulate a pedicellaria by drawing a thick thread across it, or by poking it gently with a dissecting needle; what is the reaction? Of what value would these devices be to the animal in its natural habitat? Remove several pedicellariae to make a wet mount; examine with a compound microscope, and sketch. Are all the pedicellariae the same? (*Note:* If sea urchins are available, examine their aboral surface to find the long, stalked pedicellariae, and test their reactions to various stimuli. In some urchin species, the spines and pedicellariae are associated with poison glands.)

3. In the spaces between the skeletal plates, note fingerlike sacs, surface projections of the body cavity. These are **dermal gills;** they increase the surface area available for exchange of gases (and probably of waste materials) between the body fluids and the environment. Cilia lining the spacious coelom (body cavity) keep fluids moving through the gills, and you may also notice movements of the gills themselves. Your instructor may have injected your animal or a demonstration specimen with carmine dye particles; if so, look for red particles circulating within the gills; also remove the gill and

examine it microscopically for evidence that the dye is picked up by gill tissues and excreted into the environment.

4. The surface of a sea star is fully ciliated, as you can verify by placing a small amount of carmine or carbon particles on the surface of a submerged animal. If you have time, you may wish to trace some of the patterns of ciliary movement on the rays and disc of a small sea star, or to study similar patterns within the body cavity after you have begun dissection. The danger of colonization by plants and by sessile or parasitic animals is a very real one for any marine organism, especially for one as slow-moving and defenseless as the sea star. Many marine animals and plants release gametes or larvae into the ocean; this ensures their dispersal, but also guarantees a constant rain of organisms seeking a place for attachment. External ciliary action is an important adaptation that prevents such settlers from making a home of the sea star. What other echinoderm adaptations would serve the same function?

5. On the oral side of each ray, locate the deep **ambulacral groove;** longitudinal rows of **tube feet** (how many rows?) are located within it. The original function of tube feet in echinoderms involved the procurement of food; tube feet thus acted much like tentacles, and in a few forms, this function is retained. Most modern echinoderms, however, have developed the ability to use the tube feet in locomotion.

Each tube foot can be extended by the contraction of a balloonlike **ampulla** (FIGURE 27-2, not visible externally); a valve prevents water from re-entering other parts of the water-vascular system. When the tip of the foot contacts a smooth surface, it sticks much like a bathroom plunger. Longitudinal muscles in the shaft of the foot then contract, shortening the foot and pulling the animal toward the point of attachment. The foot is then removed by contraction of radial muscles in its tip, and reattached at another location. The action of a single tube foot does not, of course, have much effect on movement, but several hundred of them operating together in a coordinated manner serve very well for locomotion. In sea stars, the tube feet are also used in feeding; the animal mounts a clam or oyster, exerts a pull on the shells of the mollusc, and thereby tires the muscles that hold the shells together. Eventually, the shells open slightly, rendering the soft parts of the mollusc vulnerable to digestion by the sea star.

Water enters the tube feet through a water-vascular system; it comes in through the madreporite and traverses a series of canals: the **stone canal,** leading orally, directly away from the madreporite; the **ring canal,** running around the mouth; and **radial canals,** one entering each ray. Branches off the radial canals end in ampullae and tube feet.

6. At the tip of each ray, you will notice a pigmented spot that acts as a light receptor. Often the ray is upturned at the tip, exposing this region to the full light impinging on the animal.

7. The main parts of the nervous system (not visible externally) are a circumoral nerve ring and radial nerves, one passing ventrally down each ray between the tube feet. The responsibility of the nervous system for tube feet coordination can be demonstrated by cutting the radial nerve in different locations on various rays, and making careful observations of the operation of tube feet in operated animals. If you have enough specimens and time for these experiments, turn in a full report of your findings.

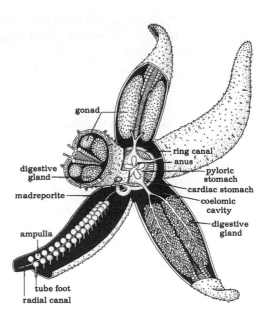

Fig. 27-3. Dissection of a sea star, showing major organs.

B. Internal anatomy
(Figure 27–3)

Injured sea stars often cast off entire rays; missing rays are then regenerated. To ensure that your specimen does not cast off rays during the dissection that follows, inject each ray (about halfway along its length on the aboral side) with 3 cc of 10% magnesium chloride solution; this serves as an anesthetic. Carry out your dissection under cold tap or sea water, and use the dissecting microscope whenever necessary.

1. Remove the aboral skeleton from rays 3, 4, and 5, and from the central disc (avoiding the anus and madreporite) in the following manner. Cut off the tips of the rays with stout scissors (*do not* use fine dissecting scissors except for internal structures) and cut up both sides of each ray toward the central disc. Then, beginning near the tip of the ray, remove the aboral skeleton in small sections. Before you cut each piece of skeleton, lift it up with forceps and free any tissues that may be hanging in mesenteries from the skeleton. When you near the central disc, join the cuts on adjacent rays with stout scissors; then remove most of the rest of the skeleton of the rays and disc, again taking it off in small sections and taking special care to free underlying tissues before you remove the skeleton. In the central disc region, you will have to cut the ducts leading from the gonads to the exterior (they empty via five pores), but leave the skeleton intact around the anus and madreporite.

2. The mouth leads into a short **esophagus** and thence into a saclike, much folded **cardiac stomach.** Above the cardiac stomach, the **pyloric stomach** receives five ducts, one from each **digestive gland.** Locate the digestive gland in each ray. Dorsal to the pyloric stomach, several brownish **rectal pouches** (probably excretory in function) will be seen. From the pyloric stomach, a short intestine leads to the small anus.

In most sea stars, the cardiac stomach is everted through the mouth during feeding, and is applied to the tissues of the prey. Since the stomach is very membranous, it can be slipped with little difficulty into a small gap between the shells of a clam or oyster. Proteolytic enzymes, secreted from the digestive glands, bring about breakdown of prey outside the sea star's body. Digestive products, along with bits of undigested food, are conveyed back into the animal by ciliary action in its digestive tract; some materials are also returned when the cardiac stomach retracts. A pair of prominent retractor muscles runs from the cardiac stomach into each ray. The cardiac stomach everts when these muscles relax and is returned into the body when they contract.

3. A pair of gonads (ovaries or testes) is located in each ray near the central disc, and opens to the exterior via a pore. Testes are chalky-white or gray; ovaries are yellow or orange.

Remove a small amount of one of the gonads and make a wet mount for microscopic observation. Both eggs and sperms are released into the ocean; fertilization and development occur outside the parent animals. Since the release of gametes must be reasonably simultaneous, sea stars (and the many other animals that reproduce in the same manner) have developed very precise biological timing mechanisms. The stimuli involved are only beginning to be studied, but all are probably related to some natural cycle such as the tides, the phase of the moon, etc. The nature of the receptors and the manner in which input becomes a signal for gamete release are poorly understood.

4. To see the **water-vascular system,** cut three or four of the digestive-gland ducts, freeing the stomachs from the rays. Cut the intestine (if not already cut) and push the stomachs gently out of the way, so that the regions lying adjacent to them are clearly visible.

Find the stone canal, a calcareous tube leading orally from the madreporite. Lying along the stone canal is the axial sinus, a part of the sea star's greatly reduced circulatory system. The stone canal leads to a perioral ring canal, five radial canals, and thence to the tube feet and ampullae.

Dispose of your specimen as directed by your instructor. Clean and dry the pan and syringe, and leave them neatly at your desk. Make sure that your dissecting instruments are well rinsed to free them of sea water, and that they are carefully dried.

PART II. OTHER ECHINODERMS

Representatives of the other echinoderm classes are on demonstration. Examine them carefully, and read any explanatory material that may accompany the displays.

A. Class Ophiuroidea

These animals are commonly called brittle stars, serpent stars, and basket stars. Like the true sea stars, they usually have five rays, but these are longer, much slenderer, more flexible, and often branched, and there is no groove on their ventral surface. The body disc is smaller and set off more distinctly from the rays than in true sea stars. The tube feet are not used in locomotion.

B. Class Echinoidea

Among the best-known members of this class are the sea urchins and sand dollars. They have no rays, but do have five bands of tube feet, which you should be able to see. The plates of the endoskeleton are fused to form a rigid box or case, which is usually spherical or flattened and oval. The body is covered by long spines.

C. Class Holothuroidea

These are the sea cucumbers. Their endoskeleton is much reduced, and the body often has a leathery texture. Unlike the other echinoderm classes discussed above, which rest on their oral surface, the sea cucumbers lie on their side. The mouth is encircled by tentacles, which are attached to the water-vascular system. Five prominent rows of tube feet run the length of the body.

D. Class Crinoidea

Most sea lilies are sessile, being attached to the substratum by a long stalk. Unlike the Asteroidea, Ophiuroidea, and Echinoidea, the mouth is on the upper surface, where it is usually surrounded by long feathery arms.

QUESTIONS FOR FURTHER THOUGHT

1. Can you suggest some reasons why echinoderms are found only in the sea?

2. Where is the skeleton of an echinoderm? What function does the skeleton serve?

3. What evidence is there that echinoderms or echinodermlike animals gave rise to chordates?

4. What method of reproduction is characteristic of echinoderms?

5. Some organisms, like nitrogen-fixing bacteria in the soil and green plants in both aquatic and terrestrial habitats, make obvious and important contributions to the economy of their habitats. What role do you think echinoderms play in the economy of the ocean?

6. How would radial symmetry be adaptive to a sea star?

7. Why does a sea star need a complicated system of channels for communication between the tube feet and the exterior? Why not just have a "sealed" water-vascular system?

8. Design the simplest possible experiment to measure the force exerted by a sea star to open a clam shell.

9. How do you think a sea lily would feed?

TOPIC 28 ADAPTATION

Biological Science, pp. 598-607.
Elements of Biological Science, pp. 370-373.

From your previous studies throughout this course, two general facts should be especially clear. First, you know that all organisms face similar and interrelated problems in coping with their environments. They must have means of exchanging the raw materials and waste products of metabolism with their environment; they must transport such materials between body cells and the sites of exchange; they must have appropriate support, coordination, and effector systems; and they must, of course, reproduce. Secondly, you know that different modes of life demand quite different means of solving the basic problems; an organism well-suited to one sort of habitat may be quite ill-suited to another. Every organism has mechanisms for coping with the peculiar problems its environment poses; such mechanisms, encompassing structural, behavioral, physiological, and other characteristics, are called **adaptations**. Adaptations, of course, are not static. Environments change, and the sets of challenges that face their inhabitants are constantly undergoing modification. If the organisms are to survive, they too must change.

While individuals die, *populations* of organisms survive change through genetic variation, which, as we have repeatedly noted, has its origin in mutation and is extended and maintained by the recombination resulting from sexual reproduction. Should environmental change occur, variation within a population provides natural selection with an array of slightly different "problem-solving" mechanisms upon which to act. Natural selection effects more rapid reproduction of individuals better fit to cope with environmental challenges and thereby encourages continual adjustment and "tuning" of organisms to shifting environmental demands.

In this laboratory, you will be asked to bring your understanding of adaptation to bear on the study of an unfamiliar organism.

Your instructor will select the "unknown" organism from among those that you have not yet studied in this course. Since you will not have time in one laboratory period to study every aspect of the organism's biology, you should limit your study to some area that particularly interests you. If you have a plant, for example, you might wish to compare its gas-exchange, supportive, vascular (if present), or reproductive systems with those of other plants you have studied; you might wish to compare (using chromatography) its pigments with those of other plants you have studied. If you have an animal, you might study its feeding behavior, locomotion, or other behavior that interests you; you might make a detailed structural comparison of some organ system (e.g., digestive system) with the same system in other animals you have already studied, and then make hypotheses about your animal's diet. Whatever study you choose, try to focus on adaptations unique to the organism you are studying, and try to deduce as much as possible about environmental pressures to which the organism is subject.

Any techniques of study (dissection, microscopy, staining, hand-sectioning, etc.) available in your laboratory are open to you; your instructor may have specific techniques to suggest. Should you wish to undertake a project that involves materials not available in your laboratory room, make sure that your instructor knows of your plans and approves them.

Your instructor may provide references; use them liberally as *guides*, but do not depend upon the observations of others in making your report. Raise questions yourself, suggest answers to as many questions as you can, and be prepared to defend your hypotheses with concrete observations.

Turn in a report as outlined by your instructor. Detail your findings, and submit appropriate drawings.

TOPIC **29** | ECOLOGICAL FIELD INVESTIGATIONS

Biological Science, Chapter 18.
Elements of Biological Science,
Chapter 18.

Ecology is the study of the relationship of organisms or groups of organisms to their environment. An organism's environment includes not only physical factors such as light, temperature, rainfall, humidity, and topography, but also all living things that directly or indirectly influence the life of the organism, such as parasites, predators, mates, and competitors. Anything not an integral part of a particular organism is part of that organism's environment.

A group of organisms of the same species in its native situation constitutes a **population.** Likewise, all the populations in a given area constitute a **community.** The community and the physical environment function together as an ecological system, or **ecosystem.** The place where an organism would normally live, or the place where one would go to find it, is defined as its **habitat.** The ecological **niche,** on the other hand, is the functional role of the organism in the ecosystem. The niche of an organism depends not only on where it lives but also on what it takes from or gives to its environment and any interactions necessary for its survival. The best way for you to appreciate the meanings of these definitions is actually to go into the field and observe the interrelationships they imply.

In this unit you will study some aspects of the relationship of organisms to their environment and record your observations in such a manner that you can effectively relay information to another individual. The studies of an ecologist are not purely descriptive; in order to be as useful as possible, they must be quantitative and analytical. For this reason, you will make some environmental measurements. The methods outlined here are, for the most part, some of the simplest and least expensive available. The measurements will not be as quantitative as those of a professional ecologist, but if they are performed carefully and consistently, they will yield reasonably accurate results.

For your laboratory work in this unit, read over the following five studies and select one that particularly interests you. You will work in pairs, or as directed by your instructor. A few materials for collection may be provided by your instructor, but you will provide most of your own collecting equipment. A list of necessary supplies precedes each study. Your instructor will ask you either to submit a written report or to prepare a short seminar on your study.

Several topics that can be regarded as aspects of ecology have been observed throughout the year. Among others, you have studied osmotic interactions between organisms and their environmental media, the problems that plants and animals face in obtaining nutrients and exchanging gases with the environment, the behavioral responses of organisms to environmental stimuli, and the role of the environment in evolution. In your report, relate your ecological field work to some specific aspect of this course.

STUDY I. SAMPLING THE BIOTIC COMMUNITY

From a distance the landscape reveals many vegetative patterns: brushy fields, grasslands, forests, streams, deserts, ponds, etc. Each of these supports a distinctive flora and fauna. The brushy fields support rabbits, quail, meadowlarks, mice, and herbs. Grasslands provide natural pasture for grazing animals such as bison and jackrabbits, and food for burrowing mammals such as prairie dogs and gophers. Deer, bears, squirrels, bobcats, grouse, and many songbirds live in the deciduous forest. The streams support algae, fish, amphibians, and other aquatic organisms.

Although casual observation of communities reveals that organisms vary from place to place, closer inspection of any one community will reveal an interdependent unit. For example, in deciduous forest communities, leaves fall off and build up in layers; the soil organisms utilize and break down this organic matter into materials that can be reused by the plants. Many plants produce pulpy fruits and nuts, such as acorns and beechnuts. These and other plant materials provide food for primary consumers, like squirrels, turkey, deer, and other herbivores; such animals in turn replace nutrients in the soil and help to disseminate the seeds. Other animals, such as fox, bobcat, and woodpecker prey upon the primary consumers. If the forest trees were cut down, many of the squirrels, birds, and deer would disappear for lack of food and protection. If the predators were eliminated, some primary consumers would multiply until their food supply ran out. If the soil organisms were destroyed, the fallen leaves would not break down, the nutrient supply in the soil would be depleted, and the trees would eventually die. All the organisms of the forest depend not only upon the large trees but also, directly or indirectly, upon one another.

Organisms living in any given area, whether it is large or small, are associated together in what we call **biotic communities.** A community can be any size chosen for study, providing it is large enough to display species interdependence.

The purpose of this study is to discover for yourself how organisms within a community depend upon each other, and to observe the variation of organisms from place to place within a community.

A. Equipment

Meter stick
Right triangle
4 wooden stakes (sticks will do)
Hammer (rock will do)
Heavy cord
200-cc bottle of alcohol
Clipboard, pad, and pencil

B. Selecting a study area

Remember that a community is a chosen area for study and can be any size. Select an area of moderate size from one of the patterns of vegetation in the landscape: a natural field, a small forest or hedgerow, a swamp, a stream edge, etc.

C. Procedure: making quadrat sample plots	If you wish to describe a particular biotic community, you must determine what organisms occur in the community and analyze their interrelationships. This is done by sampling the community. You will take two samples (called quadrat plots) from your chosen community. Quadrats are rectangular sample units or plots of known size. You will measure and analyze two 1 x 2-meter plots; both quadrats are measured in the same way.

1. Select each quadrat at random. Before entering the field, design a method for *random* selection of sample plots. Be prepared to defend your procedure.

2. Stake and rope off a 1 x 2-meter quadrat. In order to make fair comparisons, it is important that the size of both sample quadrats is identical. Use the right triangle to make the corners.

3. Describe and record the physical characteristics of your quadrat (e.g., soil texture, moisture, degree of shading, etc.).

4. After careful observation, make an effort to determine how many different species of plants are growing within the quadrat. Assign each of these species to height categories of your own choosing (e.g., 1-3 inches, 4-5 inches, 1 foot, 1-12 feet, etc.). Place each species in the category that includes its average height. Designate each species with a letter. These labels will be helpful when you compare the vegetation in your quadrats. Remember, even a professional ecologist could not identify every species he encounters. Count and record the number of individuals of each species in the quadrat. If possible, collect one or two samples of each species for comparison with those of the next quadrat.

5. Use essentially the same procedure for animals but categorize them in recognizable taxonomic groups, such as ants, beetles, turtles, grasshoppers, frogs, etc. Turn over all logs, rocks, and other cover. Remember to carefully replace all turned cover. Make notes on activities of all animals. Collect representative specimens.

Repeat Steps 1-5 with the second quadrat.

6. Record your data in the form of a bar graph that shows the relative abundance of plants and animals in the two samples of your community. Prepare your graph so that measurements from one sample can be compared easily with the corresponding measurements for the other.

7. In analyzing the data from the two quadrats, compare the total number of species of plants and animals, and the number of species in each category. Determine the percentage of species in each category that is the same for both quadrats, and compare densities of individual species. Do some plant species fall in one height class in one quadrat and in a different height class in the other? Can any differences between quadrats be explained on the basis of physical features?

8. From observation of the whole area from which you selected samples estimate how uniform or variable physical and biotic environmental features are, and then suggest how reliable your quadrats are as a basis for characterizing the larger unit.

STUDY II. MAPPING THE TERRITORY OF AN ANIMAL

Animals tend to remain within a fairly limited area for long periods of time, sometimes for their entire lives. This region is called the **home range.** Some animals actually set up a **territory,** an area that they defend from intrusion by other individuals, particularly males of the same species. Several types of territories have been recognized. One type is an area within which mating, nesting, and feeding all occur during the breeding season. Another is an area for mating and nesting only. Still other territories are used for mating or for roosting or for feeding.

Whatever its type, territoriality seems to function in spacing the individuals in such a manner that the most severe aspects of competition and individual antagonism are minimized, and social stability is improved. Territoriality also plays a role in controlling the size of the breeding population. There is usually a minimum requisite territory size, below which successful reproduction will not occur; surplus animals unable to set up such a territory are unable to reproduce.

The purpose of this study is to gain some experience working in the field with an animal. Animals are hard to find, often hard to see, and are difficult to follow because they are not stationary. Overcoming these problems can itself be an interesting challenge.

A. Equipment

9 wooden stakes
Hammer
Binoculars
Compass
Thin board 18″ x 24″
Graph paper, ruler, tape
Clipboard, pad, and pencil
Topographical map (supplied by your instructor)
Protractor

B. Finding the study area

Any animal displaying territorial behavior can be used in your study, but because birds rigidly maintain boundaries during the spring, and often in other seasons, we suggest using them. The easiest place to find territorial birds is in overgrown fields with scattered young trees 15-30 feet high, particularly fruit trees. An ideal study area would be an orchard. Stay away from areas populated by human beings and from deep forest. One or two walks across a likely area should yield a satisfactory animal. The best time of day to find birds engaged in territorial behavior is early in the morning. Your instructor may suggest some particular species of bird available in your area.

Territorial boundaries of birds can be mapped by careful observation, by plotting singing perches, and by plotting locations of territorial boundary disputes. A territorial bird will usually protect its boundaries from intrusion by other birds of the same species. When chasing an intruder, some birds will cross their territorial boundaries and double back, while other bird species will not cross their boundaries.

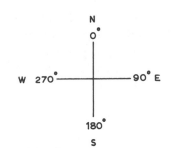

FIG. 29-1. Axes for preparing mapping graph.

C. Procedure   1. Once you have selected your study area, mark this area carefully on the topographical map supplied by your instructor. Be sure to indicate some permanent landmarks on the map. This information will enable someone else to locate and repeat your study, in order to verify it.

2. (*Note:* This step may be carried out before entering the field.) Tape four sheets of 8″ x 11″ graph paper together on the large board. Draw a line down the middle of the graph; draw another line across the middle, perpendicular to the first. Mark north at the top, south at the bottom, then east and west at the sides. Using north as 0, your protractor will give you any necessary readings around the map. Mark east 90°, south 180°, and west 270° (see FIGURE 29-1). This information will be important later when one partner calls his compass bearing from stakes driven into the study area, while the other plots with a protractor at the mapping station.

3. With stumps, rocks, or other materials available in the field, set up a temporary platform at some high point outside the study area, to serve as a table for your mapping board. Align the "north" indicator on your map with compass north. This area will be your mapping station.

4. Drive a stake at the arbitrary center of your study area. By watching the activities of the bird with your binoculars from a position *outside* the study area, observe the boundaries of the territory. Drive stakes at these boundary marks. Often, exact boundaries cannot be determined; they may depend, for example, on where trees are. Although most birds will chase an intruder beyond territorial boundaries before doubling back, they will not physically dispute an intruder until he reaches the territory. Continue your observations until you have driven a sufficient number of stakes to indicate the territorial boundaries.

5. You can measure distance by pacing, as long as you know the length of your average step or pace. Pace consistently. Once both partners agree on how many paces each square on the map represents, one of the two may step off the distance of each boundary stake from the arbitrary center stake and call this distance and the compass bearing to the other partner, who plots the information on the map.

6. Use the rest of your time making observations of this bird from outside the study area. Back off far enough so that you do not disturb the bird's normal activities; use the binoculars. Try to determine as many aspects of the animal's biology as possible. For example: What does the bird eat? Are there drinking places in the territory? What type of territory is this? Where is the next territory located? What type of nest does the bird have?

STUDY III. SELECTING INDICATOR PLANTS

Why are organisms restricted in their distribution? Why, for example, are orchids a prominent part of the flora only in tropical rain forests? A combination of environmental factors determines rather precisely what kinds of plants will be present; likewise, an environment is characterized by the presence of particular plants. The plant population of any environment reflects the amount of soil moisture, the soil chemistry, the climatic changes, the amount of light, humidity, etc. When the relationship of a particular plant species to its environment is well known, that plant becomes an **indicator** that can be used to deduce information about the environment.

This study will give you some field experience and will demonstrate the manner in which the environment influences the distribution of plants.

A. Equipment

    3 one-quart Mason jars with lids
    3 plastic bags
    2-foot pipe (½″-¾″ diameter), rod to fit as plunger
    Hammer
    3 index cards (3″ x 5″)
    Clipboard, pad, and pencil

B. Choosing the study area

Choose three different communities for study. Select one community with an abundance of water and moisture (**hydric** environment), another with very little moisture (**xeric** environment), and a third with an intermediate amount of moisture (**mesic** environment). Try to choose the study areas as close together as possible. Since water runs downhill, the rougher the terrain the better the chance of finding these different conditions close together. It is unnecessary to choose ideal hydric or xeric environments, but look for as great a difference as possible between the two you do choose. A stream side or spring in deep forest is an appropriate hydric environment, and an open high spot will probably be low in moisture. For the mesic environment choose an area between the other two.

C. Procedure

You will go to three different environments, observe the vegetation, take any necessary notes, notice any distinctive interrelationships, collect indicator herbs and soil samples, and compare the communities, using the data thus gained. Proceed as follows for each of the three selected communities.

1. Carefully observe the vegetation of the chosen area. Decide which plant of the community is best suited to serve as an indicator for the integrated effects of the environmental factors important in the habitat. Be prepared to defend your decision. Remember the chosen plant does not have to be the dominant species in the community, but certainly should be one that you have not seen outside this habitat. Collect a sample of this plant.

2. Put a mark on the pipe, one foot from the bottom, and drive it into the ground to this mark. Carefully pull the pipe out, and use the plunger to push the soil core into one of the collecting jars. Label each jar. If you can't drive the pipe into the ground anywhere near the indicator specimen, you can assume that the rest of the substrate is rock.

3. Estimate the amount of sunlight received per day (as compared with full sunlight) by each indicator. Be consistent, and be prepared to explain the method you used to make your estimation.

4. Estimate the size and shape of the area that the indicator could possibly inhabit. With what features of the terrain is the habitat correlated?

5. Observe and record any interrelationships between the indicator plant and other organisms in the environment. Would the indicator survive without these relationships (e.g., without the overhead vegetation)?

6. Take the soil cores back to your room and carry out the following procedure with each.

    a. Add water to a jar of soil, up to the shoulder.

    b. Place the lid on it and shake the jar for three minutes.

    c. Let the soil settle out overnight. The heavier particles will settle on the bottom, followed by layers of lighter soil (i.e., gravel, coarse sand, fine sand, silt, and clay). Hold a 3" x 5" index card up to the jars and draw a diagram showing the different layers. Label each layer.

    d. Compare the three environments in terms of the relative amounts of the different types of soil particles.

7. Take the indicator plants back to your laboratory for closer examination. Combine microscopy and your knowledge of plant structure to compare these plants. What differences in structure can you notice between the indicator plants that reflect conditions of their habitats?

8. (Optional) Attempt to grow the three indicator plants in terraria under conditions like those in which you found them. What are your results?

9. Prepare a list of environmental factors necessary for the survival of each of the collected indicators. From your observations and measurements, how indicative of certain environmental conditions would an indicator be? What value would plant indicators have to an ecologist?

From your field observations, what reason can you offer to explain the diversity of organisms in the plant kingdom and their distribution from place to place?

STUDY IV.
STUDYING
THE HYDRIC
ENVIRONMENT

In ecological studies, the field becomes a laboratory. In becoming acquainted with field study, attention should be focused on close examination of particular organisms in their specific habitat. Thus, isopods such as sow bugs inhabit a deciduous forest, and it is to this habitat that you would go to collect them. However, once you are in this habitat, you are not surprised to find isopods restricted to particular areas, such as under rocks and logs. Even within a given community the distribution of certain organisms may be sharply localized by small variations in moisture, light, and other conditions. A **microhabitat** is that special area within a major habitat (e.g., forest) that satisfies the living requirements of a given organism.

Any single factor can have a strong effect upon where organisms live in a community. Within a hydric environment, graded changes in moisture content are reflected by graded changes in the vegetation. As you recall from your previous studies, plants have various adaptations for obtaining and storing water; plants such as liverworts and mosses require higher percentages of water and moisture than plants with vascular systems such as lycopods and ferns.

This study will give you some experience in finding microhabitats, and will demonstrate how influential one factor can be in affecting distribution of plants within a community.

A. Equipment

2 plastic bags
Centimeter ruler
Watch
Cotton gauze, tape
Clipboard, pad, and pencil
Light meter (optional)
Two thermometers
Container of distilled water

B. Finding the microhabitats

The following five plant groups exist on the water gradient of a hydric environment. Review your understanding of their structure with reference to their need for water. The nonvascular plants require abundant water, while the vascular ones can survive with less water.

1. *Terrestrial algae* are frequently found clinging to rocks and tree trunks where there is a high percentage of moisture.

2. *Liverworts* are common around brook beds, near small waterfalls.

3. *Moss* is abundant and easily located in any hydric community.

4. *Lycopods* (FIGURE 29-2) can be found in moist areas protected from the sun.

5. *Ferns* are abundant, and most species have wider limits of tolerance than the other plants listed. They can usually be found throughout a terrestrial hydric community.

FIG. 29-2. Two common types of *Lycopodium*. (A: *Courtesy E. M. Raffensperger, Cornell University. B: Courtesy Carolina Biological Supply Company.*)

C. Procedure: measuring the environmental factors

The objectives of this study are to seek out the microhabitat of some known plant types and to measure these environments for amount of sunlight, relative humidity, and competition for land; and then to compare these environments with each other in order to gain a clear understanding of some of the reasons for species diversity.

Remember when conducting your study that each plant group, such as mosses or ferns, is made up of several species, each requiring a different interplay of factors and each with different limits of tolerance of these factors. Select for the study one typical species from each group, and make measurements. Proceed as follows:

1. Go into the hydric environment and find an example of each of the five plant types. Collect a specimen of each, place it gently into the plastic bag to be taken back to the laboratory. How large is the stand of each specimen that you have selected? How abundant is each selected type in the community?

2. Using a photographic light meter, compare the amount of light in the microhabitat of each of the specimens. How do these measurements compare with each other? How do they compare with the amount of light in an open field? If a light meter is not available, estimate the amount of light each specimen receives. Be consistent, and be prepared to explain your method of estimation.

3. Measure in millimeters the minimum amount of soil required for survival of each selected specimen. Remember that the plant may use more than the minimal amount of soil when competing plants are absent. Estimate the minimum amount of soil when there is too little to measure.

4. Measure the relative humidity at each collection point in the following manner:

> a. Cover the bulb of a thermometer with two layers of gauze and tape the gauze securely to the body of a thermometer.
>
> b. Soak the gauze in distilled water; fan it with the clipboard for three minutes directly before reading the temperature. Read this and a second thermometer at the same time, after the temperatures have remained constant (4-6 minutes). Record wet- and dry-bulb readings. Make all your recordings at 15 cm from the ground, and always shield the thermometer from the sun.
>
> c. Convert these two readings to relative humidity, using Table 1.
>
> d. When leaving the woods, measure the relative humidity of an open, uninhabited area.
>
> Relative humidity is a measure of the moistness of an environment. It is the ratio between the water vapor actually in the air at a given temperature and the maximum amount of water vapor the air could hold at the same temperature.

5. In analyzing your data and observations, what evidence can you give for variation of organisms from place to place within the community? What structural and reproductive features adapt these plants to a terrestrial environment? Is there any correlation between the height of an organism and its place on the soil moisture gradient? From your observations have you

noticed that not only too little but too much of one factor can exclude some organism? Why do some plants, such as ferns, exist throughout the community? According to your data, what are the most important environmental factors contributing to the existence of each of the species collected? Is there any correlation between the amount of water in the soil and the relative humidity where you measured it? What factors would contribute to differences in relative humidity in the environments studied? Why is relative humidity so important to the survival of some plants?

TABLE 1.

Percent relative humidity table, based on centigrade temperatures.

To find relative humidity, take dry-bulb and wet-bulb temperatures as described in text; then find intersection of dry-bulb temperature (left, vertical column) and differences between wet-bulb and dry-bulb temperatures (top, horizontal columns) and read off relative humidity in percent. (Taken from "Relative Humidity — Psychrometric Table Celsius Temperature," Revised Edition. Courtesy Technical Investigations Section, U.S. Department of Commerce, Weather Bureau, Washington, D.C. 1953.)

(Pressure = 29.24 in.)

DEPRESSION OF WET-BULB THERMOMETER (DRY MINUS WET)

DRY BULB	.5	1.0	1.5	2.0	2.5	3.0	3.5	4.0	4.5	5.0	5.5	6.0	6.5	7.0	7.5	8.0	8.5	9.0	9.5	10.0
0	91	81	73	64	55	46	38	29	21	13	5									
1	92	83	75	66	58	49	41	33	25	17	10									
2	92	84	76	68	60	52	44	37	29	22	14	7								
3	92	84	77	70	62	55	47	40	33	26	19	12	5							
4	93	85	78	71	64	57	50	43	36	29	22	16	9							
5	93	86	79	72	65	58	52	45	39	33	26	20	13	7						
6	93	86	80	73	67	60	54	48	41	35	29	24	17	11	5					
7	93	87	80	74	68	62	56	50	44	38	32	26	21	15	10					
8	94	87	81	75	69	63	57	51	46	40	35	29	24	19	14	8				
9	94	88	82	76	70	64	59	53	48	42	37	32	27	22	17	12	7			
10	94	88	82	77	71	66	60	55	50	44	39	34	29	24	20	15	10	6		
11	94	89	83	78	72	67	61	56	51	46	41	36	32	27	22	18	13	9	5	
12	94	89	84	78	73	68	63	58	53	48	43	39	34	29	25	21	16	12	8	
13	95	89	84	79	74	69	64	59	54	50	45	41	36	32	28	23	19	15	11	7
14	95	90	85	79	75	70	65	60	56	51	47	42	38	34	30	26	22	18	14	10
15	95	90	85	80	75	71	66	61	57	53	48	44	40	36	32	27	24	20	16	13
16	95	90	85	81	76	71	67	63	58	54	50	46	42	38	34	30	26	23	19	15
17	95	90	86	81	76	72	68	64	60	55	51	47	43	40	36	32	28	25	21	18
18	95	91	86	82	77	73	69	65	61	57	53	49	45	41	38	34	30	27	23	20
19	95	91	87	82	78	74	70	65	62	58	54	50	46	43	39	36	32	29	26	22
20	96	91	87	83	78	74	70	66	63	59	55	51	48	44	41	37	34	31	28	24
21	96	91	87	83	79	75	71	67	64	60	56	53	49	46	42	39	36	32	29	26
22	96	92	87	83	80	76	72	68	64	61	57	54	50	47	44	40	37	34	31	28
23	96	92	88	84	80	76	72	69	65	62	58	55	52	48	45	42	39	36	33	30
24	96	92	88	84	80	77	73	69	66	62	59	56	53	49	46	43	40	37	34	31
25	96	92	88	84	81	77	74	70	67	63	60	57	54	50	47	44	41	39	36	33
26	96	92	88	85	81	78	74	71	67	64	61	58	54	51	49	46	43	40	37	34
27	96	92	89	85	82	78	75	71	68	65	62	58	56	52	50	47	44	41	38	36
28	96	93	89	85	82	78	75	72	69	65	62	59	56	53	51	48	45	42	40	37
29	96	93	89	86	82	79	76	72	69	66	63	60	57	54	52	49	46	43	41	38
30	96	93	89	86	83	79	76	73	70	67	64	61	58	55	52	50	47	44	42	39
31	96	93	90	86	83	80	77	73	70	67	64	61	59	56	53	51	48	45	43	40
32	96	93	90	86	83	80	77	74	71	68	65	62	60	57	54	51	49	46	44	41
33	97	93	90	87	83	80	77	74	71	68	66	63	60	57	55	52	50	47	45	42
34	97	93	90	87	84	81	78	75	72	69	66	63	61	58	56	53	51	48	46	43
35	97	94	80	87	84	81	78	75	72	69	67	64	61	59	56	54	51	49	47	44
36	97	94	90	87	83	81	78	75	73	70	67	64	62	59	57	54	52	50	48	45
37	97	94	91	87	84	82	79	76	73	70	68	65	63	60	58	55	53	51	48	46
38	97	94	91	88	84	82	79	76	74	71	68	66	63	61	58	56	54	51	49	47
39	97	94	91	88	85	82	79	77	74	71	69	66	64	61	59	57	54	52	50	48

**STUDY V.
COLLECTING
INSECTS FROM
ONE PLANT**

In the study of communities, there is usually a definite correlation between the size of the community and the balance or independence of that community. *Major* communities are those of sufficient size and completeness of organization to make them nearly independent of adjoining communities. In this type of community, some fluctuation in the organisms present usually does not affect its overall balance. *Minor* communities, on the other hand, are more or less dependent on neighboring communities. In minor communities there is a constant fluctuation of organisms in and out. These visitors have a strong effect on the stability of that community. In this unit you will have an opportunity to study a minor community and to observe the fluctuation of organisms in it.

A. Equipment

Large plastic bag, rubber band
3-inch funnel
Magnifying glass
Scissors
Clipboard, pad, and pencil
100-cc jar of alcohol
Small vial of ether, cotton

**B. Selecting the
study area**

Choose a stand of dominant deciduous herbs in almost any type of field. The instructor will suggest plants available in your vicinity. Carefully check the chosen plant to be sure it has a large population of insects; examine the inside of flowers, underside of leaves, and elsewhere on the plant where insects might be found. Some herbs are quite populated with insects, while other herbs seem to be repellent.

**C. Procedure:
collecting the insects**

In this study you will observe species diversity in a small community, classify these specimens by means of a key, and observe organisms entering and leaving the community.

1. Once you have located an appropriate plant, make very careful observations of the insect population with your magnifying glass. Can you determine what each species is eating (e.g., which ones are herbivores, and which are carnivores)? What are the ecological interrelationships between the insect species? Which insects are harmful to the plant, and which are beneficial? Record as much information of this kind as possible. At this point in your study, refer to each species of insect as A, B, C, etc.

2. Leave the community, without disturbing it in any way. Find a comfortable site close enough for observation but far enough away so that you don't prevent flying fauna from visiting the plant. Spend your time watching and recording the movement of animals into and out of the community. What are the visitors doing? Do they have any effect on the community population count? Are any of the visitors beneficial to the plant? Explain all these observations.

3. When you have obtained considerable information from your observations, carefully place a plastic bag over the plant so as to enclose all

the insects. Cut the plant and place a rubber band around the end of the plastic bag.

4. Take this bag containing the community back to your room or to the laboratory. Place an ether-filled wad of cotton into the bag. Once the insects have been anesthetized and shaken from the plant, funnel them into the container of alcohol.

5. Count and record the number of individuals of each species on the plant.

6. Using your data, explain how dependent this community is upon surrounding communities. Do you think this community could exist independent of adjoining aggregations? Explain. Tie together the observed interrelationships; which insects are primary consumers, and which are secondary consumers?

7. Run each species through the following insect-order key, after reading the directions below.

A key provides an orderly and systematic method for identifying organisms in a given group. It is usually constructed on the basis of easily recognizable structural characteristics by a specialist on the group in question. To identify a given organism, one proceeds through a series of choices, starting at the beginning of the key and working forward toward more and more specific identifying features. For example, if the insect you are identifying has wings, you would proceed to choice 16. If it has two pairs of wings you next go to choice 17, etc. If you have trouble interpreting the meaning of certain choices, ask your instructor for help; error in interpretation is an important source of error in identification. Observe carefully; errors in observation will also throw you off the correct sequence.

DICHOTOMOUS KEY TO THE ORDERS OF SOME COMMON ADULT INSECTS*

(Common names are in parentheses)

1a.	Wingless	go to 2
1b.	Winged	go to 16
2a.	"Spring" present on fourth segment of abdomen, held by a catch on the ventral side; antennae shorter than body and usually bent like an elbow near the middle (springtails)	Order Collembola
2b.	"Spring" absent	go to 3
3a.	Antennae absent	Order Protura
3b.	Antennae present (may be difficult to see)	go to 4
4a.	Antennae shorter than, or about as long as, head	go to 5
4b.	Antennae clearly longer than head	go to 7
5a.	Body compressed (flattened from side to side); antennae about as long as head (fleas)	Order Siphonaptera
5b.	Body not compressed; antennae much shorter than head	go to 6
6a.	Sucking mouthparts (sucking lice)	Order Anoplura
6b.	Biting mouthparts (biting lice)	Order Mallophaga

* Taken from *Student's Manual — Laboratory and Field Investigations, BSCS Green Version* (Rand McNally & Company, Chicago, 1963), and used by permission of the Biological Sciences Curriculum Study.

7a.	Abdomen with tubular, pincherlike, or threadlike extension behind	go to 8
7b.	Abdomen without extensions behind	go to 11
8a.	Abdomen with a pair of short tubes; mouthparts in form of a tube, for sucking (plant lice)	Order Homoptera
8b.	Abdomen without tubes; mouthparts for biting or chewing (may be hidden)	go to 9
9a.	Abdomen with 3 filamentous (threadlike) appendages as long as the body; body covered with scales (silverfish and firebrats)	Order Thysanura
9b.	Abdomen with pincherlike structures or no more than 2 filamentous appendages; body not covered with scales	go to 10
10a.	Eyes absent	Order Entotrophi
10b.	Eyes present and conspicuous (earwigs)	Order Dermaptera
11a.	Mouthparts tubular, for sucking	go to 12
11b.	Mouthparts not tubular, for biting and chewing	go to 14
12a.	Body covered with scales; mouth in form of a coiled tube (some female moths)	Order Lepidoptera
12b.	Body not covered with scales; mouth in form of a straight tube	go to 13
13a.	Sucking tube long, straight, and beaklike; body flattened from top to bottom; tips of feet with claws (water striders, bat bugs, bedbugs)	Order Hemiptera
13b.	Sucking tube short and conical; body not flattened from top to bottom; tips of feet with pads (thrips)	Order Thysanoptera
14a.	Antennae and legs long and slender; body twiglike (stick insects)	Order Orthoptera
14b.	Antennae not unusually long and slender, body not twiglike	go to 15
15a.	Abdomen broadly joined to thorax (termites)	Order Isoptera
15b.	Abdomen constricted and with a beadlike enlargement at connection to thorax (ants)	Order Hymenoptera
16a.	Wings, 1 pair (flies)	Order Diptera
16b.	Wings, 2 pairs	go to 17
17a.	Front wings and hind wings similar in texture	go to 18
17b.	Front wings and hind wings not alike in texture	go to 27
18a.	Wings long and narrow, with a fringe of fine hair; tips of feet with pads (thrips)	Order Thysanoptera
18b.	Wings without fringe of fine hair; tips of feet with claws	go to 19
19a.	Wings and body covered with scales; mouthparts in form of a coiled sucking tube (butterflies and moths)	Order Lepidoptera
19b.	Wings and body not covered with scales; mouthparts not in form of a coiled tube	go to 20
20a.	Front wings distinctly narrower than hind wings; abdomen may have 2 short, prominent filaments (stoneflies)	Order Plecoptera
20b.	Front wings larger than, the same size as, or only slightly narrower than hind wings; filaments usually absent (if present, front wings much larger than hind wings)	go to 21
21a.	Front wings and hind wings similar size and shape, or front wings much larger	go to 22

21b.	Front wings longer and wider than hind wings	go to 25
22a.	Wings with a few longitudinal veins but no crossveins; body usually only 2 to 4 mm in length (termites)	Order Isoptera
22b.	Wings with crossveins; body usually more than 4 mm in length	go to 23
23a.	Wings with many crossveins and longitudinal veins, forming a network; front wings may be slightly narrower than hind wings; eyes very large (at least half of head area) (dragonflies and damselflies)	Order Odonata
23b.	Wings with veins not forming a network; front wings about the same size as hind wings; eyes small (less than half of head area)	go to 24
24a.	Wings much longer than body, smooth, with numerous crossveins, and held at an angle (rooflike) over body when at rest (dobsonflies, lacewings, snakeflies)	Order Neuroptera
24b.	Wings not much longer than body, smooth, with hairs or patches of scales, many longitudinal veins but few crossveins, and held flat on back when at rest (caddisflies)	Order Trichoptera
25a.	Abdomen with 2 or 3 filaments longer than body; hind wings about one-third the size of front wings (mayflies)	Order Ephemeroptera
25b.	Abdomen without filaments; hind wings at least half as large as front wings	go to 26
26a.	Wings held at an angle (rooflike) over body when at rest; mouthparts absent or form a jointed sucking tube (cicadas, leafhoppers, aphids, etc.)	Order Homoptera
26b.	Wings held flat on back when at rest; mouthparts for biting, chewing, lapping, or sucking; abdomen often has a stinger (ants, bees, wasps)	Order Hymenoptera
27a.	Front wings leathery at base, membranous at tip; mouthparts in the form of a long sucking tube ("true" bugs)	Order Hemiptera
27b.	Front wings leathery or parchmentlike throughout; mouthparts for biting or chewing	go to 28
28a.	Abdomen with 2 pincherlike or forcepslike extensions behind; wings short, not covering all of abdomen (earwigs)	Order Dermaptera
28b.	Abdomen without pincherlike extensions behind; wings usually cover all of abdomen	go to 29
29a.	Front wings with veins, long, narrow, and parchmentlike; hind wings broad, often fan-shaped (grasshoppers, katydids, crickets, cockroaches, mantises, etc.)	Order Orthoptera
29b.	Front wings veinless, usually not parchmentlike (if long and narrow, hind wings also long and narrow)	go to 30
30a.	Front wings and hind wings broad; front wings meet in center of back, forming hard, veinless cover for hind wings (beetles)	Order Coleoptera
30b.	Front wings and hind wings long and narrow; front wings do not form hard cover for hind wings; tip of abdomen may be curved forward over back (scorpionflies)	Order Mecoptera

TOPIC 30 MICROBIAL ECOLOGY

Biological Science, pp. 640-642, 711-718.
Elements of Biological Science, pp. 394-397, 452-457.

by Steven C. Carlin

Although many people associate bacteria solely with disease, you, as students of biology, doubtless know that the pathogenic bacteria are no more representative of the bacterial community than are the criminals of the human community. In both cases, the villains are so labeled because they are parasites, harming other organisms or individuals in providing for their own well-being. Most bacteria, however, are saprophytes, feeding on dead or decaying matter. These and other bacteria are crucial in the cycling of raw materials between living and nonliving systems; without them, the building blocks of protoplasm would remain permanently bound in individual plants and animals after their death, and would then not be available for the creation of successive new generations of organisms. How an organism the size of a bacterium — roughly the same size as a chloroplast or mitochondrion of higher organisms — can make such important contributions to the cycling of raw materials may be difficult to comprehend until you realize that it is just their small size and simplicity that gives bacteria the enormous metabolic and reproductive potential such cycling requires. The small size of bacteria makes their surface-to-volume ratio very large; the higher this ratio, the faster an organism can exchange materials with its environment and the greater its potential metabolic rate. The simplicity of bacteria allows them to reproduce at an enormous rate, while permitting simultaneous specialization and adaptability. Thus, what bacteria lack in mass they more than make up by their activity, numbers, and versatility.

The small size of bacteria, however, makes it very difficult to study them individually. Higher organisms usually can be classified on the basis of structure, but as seen in FIGURE 30-1 many bacteria are simply single cells shaped like rods (bacilli), dots (cocci), or helical coils (spirilla), although

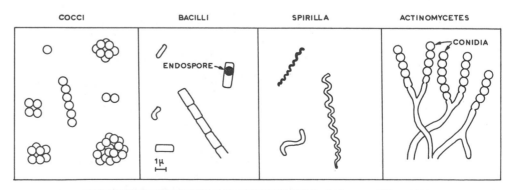

FIG. 30-1. Some of the common morphological classes of bacteria.

others such as the mycelial bacteria (actinomycetes) are more complex. Therefore, what distinguishes one bacterial species from another is often physiological rather than morphological characteristics. In studying the chemical changes brought about by bacteria, it is necessary to measure the activity of an entire population of cells, since each individual cell makes a contribution too small to detect. In consequence, much of what is known about bacteria has been discovered by isolating them from nature and studying the changes they bring about when grown in pure culture under laboratory conditions.

To isolate a specific organism from a mixture of different species such as would be found in soil, water, or air, a culture solution (medium) that favors the growth of the desired organism over all its natural competitors is devised on the basis of some known physiological peculiarity of the desired organism. This so-called **enrichment-culture technique** is outlined and described in FIGURE 30-2. If one wishes to isolate a given bacterial species, free from contaminants, this technique must be used in conjunction with one of the **pure-culture methods** outlined in Appendix 7. Studying microorganisms in pure culture, however, is somewhat artificial because it neglects the interactions that normally occur among the **populations,** groups of individuals of the same species, that make up the **biotic community** of a given area.

Relationships between populations of any two species in a microbial community may have positive, negative, or neutral survival value for the populations involved. The effects of beneficial or harmful relationships may, in turn, be either mutual or unilateral. Associations between microorganisms usually involve food, but they may also involve shelter or transport; they may be continuous or transitory, obligate (mandatory) or facultative (optional), and may involve either direct or indirect contact. It is not sufficient, however, to consider any given interspecific interaction in isolation, since the two interacting populations are each subject to many other external influences. Each population is affected by both its physical surroundings (**abiotic environment**) and its associations with the total collection of populations in the community (**biotic environment**). Together, all the abiotic and biotic factors that impinge upon a given individual constitute its environment; how organisms interact with their environment is the subject matter of the science called **ecology.**

The species composition of a biotic community and the density of any given population within that community is a direct reflection of the environment. The density of any bacterial population continues to increase as long as the environmental conditions for its growth remain favorable, but even in the absence of other organisms this density will soon reach a maximum value and then level off due to such factors as crowding and the accumulation of toxic waste products. The presence of other organisms influences the value of this steady state population density; there is, in fact, a constant interplay between the various populations in a community. So long as no external physical influences or internal evolutionary influences disturb this interplay, it will lead to a more or less stable community composition in which each bacterial population has a certain steady state density; this *stabilization* of the community composition by the balancing of the reciprocal influence of populations upon one another and upon the physical surroundings is called the **biotic equilibrium.**

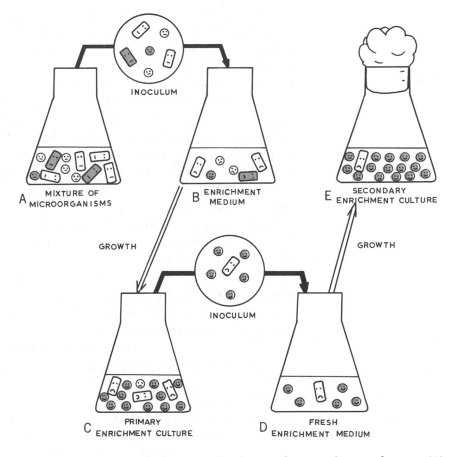

FIG. 30-2. Isolation of bacteria by the enrichment-culture technique. (A) A suspension containing a mixture of microorganisms is obtained from a natural source such as soil or polluted water. (B) A drop of this mixed suspension is used as an inoculum for a flask containing an enrichment culture solution (medium). The composition of the medium and the conditions under which it is incubated are chosen to make the environment favorable for the growth of the desired organism, but unfavorable for its natural competitors; only the desired organism is "happy" in this artificial environment. Note the animated expressions of cells. (C) The organisms are allowed to grow. Since the desired organism can grow much faster than any of its competitors under these conditions, its relative abundance in the population is increased. (D) A drop of this primary enrichment culture is now used to inoculate a flask of fresh enrichment medium. This dilutes out any contaminating organisms and nutrients capable of supporting the growth of microorganism contaminants; these nutrients may have been introduced with the sample or produced by the desired organisms itself. (E) The growth of this inoculum leads to a secondary enrichment culture that contains a great preponderance of the desired organisms, even if that organism only made up a small fraction of the original sample.

Whether or not a species is present in a community at equilibrium depends upon the ability of the species to fit into the ecosystem defined by that community and its physical surroundings. Whenever a species is present in a particular ecosystem, it assumes a certain functional role and position in that ecosystem; this **ecological niche** encompasses all aspects of the organism's existence. If the environment fulfills the conditions necessary for the establishment of a bacterial species' niche and if there are no barriers preventing the species' dispersal, then this species can usually be assumed to be present. Conversely, the absence of a species from a given environment usually means either that the environment is unfavorable or that there is a barrier to dispersal. Since attempts to introduce a desired microorganism into a habitat where it is absent usually result in failure, it is clear that unfavorable environmental conditions are a more important factor than barriers to dispersal in limiting microbial distribution; the latter factor, however, is quite important in limiting the distribution of higher organisms. Therefore, only if the unfavorable environmental conditions are corrected can the presence of that microorganism be assured. Of course, changing the environment to conform to the requirements of one organism's niche disrupts the delicate balance of biotic and abiotic factors responsible for maintaining the biotic equilibrium and leads to a series of changes in the composition of the community until a new equilibrium is established. Especially in the case of microorganisms, where some populations have the potential of doubling every twenty minutes, and where organisms are constantly adapting to survive in previously unfavorable environments, biotic communities are constantly changing, dynamic entities.

The response of a community to slight environmental changes is often marked; the sequence of changes initiated when an unwary farmer ploughs under a nitrogen-poor crop residue, such as straw, provides a particularly graphic example. Food is the main factor limiting growth of the soil microflora; when straw is added, there is a tremendous upsurge in microbial numbers in response to this new food source. Only certain organisms, however, can decompose cellulose, a major constituent of straw. These *cellulolytic* organisms usually make up only a small part of the microflora; their slow growth rate does not allow them to compete effectively with other, more active organisms. Nevertheless, the addition of straw to the soil gives them a temporary competitive advantage, and their numbers increase markedly. Soon a secondary population of organisms begins to develop at the expense of cellulose breakdown products liberated by these organisms. At first, this secondary population benefits the cellulose decomposers by continuously removing inhibitory metabolic wastes, but eventually the secondary population begins to compete with the primary one for limiting nutrients. There results a **succession** in the microbial community: the secondary population becomes dominant while the primary cellulolytic organisms decrease in numbers. A series of such successions will occur until the original biotic equilibrium is restored. What our unsuspecting farmer did not realize was the net effect of all these changes: the *immobilization* of the soil's plant-available nitrogen by its conversion into microbial protoplasm. In a soil containing nitrogen adequate for plant growth, the microflora usually is limited by the supply of carbon; the addition of carbon in the form of straw, however, will result in the proliferation of the microflora until the supply of nitrogen becomes limiting. This, of course, results in a marked decrease in soil fertility. The wise farmer, therefore, burns his straw stubble. (If you saw a farmer turning under the remains of a crop of legumes [pod-bearing plants that harbor symbiotic nitrogen-fixing bacteria in root nodules], would you be justified in telling him that he should consider burning the crop remains? Explain your answer.) From this example, you can see clearly that environmental change is the thread that weaves the pattern of community change.

In this laboratory, you will learn how to work with bacteria, and by studying a simulated "epidemic" you will see how easy it is for bacteria to spread to any place where the conditions are favorable for their growth; you will observe microorganisms in their natural environment by the "buried slide" technique and will observe a natural succession; finally, you will use the enrichment-culture technique to isolate nitrogen-fixing bacteria from the soil, and will study various interactions of bacterial populations in the model ecosystem of a mixed laboratory culture.

PART I. MICROBIAL DISPERSAL: AN EPIDEMIC

By observing the mechanics of the development of an epidemic in a human population, you will demonstrate the ease with which bacteria are dispersed in nature. The "epidemic" will be simulated by the spread of a bacterial population among the members of your laboratory class. The organism to be used in this demonstration is naturally not a pathogen, but one that may easily be detected in culture because of its pigment production.

A. Handshake-mediated dispersal

Each of you will be given a wetted gumdrop in a petri dish. Your instructor has wetted one of the gumdrops with a suspension of the red-pigmented bacterium *Serratia marcescens*, and has wetted the others with plain water; you will not be told which person received the "infected" gumdrop. The sugary solution from the partially dissolved candy will serve as a vehicle for passing the infecting organism from individual to individual.

1. Before starting the experiment, thoroughly wash your hands with soap and water and obtain two sterile cotton swabs in a cotton-plugged tube, and a nutrient-agar plate. Label the top half of the plate with your name and seat number; then turn the plate over and divide the bottom half into two numbered segments.

2. Handle your gumdrop in order to make your hands as sticky as possible, since the method for passing the organism through the group will be handshaking.

3. Now, under the instructor's supervision, each person in turn will shake hands with some other person selected at random from the other members of the class. To keep the proceedings orderly, each person will get his turn at choosing a handshaking partner, according to his number in the seating chart. When it is your turn, you may choose any other person in the class as a handshaking partner, even if that person has already shaken hands with someone else. Your instructor will record all contacts between individuals as they occur, so that the exact sequence of contacts will be known when you try to trace the source of the infection and the course of its spread through the group.

4. When you return to your seat, plate out the material on your contact hand by stroking the contact area with a sterile swab and then stroking the swab over that half of the nutrient-agar plate previously labeled "1." Be sure *not* to wipe or wash the remaining sticky material off your hands until the entire experiment has been completed.

5. When everyone in the class has had his turn at choosing a handshaking partner, repeat the *same* series of handshaking operations.

6. When you return to your seat, transfer the bacteria from your hand to the second half of the nutrient-agar plate, making sure that you use a new sterile swab. Do not wash your hands until the second handshaking round has been completed.

B. Tracing the epidemic

At the next laboratory session, plate readings for the entire class will be recorded. From these data, determine the culprit responsible for the epidemic and trace the course of its spread. Remember that other bacteria may be present on your hands, but that only the infecting organism will produce red colonies. Compare the extent of the epidemic for the two rounds of handshaking. If you changed the experimental protocol, swabbing your hands onto non-nutritive agar and washing your hands with a disinfectant at the end of the handshaking rounds, would the handshake-mediated dispersal of *Serratia marcescens* still lead to its effective colonization of new habitats, or would it die out because the requirements for its niche had not been fulfilled? If your instructor liked yellow better than red and used the golden-colored organism *Staphylococcus aureus* without realizing that this organism is the causative agent of boils (localized swellings of the skin caused by infection of the sebaceous glands that empty into the hair follicles), would every student whose plates showed yellow colonies after the second round of handshaking come down with boils on his hands? Explain your answer. Do you think that the number of students getting boils would be different if everyone in the class had been overworking himself the week of the handshake experiment?

PART II. MICROBIAL HABITAT: THE SOIL MICRO-ENVIRONMENT

The purpose of this study is to give you an appreciation of the physical nature of the soil microenvironment and of the relationship between the microbial community and the inert soil particles, microscopic pieces of organic matter, and water films that make up this microbial habitat. You will also have an opportunity to observe changes in the microflora when straw is added to the soil. So that you can actually look into the soil microenvironment, you will bury microscope slides in a container of soil, removing and staining them after periods of one and two weeks. This will enable you to observe microscopically the microorganisms and soil particles adhering to the slides.

A. Burying slides in soil

You will be given two containers filled with dry soil. Your instructor may give different types of soil (e.g., sandy, loamy, or clayey) to different students and, if so, note the type of soil you have received. Mix about 1.2 grams of finely pulverized straw per 100 grams of soil in one of the containers provided. (If balances are not available, your instructor may provide you with a weighed packet of straw and soil.) Place four clean microscope slides vertically into each of the containers, and add water until the moisture content of the soil is about 15 percent (i.e., about 15 ml water per 100 grams soil). Put these containers in your drawer until the next laboratory period.

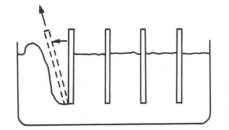

Fig. 30-3. Method of removing "buried slide" without disturbing the "soil print" on one side. See text for full description.

B. Examining "soil print" on slides

1. After one week, remove two slides from each container as follows. Carefully dig the soil away from one side of each slide (but try not to disturb the others), tilt the slide toward the excavation, and lift it out, as shown in FIGURE 30-3. Done correctly, this will leave a "soil print," consisting of a film of soil particles and microorganisms, on the side of the slide opposite the excavation. Place the containers back in your drawer, making sure to add more water if the soil is drying out.

2. After labeling the slides with a wax pencil, fix them by passing them once or twice through a flame; be sure not to overheat them.

3. Stain one slide from each container with erythrosine (15 minutes) and one with gentian violet (1-2 minutes).

4. Examine the slides with low and high power of a compound microscope and make representative drawings. What types of organisms can you see? Can you find representatives of the various morphological classes of bacteria (FIGURE 30-1) in the soil community? Can you distinguish thick fungal mycelia (Topic 18) from thin actinomycete filaments (FIGURE 30-1)? What is the physical relationship of the microorganisms to the inert and organic soil particles? Do the two stains reveal different types of organisms? Are there any differences in numbers or types of organisms present in soil with and without straw?

5. If several different types of soil were available, compare your results with those of students who had different types of soil. Does the microflora of different types of soil seem to be different? If so, is the difference only quantitative, or is it also qualitative? Was the effect of straw the same for the various types of soil?

6. At the end of two weeks, remove the remaining slides and follow the same procedure for staining and observing. Has the microflora in the straw-treated soil changed in composition or in numbers? If so, can you account for these changes? Has the microflora in the untreated soil changed significantly? If not, why do you think this is so?

PART III. MICROBIAL ISOLATION: AN ENRICHMENT CULTURE

In order to demonstrate the enrichment-culture technique, you will isolate nitrogen-fixing bacteria from the soil; before proceeding, study FIGURE 30-2 until you thoroughly understand the general principles involved in this technique. Most microorganisms are unable to utilize the vast reservoir of nitrogen gas in the atmosphere; they depend upon inorganic nitrogen compounds like ammonia and nitrate or organic compounds like amino acids and vitamins. The nitrogen-fixing bacteria, however, are able to utilize atmospheric nitrogen, and they can therefore grow in environments where chemically combined nitrogen is scanty or lacking. Organisms in the genus *Azotobacter*, the most common genus of free-living nitrogen-fixing bacteria, cannot tolerate acid environments; they also require unusually large amounts of inorganic phosphate for growth. The mere addition of either calcium carbonate, phosphate, or both to a soil devoid of nitrogen-fixing bacteria often leads to their appearance; can you explain these effects? Where do you think the *Azotobacter* came from? Do you think that a farmer could improve the fertility of his land merely by inoculating the soil with *Azotobacter*? If not, what measurements should the farmer make in order to determine what might be wrong with his soil? To obtain the energy required for nitrogen fixation, these organisms must metabolize large quantities of carbonaceous material. Although many soil organisms cannot utilize mannitol (a sugar alcohol) as a carbon source, *Azotobacter* can readily metabolize this compound. You now know four facts about the physiology of *Azotobacter* that should enable you judiciously to change the soil environment in ways that should selectively stimulate the growth of this organism. What are these four facts? In this study you will be able to observe the formation of macroscopically visible colonies of *Azotobacter* on the surface of a soil paste that you have previously supplemented with calcium carbonate, mannitol, and phosphate.

A. Preparing the soil paste

1. Your instructor will give you, or ask you to bring to the laboratory, a small container of soil; as in the previous section, different students in the laboratory may use different types of soil. For each 50 grams of soil add 0.75 grams of mannitol, 0.1 grams of calcium carbonate ($CaCO_3$), and 5 ml of 3% dibasic sodium phosphate (Na_2HPO_4) solution.

2. After making these additions, mix well, and add enough water to make a *thick* paste; avoid adding too much water. Put the soil paste into small plastic petri dishes and smooth off the surface with a wet glass slide. The top should be made as smooth and even as possible; the paste should be wet enough to make it glisten after it has been smoothed over, but not so wet as to make it watery. Why? (Remember that *Azotobacter* requires oxygen.)

3. Place your soil plate in a moist chamber, such as a large plastic petri dish with some water in the bottom half, and label the plate or chamber as "treated."

4. Now make a paste of *untreated* soil, place it in a moist chamber, and label appropriately.

5. Put the moist chambers in the storage area designated by the instructor; he will incubate them at room temperature until colonies form, and then he will place them in the refrigerator until the next laboratory period.

B. Observing Azotobacter enrichment

1. The following week you should be able to see clear or opaque raised colonies on the soil surface; most of these glistening colonies are *Azotobacter*. If colonies failed to develop on your mannitol-treated plate, get a fresh sample of the soil you used and some pH paper from the instructor; wet the soil and test the pH. Is the pH value 6 or below? If not, did your soil sample come from a salty area? Or did the soil plate dry out? How

would you explain the development of colonies on your mannitol plate, but not on your untreated plate? Could you explain the development of colonies on both plates?

2. With a bacteriological loop, transfer a portion of a colony to a drop of water on a microscope slide, pass the slide through a flame several times to fix the bacteria onto the slide, and stain 1-2 minutes with gentian violet. If you see several morphologically different types of colonies, make and stain a slide for each one. You can conserve glass slides by putting several drops of water on each slide and transferring a different colony type to each drop; devise some labeling code that will enable you to correlate what you see under the microscope with colony morphology.

3. Observe the stained slide(s) microscopically. *Azotobacter* cells are much larger than most bacterial cells; they are usually round or oval and look very much like yeast. Make drawings of the microscopic appearance of the colonies you stained. What other types of microorganisms, if any, appeared on your soil paste? Explain why these organisms could grow extensively enough to form colonies on the surface of the paste, since carbon, but no nitrogen, was added. Why do the colonies form on the surface of the plate, rather than within it? If you wanted to obtain a pure culture of *Azotobacter*, what might your next step be?

4. Do you think that *Azotobacter* colonies still would have developed if you had added ammonium chloride (NH_4Cl) to the soil? Explain your answer. You may want to try this as an optional experiment.

5. If different types of soil were used by some of your classmates, compare your results with theirs. How might you explain any disparity between your results and theirs?

6. The soil paste in your plate might have "puffed up" during the week; how could you explain this? If you cannot answer this question, dig up some of the soil and examine the lower layers.

PART IV. MICROBIAL INTERACTIONS: A LABORATORY MODEL

The purpose of this study is to demonstrate some of the interactions that can occur among populations of different microorganisms when they are placed in a common physical environment, and to show how such interactions can lead to a functional community organization. It is, however, very difficult to observe and understand microbial interactions under natural conditions; the primary difficulty is the morphological similarity of many of the interacting species. Although each species can be identified by studying its physiology in the laboratory, such physiological distinctions are usually of little help in identifying organisms in their natural habitats. A further difficulty is the superimposition of one interspecific interaction upon another; this complexity, while not unique to microbial communities, may obscure or even overrule any one interaction.

In order to overcome these difficulties, you will study interspecific interactions and community regulation by setting up a simple and well-defined ecosystem in the laboratory. The substratum for this ecosystem will be agar; this will allow populations of different organisms to be distinguished on the basis of colony morphology. The method of streaking and cross-streaking (described below) will allow further colony distinctions on the basis of location. Finally, the use of a medium lacking compounds that are essential nutrients for some organisms but that are synthesized and even excreted by others, will permit organisms to be distinguished physiologically. The experimental design will thus readily enable you to distinguish the interacting organisms, even if their cell morphology is identical. Moreover, by either eliminating or controlling any natural interactions other than the one under study, you will be able to observe various regulatory processes either in isolation or in controlled superimposition. It must be emphasized, however, that this laboratory ecosystem can only serve as a *model* for bringing out the principles underlying community regulation in nature; the validity of this model depends upon how closely it simulates the essential features of the natural ecosystem.

The model you will use is based upon the production, by nitrogen-fixing bacteria, of the fixed nitrogen compounds that are essential nutrients for other soil microorganisms. You will streak various nitrogen-dependent organisms onto a nitrogen-free medium and will then make a single cross-streak of a nitrogen-fixing organism, such as the *Azotobacter* you have isolated. The *Azotobacter* will grow rapidly at the expense of atmospheric nitrogen; some of the nitrogen that it fixes will be excreted into the medium as amino acids, ammonia, and vitamins. The excreted nitrogen will diffuse through the agar away from the *Azotobacter* streak, forming a gradient in which the nitrogen concentration decreases with increasing distance from the streak. The excreted nitrogen will then allow the nitrogen-dependent organism(s) to grow as "satellite" colonies around the "feeder" streak of *Azotobacter* (FIGURE 30-4); such interspecific interactions, in which one population of organisms benefits without affecting the other, are called **commensalisms,** and the dependent organisms are called **commensals.**

This satellitic relation will serve as the central element of the model; by changing the feeder, the commensal, or the carbon source, you will be able to demonstrate some of the principles of community regulation. Thus, using a mixture of soil microorganisms as the commensal will allow you to observe competition between organisms for the excreted nitrogen. Using an antibiotic-producing organism (*Streptomyces griseus,* which produces streptomycin) will permit you to demonstrate how superimposing an antibiosis on the commensal relation can lead to a succession in the model community. Using different *Azotobacter* species as feeders will enable you to show how the species differ in their capacity to support satellite organisms. Finally, by varying an abiotic factor, the carbon source, you will be able to observe changed associations between organisms. These studies will illustrate the concepts of community and equilibrium in microbial ecology.

A. Streaking satellitism plates

To minimize the number of plates that must be made for you, the various combinations of nitrogen-dependent commensals, feeder organisms, and media required for this study will be apportioned among groups of students; your instructor will divide the class into groups of appropriate size and will give you the proper cultures and media. Be sure to familiarize yourself with microbiological technique (Appendix 7) before coming to class.

The various combinations of interacting organisms and media required

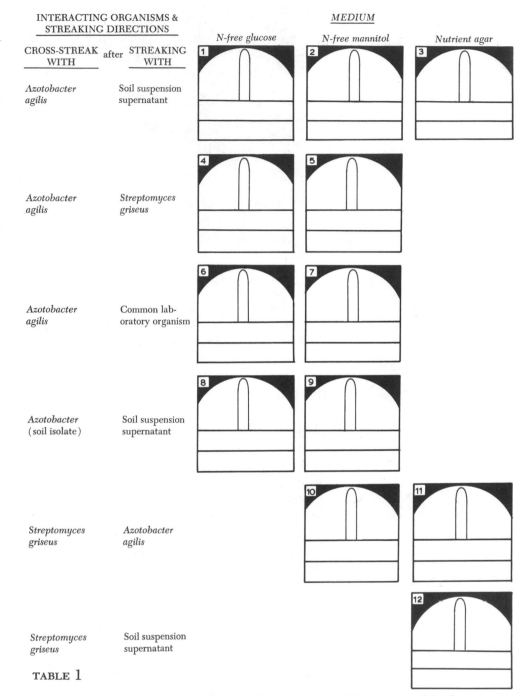

INTERACTING ORGANISMS & STREAKING DIRECTIONS		*MEDIUM*		
CROSS-STREAK WITH	after STREAKING WITH	*N-free glucose*	*N-free mannitol*	*Nutrient agar*
Azotobacter agilis	Soil suspension supernatant	1	2	3
Azotobacter agilis	*Streptomyces griseus*	4	5	
Azotobacter agilis	Common laboratory organism	6	7	
Azotobacter (soil isolate)	Soil suspension supernatant	8	9	
Streptomyces griseus	*Azotobacter agilis*		10	11
Streptomyces griseus	Soil suspension supernatant			12

TABLE 1

for each group are presented in Table 1; you will be responsible for one or more of the numbered blocks in the table.

1. After obtaining the cultures and media required to prepare the satellitism plate(s) represented by your block(s) in the table, label your plate(s) with your name, the block number, and the interacting organisms.

2. Be sure to streak the organisms in the order specified in the table; use sterile technique. You may plate the first organisms either by streaking or spreading. If you choose the streaking method, one loopful of a liquid suspension or a tiny portion of a colony will be a sufficient inoculum. If you choose the spreading method, use one drop of a liquid suspension delivered

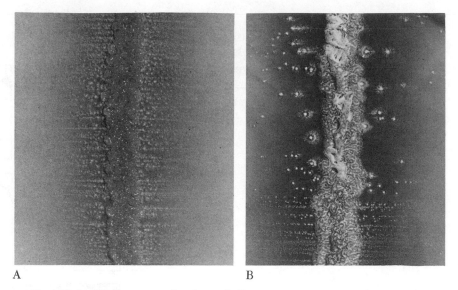

A   B

Fig. 30-4. Satellitic growth of a soil *Streptomyces* sp. around an *Azotobacter agilis* feeder streak. (A) One-week incubation. (B) Two-week incubation.

with a sterile Pasteur capillary pipette and bulb. To prepare the soil suspension, shake or stir some soil with water and allow the soil particles to settle out; use the water layer (often called a supernatant) above the layer of settled soil.

3. Make sure that the agar surface is dry before cross-streaking the plate with the second organism; use a heavy inoculum to make a single cross-streak, perpendicular to the streaks of the first organism (see FIGURE 30-4). If your block calls for the use of the *Azotobacter* you isolated previously, be sure that the colony from which you take an inoculum is indeed *Azotobacter*; your previous microscopic examination of the various colonies appearing on the soil paste should enable you to identify *Azotobacter* by its colony morphology.

4. When you have finished streaking and cross-streaking, place your plate(s) upside down (why?) in your drawer until the next laboratory period.

B. Interpretation of satellitism experiments

1. After one week, examine your plate(s) and, in the appropriate block(s) of the table, record the following three types of observation in the form of drawings; each block has been divided into three parts for this purpose.

a. The size and density of the satellite colonies as a function of distance from the cross-streak (make your drawings on the diagrammatic satellitism plate in the correct block).

b. The morphological characteristics of the satellite colonies and of the cross-streak (indicate such characteristics as border, color, elevation, shape, texture, etc.).

c. The microscopic appearance of cells from each colony type, including the feeder colonies (prepare and fix a smear from each colony type and stain with gentian violet; make your drawings of cell morphology directly below the drawings of colony morphology to which they correspond).

2. When you have completed these observations, trade plates with your neighbor and make similar observations; continue trading plates until you have filled in all the blocks in the table. There will be some repetition in the types of colony you see. You need to prepare a stained mount only the first time you encounter a new colony type; in subsequent encounters, rely on the drawings you made from the slide. If you are not sure whether a colony is different, prepare a stained mount to make sure. Whenever a plate is streaked with an organism that you have not previously examined micro-scopically, prepare a stained mount for observation; do this even if the colonies of this organism look similar to other colonies you have observed. Remember that you can conserve glass slides by making smears of several different organisms on one slide; don't forget to use a wax pencil to code the smears.

3. Examine the results in Blocks 1, 2, and 3. Can you explain why there is a much greater diversity of organisms present on nutrient agar than on either of the nitrogen-free media? (*Hint:* Remember that nutrient agar contains nitrogen compounds.) What ecological phenomenon does this illustrate?

On the nitrogen-free plates, are the predominant types of satellitic colonies hard, compact, and fuzzy (i.e., do they look like the satellitic colonies in FIGURE 30-4)? Microscopically, are they thin filaments (as opposed to relatively thick fungal hyphae)? Do your nitrogen-free plates have the musty odor of soil? If your answers to the preceding three questions are "yes," then your satellite colonies are probably actinomycetes (mycelial bacteria). Are there many actinomycetes on your nutrient-agar plate? Can you explain why the other colony types that grew on nutrient agar failed to grow satellitically around the *Azotobacter agilis* feeder streak?

(*Hint:* This selective stimulation of certain groups of bacteria in the microflora has an analogy in the so-called **rhizosphere effect:** the total number of bacteria is much greater in the rhizosphere [the soil adjacent to the root systems of growing plants] than in soil more distant from the plant. Some nutritional groups of bacteria, however, are stimulated to a greater extent than others; thus, organisms requiring amino acids are stimulated more than those requiring vitamins or those requiring no supplementary growth factors. This differential increase in the proportion of amino-acid-requiring bacteria is accompanied by a decrease in the proportion of bacteria with more complex nutritional requirements. The overall stimulation of the microflora can readily be explained by the secretion of amino acids by the roots. Since these compounds can serve as both carbon and nitrogen sources, even organisms that do not actually require amino acids would be expected to be stimulated; some other factor, therefore, must be responsible for the differential stimulation of the amino-acid-requiring bacteria. The explanation lies in the excretion [by those organisms without special nutritional requirements] of substances that are toxic to many vitamin-requiring bacteria, but that are often stimulating to amino-acid-requiring bacteria.)

4. Look again at Blocks 1-3. Are there any large colonies away from the feeder streak that have satellite colonies around them? What might this organism be?

Does the feeder streak on your nitrogen-free plates have a ropelike appearance? Can you explain why? (*Hint:* Compare the two photographs in Figure 30-4; the first is after one week, the second after two weeks.)

In light of your answer to the previous question, do you think that the commensal organism eventually becomes an antagonist of the feeder organism in this model ecosystem?

Do you think that the model ecosystem accurately represents the situation in nature? (*Hint:* In soil, the microenvironment is discontinuous; microcolonies of different organisms usually occupy separate soil particles.)

Do the number and size of the satellite colonies in Block 2 increase in a gradient from the edge of the plate toward the feeder streak? In Block 1? Explain any difference between the two blocks. (*Hints:* The amount of nitrogen excreted by the *Azotobacter* may be greater on one carbon source than on the other; maximal growth of the satellite organism is possible once the nitrogen concentration at any particular distance from the feeder streak reaches a certain level.)

5. The following questions refer to Blocks 4, 5, 10, 11, and 12.

Is there a zone of clearing around the *Streptomyces griseus* cross-streak in Blocks 11 and 12? If so, how is the *S. griseus* inhibiting the other organisms?

In Block 12 are there some types of colonies that still grow near the cross-streak? Explain why this could happen.

Are the results in Block 10 different from those in Block 11? If so, explain why.

Can you explain your results in Block 5 in terms of a succession? (*Hint:* Your results in Blocks 10 and 11 may help you explain what happened in Block 5.)

Did changing the abiotic environment by using different carbon sources lead to different results in Blocks 4 and 5? If so, explain how changing the carbon source could cause the difference observed. (*Hint:* The same principle can be used to explain both this difference and that between Blocks 1 and 2.)

6. Did the common laboratory organism you used in Blocks 6 and 7 grow satellitically around the *Azotobacter* feeder streak? If growth was poor, suggest a possible explanation. (*Hint:* The rhizosphere effect described above may give you a clue.)

Are there any differences between Blocks 6 and 7? If so, explain why.

7. Compare Block 1 with 8, and 2 with 9; was the *Azotobacter* you isolated from soil as good a nitrogen feeder as *Azotobacter agilis?* (If your soil *Azotobacter* grew very copiously and produced a brown or black pigment, it is probably *Azotobacter chroococcum.*) If not, do you think the explanation is that the soil *Azotobacter* excretes less nitrogen? If you don't think so, what other explanation might you invoke? (*Hint:* Remember the rhizosphere effect.)

Explain any differences between Blocks 8 and 9.

8. If you were designing a laboratory exercise to demonstrate succession, why might you choose a microbial system? Do you think you could actually observe a complete succession in a plant or an animal community?

APPENDICES

APPENDIX 1 DRAWING

There are many good reasons for making drawings in biology. One of them, and perhaps the most important, is that drawing forces you to observe closely and accurately. You may not believe this, but if you accept it and approach the necessity of drawing with intent to profit by using it as a tool to learn the techniques of close observation, you will never regret the effort expended.

Secondly, drawings form an important record of things you see. While photographs are limited to representing appearance in a single moment in time, drawings can be composites of the observer's cumulative experience with many different specimens of the same material. The total picture of an object thus represented can often communicate information much more effectively than a photograph; communication is extremely important to a scientist who wishes to have his published work clearly understood by others.

Your attention to the outline of suggestions below will help you to make effective drawings. If you are careful to follow the suggestions at the beginning they will soon become habitual.

1. Your drawing should be a complete, accurate representation of the material you have observed, and should communicate your understanding of the material to anyone who looks at it. Avoid making "idealized" drawings; your drawing should be a picture of what you actually see, not what you imagine should be there.

2. A drawing should be large enough to represent easily all the details you see *without crowding* them. Rarely, if ever, are drawings too large, but they are often much too small. Show only as much as is necessary for an understanding of the structure; a small section shown in detail will often suffice. It is time-consuming and unnecessary, for example, to reproduce accurately the entire contents of a microscope field.

3. A sharp, hard pencil used on hard, smooth-surfaced paper enables you to produce maximum resolution with minimum effort. Use a 3H or 4H pencil, good hard paper, and a clean eraser.

4. Use only simple, narrow lines. Represent depth by stippling (dots close together). Indicate depth only when it is essential to your drawing — usually it is not. The ratio of observations to lines drawn should be at least 10:1, not the reverse.

5. Leave plenty of margin for labels. Label the drawing with a) source of material, b) magnification under which it was observed, c) structures, and in the case of living materials, d) movements you have seen. Try to keep label lines parallel to one side of the drawing paper (usually the bottom), and avoid crossing label lines.

Remember that drawings are intended as records for you, and as a means of encouraging close observation; artistic ability is unnecessary. Before you turn in a drawing, ask yourself if you know what every line represents. If not, look more closely at the material you are working on.

APPENDIX 2 CLASSIFICATION

To deal with the vast array of life, biologists need an orderly system by which species can be classified in a meaningful manner. The classification system used in biology today consists of a hierarchy of graded *taxonomic* ranks; **taxonomy** is the specialization within biology that deals exclusively with bringing order into the classification system. Each taxonomic rank automatically encodes information about every aspect of the organisms it includes, and thereby conveys information useful in almost every other area of biology. Some of the ranks, in order of decreasing inclusiveness, are as follows:

Kingdom
    Phylum or Division
        Class
           Order
              Family
                Genus
                  Species

Each category in this hierarchy is a collective unit containing one or more groups from the next-lower level in the hierarchy. Thus, a genus is a group of closely related species, unified by some characteristic that is considered more fundamental than the differences which define separate species of that genus. A family is a group of related genera, etc. Such a scheme represents an effort to group organisms according to relationship. Often relationships can be determined from fossils, but where fossils are not available, we are often forced to guess relationships by studying similarities in anatomy, development, behavior, and other characteristics of living organisms.

Because this system of hierarchy presumably reflects degree of kinship, we can set up "phylogenetic trees" to show probable evolutionary histories. Using the trees will help you to understand the hypothetical relationships.

The first time you encounter an organism, look up its genus in the adjoining table, and place its scientific name in the proper position on its phylogenetic tree. A few genera you use will not be found in the adjoining list; if you encounter such organisms and if their position is not obvious from context, ask your instructor to help. Each species is placed at the tip of a branch which you must draw on the tree yourself.

A Classification of Living Things

KINGDOM MONERA

DIVISION SCHIZOMYCETES. Bacteria

Class Myxobacteria. *Myxococcus, Chondromyces, Cytophaga, Sporocytophaga*

Class Spirochetes. *Leptospira, Cristispira, Spirocheta, Treponema*

Class Eubacteria. *Azotobacter, Staphylococcus, Escherichia, Salmonella, Pasteurella, Streptococcus, Bacillus, Spirillum, Caryophanon, Actinomyces*

Class Rickettsiae. *Rickettsia, Coxiella*

DIVISION CYANOPHYTA. Blue-green algae. *Gloeocapsa, Microcystis, Oscillatoria, Nostoc, Scytonema*

KINGDOM PLANTAE

DIVISION EUGLENOPHYTA. Euglenoids. *Euglena, Eutreptia, Phacus, Colacium*

DIVISION CHLOROPHYTA. Green algae

Class Chlorophyceae. True green algae. *Chlamydomonas, Volvox, Ulothrix, Spirogyra, Oedogonium, Ulva*

Class Charophyceae. Stoneworts. *Chara, Nitella, Tolypella*

DIVISION CHRYSOPHYTA

Class Xanthophyceae. Yellow-green algae. *Botrydiopsis, Halosphaera, Tribonema, Botrydium*

Class Chrysophyceae. Golden-brown algae. *Chrysamoeba, Chromulina, Synura, Mallomonas*

Class Bacillariophyceae. Diatoms. *Pinnularia, Arachnoidiscus, Triceratium, Pleurosigma*

DIVISION PYRROPHYTA. Dinoflagellates. *Gonyaulax, Gymnodinium, Ceratium, Gloeodinium*

DIVISION PHAEOPHYTA. Brown algae. *Sargassum, Ectocarpus, Fucus, Laminaria*

DIVISION RHODOPHYTA. Red algae. *Nemalion, Polysiphonia, Dasya, Chondrus, Batrachospermum*

DIVISION MYXOMYCOPHYTA. Slime molds

Class Myxomycetes. True slime molds. *Physarum, Hemitrichia, Stemonitis*

Class Acrasiae. Cellular slime molds. *Dictyostelium*

Class Plasmodiophoreae. *Plasmodiophora*

Class Labyrinthuleae. *Labyrinthula*

DIVISION EUMYCOPHYTA. True fungi

Class Phycomycetes. Algal fungi. *Rhizopus, Saprolegnia, Phytophthora, Albugo*

Class Ascomycetes. Sac fungi. *Neurospora, Aspergillus, Penicillium, Saccharomyces, Morchella, Ceratostomella*

Class Basidiomycetes. Club fungi. *Ustilago, Puccinia, Coprinus, Lycoperdon, Psalliota, Amanita*

DIVISION BRYOPHYTA

Class Hepaticae. Liverworts. *Marchantia, Conocephalum, Riccia, Porella*

Class Anthocerotae. Hornworts. *Anthoceros*

Class Musci. Mosses. *Polytrichum, Sphagnum, Mnium*

DIVISION TRACHEOPHYTA. Vascular plants

Subdivision Psilopsida. *Psilotum, Tmesipteris*

Subdivision Lycopsida. Club mosses. *Lycopodium, Phylloglossum, Selaginella, Isoetes, Stylites*

Subdivision Sphenopsida. Horsetails. *Equisetum*

Subdivision Pteropsida. Ferns. *Polypodium, Osmunda, Dryopteris, Botrychium, Pteridium*

Subdivision Spermopsida. Seed plants

Class Pteridospermae. Seed ferns. No living representatives

Class Cycadae. Cycads. *Zamia*

Class Ginkgoae. *Ginkgo*

Class Coniferae. Conifers. *Pinus, Tsuga, Taxus, Sequoia*

Class Gneteae. *Gnetum, Ephedra, Welwitschia*

Class Angiospermae. Flowering plants

Subclass Dicotyledoneae. Dicots. *Magnolia, Quercus, Acer, Pisum, Taraxacum, Rosa, Chrysanthemum, Aster, Primula, Ligustrum, Ranunculus*

Subclass Monocotyledoneae. Monocots. *Lilium, Tulipa, Poa, Elymus, Triticum, Zea, Ophyrys, Yucca, Sabal*

KINGDOM ANIMALIA

SUBKINGDOM PROTOZOA

PHYLUM PROTOZOA. Acellular animals

Subphylum Plasmodroma

Class Flagellata (or Mastigophora). Flagellates. *Trypanosoma, Calonympha, Chilomonas* (also *Euglena, Chlamydomonas,* and other green flagellates included in Plantae as well)

Class Sarcodina (or Rhizopoda). Protozoans with pseudopods. *Amoeba, Globigerina, Textularia, Acanthometra*

Class Sporozoa. *Plasmodium, Monocystis*

Subphylum Ciliophora

Class Ciliata. Ciliates. *Paramecium, Opalina, Stentor, Vorticella, Spirostomum*

SUBKINGDOM PARAZOA

PHYLUM PORIFERA. Sponges

Class Calcarea. Calcareous (chalky) sponges. *Scypha, Leucosolenia, Sycon, Grantia*

Class Hexactinellida. Glass sponges. *Euplectella, Hyalonema, Monoraphis*

Class Demospongiae. *Spongilla, Euspongia, Axinella*

SUBKINGDOM MESOZOA

PHYLUM MESOZOA. *Dicyema, Pseudicyema, Rhopalura*

SUBKINGDOM METAZOA

SECTION RADIATA

PHYLUM COELENTERATA (or Cnidaria)

CLASS HYDROZOA. Hydrozoans. *Hydra, Obelia, Gonionemus, Physalia*

CLASS SCYPHOZOA. Jellyfishes. *Aurelia, Pelagia, Cyanea*

CLASS ANTHOZOA. Sea anemones and corals. *Metridium, Pennatula, Gorgonia, Astrangia*

PHYLUM CTENOPHORA. Comb jellies

CLASS TENTACULATA. *Pleurobrachia, Mnemiopsis, Cestum, Velamen*

CLASS NUDA. *Beroe*

SECTION PROTOSTOMIA

PHYLUM PLATYHELMINTHES. Flatworms

CLASS TURBELLARIA. Free-living flatworms. *Planaria, Dugesia, Leptoplana*

CLASS TREMATODA. Flukes. *Fasciola, Schistosomum, Prosthogonimus*

CLASS CESTODA. Tapeworms. *Taenia, Dipylidium, Mesocestoides*

PHYLUM NEMERTINA (or Rhynchocoela). Proboscis worms. *Cerebratulus, Lineus, Malacobdella*

PHYLUM ACANTHOCEPHALA. Spiny-headed worms. *Echinorhynchus, Gigantorhynchus*

PHYLUM ASCHELMINTHES

CLASS ROTIFERA. Rotifers. *Asplanchna, Hydatina, Rotaria*

CLASS GASTROTRICHA. *Chaetonotus, Macrodasys*

CLASS KINORHYNCHA (or Echinodera). *Echinoderes, Semnoderes*

CLASS PRIAPULIDA. *Priapulus, Halicryptus*

CLASS NEMATODA. Round worms. *Ascaris, Trichinella, Necator, Enterobius, Ancylostoma, Heterodera*

CLASS NEMATOMORPHA. Horsehair worms. *Gordius, Paragordius, Nectonema*

PHYLUM ENTOPROCTA. *Urnatella, Loxosoma, Pedicellina*

PHYLUM ECTOPROCTA (or Bryozoa). Bryozoans, moss animals

CLASS GYMNOLAEMATA. *Paludicella, Bugula*

CLASS PHYLACTOLAEMATA. *Plumatella, Pectinatella*

PHYLUM PHORONIDA. *Phoronis, Phoronopsis*

PHYLUM BRACHIOPODA. Lamp shells

CLASS INARTICULATA. *Lingula, Glottidia, Discina*

CLASS ARTICULATA. *Magellania, Neothyris, Terebratula*

PHYLUM MOLLUSCA. Molluscs

CLASS AMPHINEURA. Chitons. *Chaetopleura, Ischnochiton, Lepidochiton, Amicula*

CLASS MONOPLACOPHORA. *Neopilina*

CLASS GASTROPODA. Snails and their allies (univalve molluscs). *Helix, Busycon, Crepidula, Haliotis, Littorina, Doris, Limax*

CLASS SCAPHOPODA. Tusk shells. *Dentalium, Cadulus*

CLASS PELECYPODA. Bivalve molluscs. *Mytilus, Ostrea, Pecten, Mercenaria, Teredo, Tagelus, Unio, Anodonta*

CLASS CEPHALOPODA. Squids, octopuses, etc. *Loligo, Octopus, Nautilus*

PHYLUM SIPUNCULIDA. *Sipunculus, Phascolosoma, Dendrostomum*

PHYLUM ECHIURIDA. *Echiurus, Urechis, Thalassema*

PHYLUM ANNELIDA. Segmented worms

CLASS POLYCHAETA (including Archiannelida). Sandworms, tubeworms, etc. *Nereis, Chaetopterus, Aphrodite, Diopatra, Arenicola, Hydroides, Sabella*

CLASS OLIGOCHAETA. Earthworms and many freshwater annelids. *Tubifex, Enchytraeus, Lumbricus, Dendrobaena*

CLASS HIRUDINEA. Leeches. *Trachelobdella, Hirudo, Macrobdella, Haemadipsa*

PHYLUM ONYCHOPHORA. *Peripatus, Peripatopsis*

PHYLUM TARDIGRADA. Water bears. *Echiniscus, Macrobiotus*

PHYLUM PENTASTOMIDA. *Cephalobaena, Linguatula*

PHYLUM ARTHROPODA

Subphylum Trilobita. No living representatives

Subphylum Chelicerata

CLASS EURYPTERIDA. No living representatives

CLASS XIPHOSURA. Horseshoe crabs. *Limulus*

CLASS ARACHNIDA. Spiders, ticks, mites, scorpions, whipscorpions, daddy longlegs, etc. *Archaearanea, Latrodectus, Argiope, Centruroides, Chelifer, Mastigoproctus, Phalangium, Ixodes*

CLASS PYCNOGONIDA. Sea spiders. *Nymphon, Ascorhynchus*

Subphylum Mandibulata

CLASS CRUSTACEA. *Homarus, Cancer, Daphnia, Artemia, Cyclops, Balanus, Porcellio, Cambarus*

CLASS CHILOPODA. Centipeds. *Scolopendra, Lithobius, Scutigera*

CLASS DIPLOPODA. Millipeds. *Narceus, Apheloria, Polydesmus, Julus, Glomeris*

CLASS PAUROPODA. *Pauropus*

CLASS SYMPHYLA. *Scutigerella*

CLASS INSECTA. Insects

ORDER COLLEMBOLA. Springtails. *Isotoma, Achorutes, Neosminthurus, Sminthurus*

ORDER PROTURA. *Acerentulus, Eosentomon*

ORDER DIPLURA. *Campodea, Japyx*

ORDER THYSANURA. Bristletails, silverfish, firebrats. *Machilis, Lepisma, Thermobia*

ORDER EPHEMERIDA. Mayflies. *Hexagenia, Callibaetis, Ephemerella*

ORDER ODONATA. Dragonflies, damselflies. *Archilestes, Lestes, Aeshna, Gomphus*

ORDER ORTHOPTERA (including Isoptera). Grasshoppers, crickets, walking sticks, mantids, cockroaches, termites, etc. *Schistocerca, Romalea, Nemobius, Megaphasma, Mantis, Blatta, Periplaneta, Reticulitermes*

ORDER DERMAPTERA. Earwigs. *Labia, Forficula, Prolabia*

ORDER EMBIARIA (or Embiidina or Embioptera). *Oligotoma, Anisembia, Gynembia*

ORDER PLECOPTERA. Stoneflies. *Isoperla, Taeniopteryx, Capnia, Perla*

ORDER ZORAPTERA. *Zorotypus*

ORDER CORRODENTIA. Book lice. *Ectopsocus, Liposcelis, Trogium*

ORDER MALLOPHAGA. Chewing lice. *Cuclotogaster, Menacanthus, Menopon, Trichodectes*

ORDER ANOPLURA. Sucking lice. *Pediculus, Phthirius, Haematopinus*

ORDER THYSANOPTERA. Thrips. *Heliothrips, Frankliniella, Hercothrips*

ORDER HEMIPTERA (including Homoptera). Bugs, cicadas, aphids, leafhoppers, etc. *Belostoma, Lygaeus, Notonecta, Cimex, Lygus, Oncopeltus, Magicicada, Circulifer, Psylla, Aphis*

ORDER NEUROPTERA. Dobsonflies, alderflies, lacewings, mantispids, snakeflies, etc. *Corydalus, Hemerobius, Chrysopa, Mantispa, Agulla*

ORDER COLEOPTERA. Beetles, weevils. *Copris, Phyllophaga, Harpalus, Scolytus, Melanotus, Cicindela, Dermestes, Photinus, Coccinella, Tenebrio, Anthonomus, Conotrachelus*

ORDER HYMENOPTERA. Wasps, bees, ants, sawflies. *Cimbex, Vespa, Glypta, Scolia, Bembix, Formica, Bombus, Apis*

ORDER MECOPTERA. Scorpionflies. *Panorpa, Boreus, Bittacus*

ORDER SIPHONAPTERA. Fleas. *Pulex, Nosopsyllus, Xenopsylla, Ctenocephalides*

ORDER DIPTERA. True flies, mosquitoes. *Aedes, Asilus, Sarcophaga, Anthomyia, Musca, Chironomus, Tabanus, Tipula, Drosophila*

ORDER TRICHOPTERA. Caddisflies. *Limnephilus, Rhyacophila, Hydropsyche*

ORDER LEPIDOPTERA. Moths, butterflies. *Tinea, Pyrausta, Malacosoma, Sphinx, Samia, Bombyx, Heliothis, Papilio, Lycaena*

SECTION DEUTEROSTOMIA

PHYLUM CHAETOGNATHA. Arrow worms. *Sagitta, Spadella*

PHYLUM ECHINODERMATA

CLASS CRINOIDEA. Crinoids, sea lilies. *Antedon, Ptilocrinus, Comactinia*

CLASS ASTEROIDEA. Sea stars. *Asterias, Ctenodiscus, Luidia, Oreaster*

CLASS OPHIUROIDEA. Brittle stars, serpent stars, basket stars, etc. *Asteronyx, Amphioplus, Ophiothrix, Ophioderma, Ophiura*

CLASS ECHINOIDEA. Sea urchins, sand dollars, heart urchins. *Cidaris, Arbacia, Strongylocentrotus, Echinanthus, Echinarachnius, Moira*

CLASS HOLOTHUROIDEA. Sea cucumbers. *Cucumaria, Thyone, Caudina, Synapta*

PHYLUM POGONOPHORA. Beard worms. *Siboglinum, Lamellisabella, Oligobrachia, Polybrachia*

PHYLUM HEMICHORDATA

CLASS ENTEROPNEUSTA. Acorn worms. *Saccoglossus, Balanoglossus, Clossobalanus*

CLASS PTEROBRANCHIA. *Rhabdopleura, Cephalodiscus*

PHYLUM CHORDATA. Chordates

Subphylum Urochordata (or Tunicata). Tunicates

CLASS ASCIDIACEA. Ascidians or sea squirts. *Ciona, Clavelina, Molgula, Perophora*

CLASS THALIACEA. *Pyrosoma, Salpa, Doliolum*

CLASS LARVACEA. *Appendicularia, Oikopleura, Fritillaria*

Subphylum Cephalochordata. Lancelets, amphioxus. *Branchiostoma, Asymmetron*

Subphylum Vertebrata. Vertebrates

CLASS AGNATHA. Jawless fishes. *Cephalaspis,** *Pterasis,** *Petromyzon, Entosphenus, Myxine, Eptatretus*

CLASS PLACODERMI. No living representatives

CLASS CHONDRICHTHYES. Cartilaginous fishes. *Squalus, Hyporion, Raja, Chimaera*

CLASS OSTEICHTHYES. Bony fishes

SUBCLASS SARCOPTERYGII

ORDER CROSSOPTERYGII (or Coelacanthiformes). Lobe-fins. *Latimeria*

ORDER DIPNOI (or Dipteriformes). Lungfishes. *Neoceratodus, Protopterus, Lepidosiren*

SUBCLASS BRACHIOPTERYGII. Bichirs. *Polypterus*

SUBCLASS ACTINOPTERYGII. Higher bony fishes. *Amia, Cyprinus, Gadus, Perca, Salmo*

CLASS AMPHIBIA

ORDER ANURA. Frogs and toads. *Rana, Hyla, Bufo*

ORDER URODELA. Salamanders. *Necturus, Triturus, Plethodon, Ambystoma*

ORDER APODA. *Ichthyophis, Typhlonectes*

CLASS REPTILIA

ORDER CHELONIA. Turtles. *Chelydra, Kinosternon, Clemmys, Terrapene*

ORDER RHYNCHOCEPHALIA. Tuatara. *Sphenodon*

ORDER CROCODYLIA. Crocodiles and alligators. *Crocodylus, Alligator*

ORDER SQUAMATA. Snakes and lizards. *Iguana, Anolis, Sceloporus, Phrynosoma, Natrix, Elaphe, Coluber, Thamnophis, Crotalus*

CLASS AVES. Birds. *Anas, Larus, Columba, Gallus, Turdus, Dendroica, Sturnus, Passer, Melospiza*

CLASS MAMMALIA. Mammals

SUBCLASS PROTOTHERIA

ORDER MONOTREMATA. Egg-laying mammals. *Ornithorhynchus, Tachyglossus*

SUBCLASS THERIA. Marsupial and placental mammals

ORDER MARSUPIALIA. Marsupials. *Didelphis, Sarcophilus, Notoryctes, Macropus*

ORDER INSECTIVORA. Insectivores (moles, shrews, etc.). *Scalopus, Sorex, Erinaceus*

* Extinct

ORDER DERMOPTERA. Flying lemurs. *Galeopithecus*

ORDER CHIROPTERA. Bats. *Myotis. Eptesicus, Desmodus*

ORDER PRIMATES. Lemurs, monkeys, apes, man. *Lemur, Tarsius, Cebus, Macacus, Cynocephalus, Pongo, Pan, Homo*

ORDER EDENTATA. Sloths, anteaters, armadillos. *Bradypus, Myrmecophagus, Dasypus*

ORDER PHOLIDOTA. Pagonlin. *Manis*

ORDER LAGOMORPHA. Rabbits, hares, pikas. *Ochotona, Lepus, Sylvilagus, Oryctolagus*

ORDER RODENTIA. Rodents. *Sciurus, Marmota, Dipodomys, Microtus, Peromyscus, Rattus, Mus, Erethizon, Castor*

ORDER CETACEA. Whales, dolphins, porpoises. *Delphinus, Phocaena, Monodon, Balaena*

ORDER CARNIVORA. Carnivores. *Canis, Procyon, Ursus, Mustela, Mephitis, Felis, Hyaena, Eumetopias*

ORDER TUBULIDENTATA. Aardvark. *Orycteropus*

ORDER PROBOSCIDEA. Elephants. *Elephas, Loxodonta*

ORDER HYRACOIDEA. Coneys. *Procavia*

ORDER SIRENIA. Manatees. *Trichechus, Halicore*

ORDER PERISSODACTYLA. Odd-toed ungulates. *Equus, Tapirella, Tapirus, Rhinoceros*

ORDER ARTIODACTYLA. Even-toed ungulates. *Pecari, Sus, Hippopotamus, Camelus, Cervus, Odocoileus, Giraffa, Bison, Ovis, Bos*

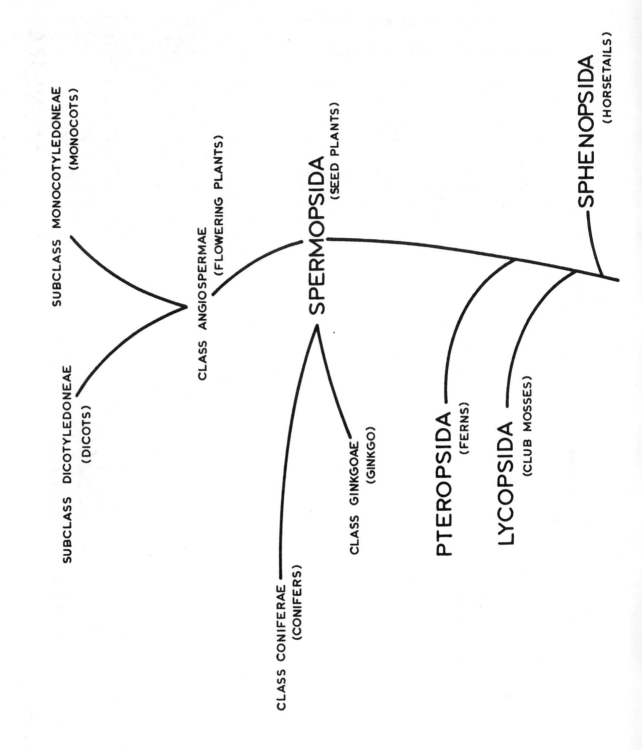

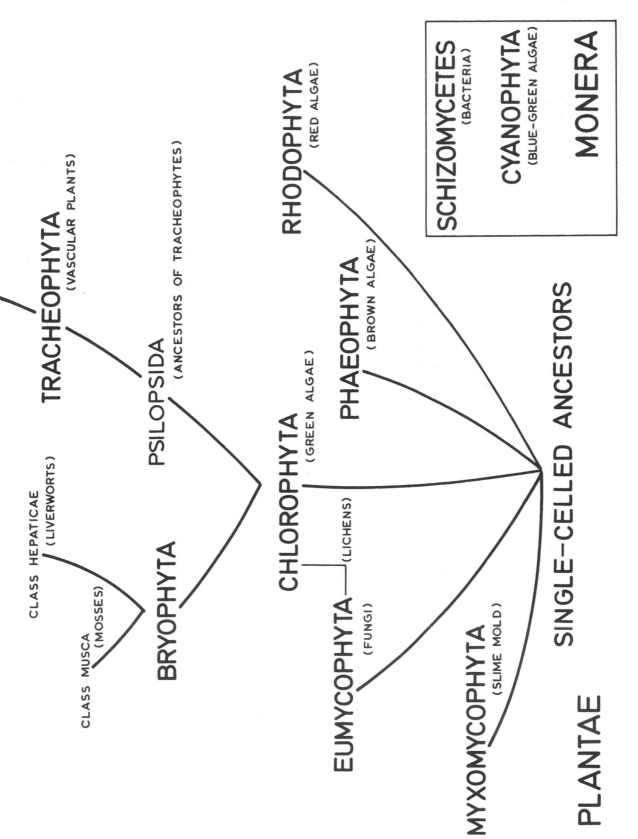

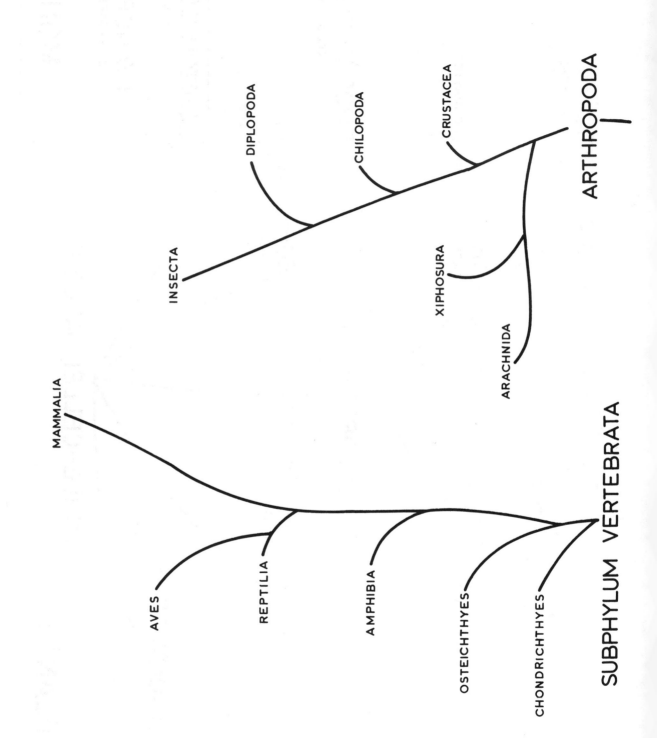

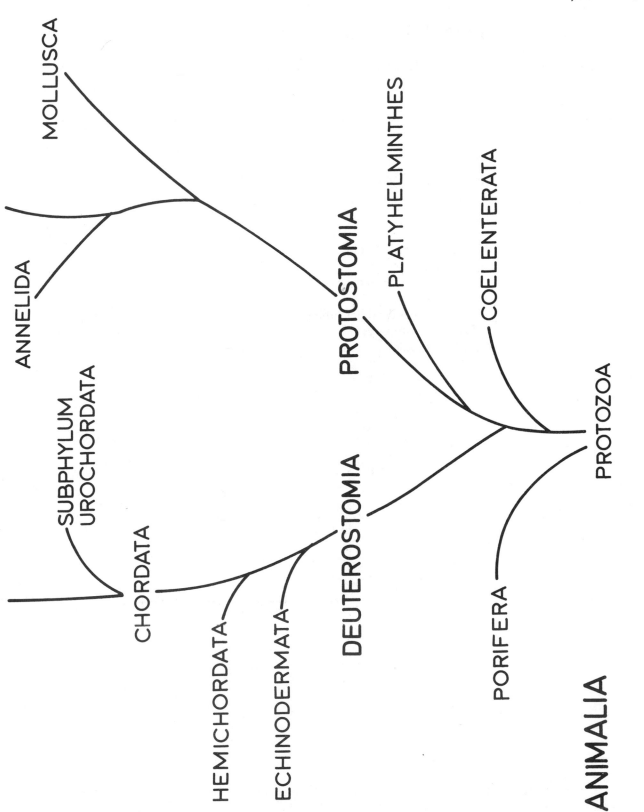

MOLLUSCA

ANNELIDA

PROTOSTOMIA

PLATYHELMINTHES

COELENTERATA

SUBPHYLUM
UROCHORDATA

CHORDATA

PROTOZOA

DEUTEROSTOMIA

HEMICHORDATA

ECHINODERMATA

PORIFERA

ANIMALIA

APPENDIX 3 THE CLOSED MANOMETER SYSTEM

PART I.
MATERIALS

Water bath (2-liter beaker)
Supports (one inside and one atop the beaker)
Test tubes
Plastic U-block
2 T-rigs, each with T-tube
       2-hole stopper
       1-cc calibrated syringe
       Luer-lok
       glass and rubber tubing
2 glass escape stoppers
1 bottle of Brodie's manometer fluid, with syringe and needle

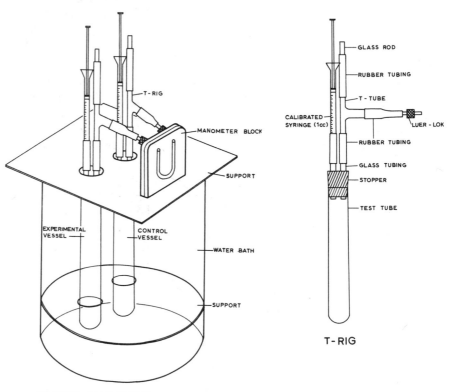

CLOSED MANOMETER SYSTEM

PART II.
ARRANGEMENT

Two test tubes are connected by glass and rubber tubing to a U-shaped capillary in a plastic block. One test tube acts as a control chamber and the other as an experimental chamber. Each test tube also has an escape tube and a second outlet closed by a calibrated syringe. When the escape tubes are plugged, any decrease in volume of gas contained in one tube will be compensated by a decrease in pressure in the line toward the U-tube, and the manometer fluid will move in the direction of the decreased volume. To measure the change in volume, air is removed or introduced to the system with one of the calibrated syringes, until the manometer fluid is returned to the starting level indicated by a line engraved on the U-tube block. Changes in gas volume caused by variations in temperature (or any other factor that is not the subject of measurement) are automatically corrected, since the control and experimental chambers are connected to either side of the U-tube. Furthermore, the two tubes rest in a constant temperature bath, and thus most of the apparatus is isolated from temperature change. The control and experimental tubes should always contain about the same volume of air, so that any pressure or temperature changes will register equally on both sides of the U-tube.

PART III. PROCEDURE

1. Introduce a column of manometer fluid (Brodie's solution) into the capillary with the needle and syringe provided for this purpose. Take care not to trap air in the column, and remove excess fluid with pieces of absorbent paper. The fluid column should be about level with the line on the front of the block.

2. With rubber stoppers and attached tubes, tightly stopper the two test tubes, and lower them into the water bath. With the escape tubes open, attach the T-rigs to the manometer block. The escape tubes may then be closed with glass-rod stoppers.

3. Test the system for leaks in the following manner: Allow the empty, closed apparatus to equilibrate for several minutes in the water bath; then raise the column of manometer fluid on one side of the U-tube by adjusting the appropriate syringe. If the column has moved after a few minutes, check for leaks and tighten all connections before proceeding.

4. When you are sure that the system is closed and that the apparatus can be assembled without leaks, proceed to the experiments.

5. During the course of an experiment when it is necessary to reset a syringe, first open the escape tubes of both chambers simultaneously; reset the syringe, close the escape tubes, and begin reading again. If properly arranged, the maximum amount of displacement from any side is 2 cc. The volume of the manometer block is about ½ cc. Therefore measurements must be made at intervals that displace less than ½ cc. When the fluid is releveled, the measurements of volume are taken from the calibrations on the syringes. Take readings at intervals frequent enough to get sufficient data for a graph.

Option to Topic 3 – MEASUREMENT OF ENZYME ACTIVITY BY SAMPLING

Read the introduction to Topic 3. Instead of measuring rate of enzyme activity by amount of oxygen evolved, we shall measure the amount of substrate (hydrogen peroxide) remaining in samples taken from the reaction at intervals. At specified intervals, samples will be taken from the reaction mixture and immediately treated to stop the reaction. At the end of the experiment, after all samples have been taken, each one will be tested to determine how much hydrogen peroxide remains in it. This measurement indicates the amount of decomposition of the substrate. The conditions under which the reaction takes place may be varied, just as in Topic 3.

PART I. PREPARATION OF SAMPLING APPARATUS AND ENZYME
A. The sampling apparatus

1. Number test tubes 1, 2, 3, and 4; with a wax pencil, mark each one 1 cm and 2 cm from the bottom.

2. Place 2N sulfuric acid (H_2SO_4) into each tube, to the 1-cm mark, using a dropper reserved for this purpose. These tubes will be used to measure the substrate remaining in the sample.

3. Label four more test tubes A, B, C, and D; mark at exactly 3 cm and 6 cm from the bottom. These tubes will be used for the reaction between enzyme and substrate.

B. Extraction of catalase

Follow the instructions in Topic 3, Part IB, for preparing catalase solution from liver.

In the following experiments, catalase will be added to a test tube containing the substrate hydrogen peroxide. The reaction will take place over a period of about 3-12 minutes. At minutes 1, 3, 5, and 7, a sample of the reaction mixture will be poured into a test tube containing sulfuric acid, which will immediately inactivate the enzyme. At the end of the reaction, the amount of hydrogen peroxide in each of the acid mixtures will be determined by adding 2% potassium permanganate solution ($KMnO_4$), until the purple color no longer disappears. The amount of permanganate added to reach this *end point* is a measure of the hydrogen peroxide present.

PART II. EXPERIMENTS WITH CATALASE
A. Effect of time on the rate of reaction

1. Put 3% hydrogen peroxide in Test Tube A, up to the 6-cm mark.

2. Add six drops of catalase stock solution; *record the time.*

3. Stir with thin glass rod.

4. Exactly one minute after adding the enzyme, pour a sample of the reaction mixture into Test Tube 1 up to the 1-cm mark; shake.

5. At 3, 5, and 7 minutes after adding the enzyme, repeat the procedure in step 4, using Test Tubes 2, 3, and 4 respectively. Shake the tube each time after adding the reaction mixture to the acid.

6. After all the samples have been taken and inactivated, use the following procedure on each of them to determine the amount of hydrogen peroxide remaining: Add KMnO₄ (potassium permanganate), using the special dropper reserved for this purpose, until the purple color of the permanganate no longer disappears. Add the potassium permanganate by dropperfuls at first, then more slowly, by drops, toward the end point. Ideally, the mixture should be a very faint pink at the end point. The reaction which occurs is as follows:

$$5\ H_2O_2 + 2\ KMnO_4 + 4\ H_2SO_4 \rightarrow 2\ KHSO_4 + 2\ MnSO_4 + 8\ H_2O + 5\ O_2$$
$$\text{(purple)} \qquad\qquad\qquad\qquad \text{(colorless)}$$

The column of fluid *above* the 2-cm mark represents the total of KMnO₄ added. The height of this column, measured to the nearest millimeter, will be used as an inverse index of catalase activity (i.e., the higher the column, the more hydrogen peroxide remained and the less was decomposed by the catalase.).

7. Record (in Table 1) the height of the potassium permanganate column (in millimeters) for the samples at 1, 3, 5, and 7 minutes. Plot the data on Graph A. Use the tables and graph grids provided at the end of Topic 3, but be sure to change the labels and captions properly.

8. Is the activity of the enzyme increasing, decreasing, or remaining constant over successive units of time? Explain your results.

B. Effect of catalase concentration

In three clean test tubes, prepare dilutions of the catalase stock solution, as follows:

CONCENTRATION	AMOUNT OF STOCK	AMOUNT OF WATER
3/4	3 cc	1 cc
1/2	2 cc	2 cc
1/4	1 cc	3 cc

With each of these diluted stock solutions, repeat Part A above. Use Test Tubes B, C, and D filled to the 6-cm mark with H₂O₂. Before adding six drops of your enzyme dilution, be sure to prepare fresh sample tubes 1, 2, 3, and 4, with sulfuric acid. Make sure that the tubes in which the reactions will occur are also appropriately labeled. Wash the stock dropper thoroughly between uses.

Using your data, also plot on Graph A the potassium permanganate added vs time, for each of the three concentrations. How does enzyme activity vary with enzyme concentration?

Before beginning more experiments, make sure that all the test tubes you have used so far are properly cleaned with detergent solution, rinsed, and dried.

C. Effect of temperature

Measure enzyme activity in the usual manner, at 37°C and 10°C. Choose a concentration of catalase that seems suitable on the basis of your previous observations. The catalase and substrate should be brought to the testing temperature before they are mixed. Use cold water and ice in a beaker to establish and maintain a 10°C water bath; mix cold and hot water from the tap to get a 37°C water bath. Check the temperature and record it at the beginning and at the end of the experiment. Record the results, and plot them on Graph B, comparing them with the room-temperature results of the selected concentration. One good way to make the comparison is to have all three plots together on the same graph.

From these data, what can you conclude about how temperature affects enzyme activity?

D. Effect of pH   Measure enzyme activity with the reaction mixture buffered at pH 5, 7, and 9. Use the same procedure as in Part A, except put the desired buffer in the testing tube to the 3-cm mark; then bring to the 6-cm mark with hydrogen peroxide and shake. Add only 12 drops of stock enzyme to maintain a reasonable rate of reaction.

Record the results and plot them together on Graph C. What is the effect of pH on enzyme activity? At what pH would catalase normally have to act?

PART III. ENZYME INHIBITION   Any treatment that changes the structure of an enzyme or interferes with the formation of the complex between enzyme and substrate, will have an inhibiting effect upon the reaction, and may eliminate it altogether. Measure the activity of catalase after it has been treated with:

A. Heat   Place a small test tube containing a few cc of catalase solution into a boiling water bath for two minutes. Cool the boiled catalase and begin your experiment. Record your results and plot them on Graph D. How does boiling affect the enzyme? the reaction?

B. Trypsin   To ten drops of enzyme solution in a small test tube, add an equal amount of 5% trypsin. (What is trypsin?) Place the test tube in a 37°C water bath. After 30 minutes or longer, cool the trypsin-treated enzyme and measure enzyme activity as usual except for using 12 drops of trypsin-treated enzyme solution to maintain a reasonable rate of reaction. Record the results, and plot them on Graph D.

C. Hydroxylamine   Hydroxylamine attaches to the iron atom (a part of the catalase molecule), and thereby interferes with the formation of the enzyme-substrate complex. To ten drops of stock catalase in a test tube, add an equal quantity of 10% hydroxylamine. Shake well and allow to stand for one minute. Measure enzyme activity in the usual manner, except for using 12 drops of treated catalase to maintain a reasonable rate of reaction. Record the results, and plot them on Graph D.

PITHING OF A FROG

In many experiments with frogs, the central nervous system must be inactivated in order to eliminate sensation and response. This may be accomplished by anesthesia, but most anesthetics wear off, and they may impair functioning of the internal organs; a relatively normal condition can be maintained for several hours with a simple operation known as **pithing.**

1. Obtain a frog that has been cooled by refrigeration or with crushed ice. Dry the animal on a paper towel.

2. Hold the frog in your fist, dorsal side up, with the head between your thumb and index finger. Now put your index finger atop the head and press down gently, bending the spinal cord at the neck.

3. With your free hand, bring the point of a *dull** dissecting needle posteriorly, down the middorsal line from the eyes, until you feel the first indentation in the spinal column, just behind the skull. Set the dull needle aside, and with a *sharp* dissecting needle, puncture the skin at this point, taking care to make a hole *only* in the skin.

4. Now insert the dull needle into the spinal cord at this point, and push forward, parallel to the external surface, as far as it will go. Move the needle to the right and left, destroying the brain. This *anterior* pithing destroys all of the brain, but leaves the spinal cord intact. Complete destruction of the CNS entails pithing of the spinal cord, described in step 5.

5. Slowly bring the needle back to its point of entrance into the body; without taking it out, turn it at a right angle to the body surface (pointing directly in), and then turn the handle back toward the head, so that the pointed end, heading posteriorly, is again about parallel to the external surface. Gently push the needle into the spinal cord, as far as it will go. If it has entered the spinal cord, the legs will jerk out straight. When you remove the needle, rotate it slowly so that all the nerves of the cord are disconnected.

* Dulled by a grindstone or on a cement floor.

APPENDIX 6 USE OF OIL-IMMERSION OBJECTIVES

The working distance of an oil-immersion (90-100X) objective is less than one millimeter, which means that the nose of the objective must almost touch the slide for the specimen to be in focus. The oil-immersion field is much smaller than the high-power field; the depth of focus with oil immersion is also very small. Due to the number of lenses in the oil-immersion objective through which light must pass, the light intensity almost always needs to be increased considerably.

In order to obtain maximum resolution from an oil-immersion objective, the layer of air between the objective nose and the specimen is replaced by a drop of oil, into which the objective nose is carefully immersed. This procedure eliminates much of the refraction and loss of light that occur when light passes from a slide into the air before reaching the lens of a dry objective.

TO USE THE OBJECTIVE

Obtain an oil-immersion objective in its case, remove the objective carefully, and put the case in a safe place. Screw the objective into the nosepiece of your microscope.

After you have located the detail under high power, center it carefully in the high-power field and then follow the directions below:

1. Raise the body tube high enough above the slide so that you can easily swing the oil-immersion objective into position. Deposit a drop or two of immersion oil on the slide, and on the condenser lens, too, if you have a condenser. Click the objective into position.

2. *While watching from the side*, slowly lower the objective until its tip makes contact with the oil. Then, looking very closely at the slide, lower the objective a little more, until you can see that it is nearly touching the slide.

3. While looking into the ocular (with both hands on the fine adjustment) very slowly raise the objective. As you do this, you should bring the object into focus. If you do not, you may not have gotten the objective close enough to the slide; try again, and then ask for help if you continue to have difficulty.

TO CLEAN UP

Raise the body tube and wipe off the slide and objective with lens paper to remove most of the oil. The remaining oil can be cleaned off with a little xylol.

APPENDIX 7 MICROBIOLOGICAL TECHNIQUE *by Steven C. Carlin*

PART I. GENERAL TECHNIQUES

Bacteria and fungal spores are everywhere in the laboratory; they are in the air, on the floors, on glassware and tabletops, and on human beings. This makes **contamination** by stray organisms a serious problem when experimenting with microorganisms. Sometimes it is possible to design an artificial environment that will allow the growth of only the desired organism, but generally the combination of nutrients and physical conditions used to cultivate one microorganism is also highly favorable for the growth of many other microorganisms. Therefore, the design of a suitable nutritive solution (**culture medium**) is a necessary, but not sufficient, condition for growing a population of cells belonging to a single species of microorganism (a **pure culture**) in the laboratory. A sufficient condition exists only after precautions have been taken to free the culture medium of unwanted microorganisms and to prevent its contamination during and after the introduction (**inoculation**) into the medium of a number of cells (the **inoculum**) of the desired organism. The process whereby any stray organisms in a freshly prepared medium, or on any containers or instruments coming in contact with the medium, are killed or removed is called **sterilization,** and the procedure by which microorganisms are manipulated without outside contamination is called aseptic or **sterile technique.**

A. Sterilization

Sterilization of media is usually effected by killing any contaminating cells or spores with superheated steam (autoclaving), unless the medium contains heat-labile components, in which case it is sterilized by filtering it through one of several special materials that prevent the passage of bacteria. In the former method, the steam is superheated by increasing the pressure in the sterilization chamber (autoclave); this is necessary because the spores of some bacteria and fungi are so resistant that they can withstand steam at 100°C or dry heat at much higher temperatures. In the filtration method, all glass or plastic ware must be previously sterilized by autoclaving or by chemical disinfection, respectively. Liquid media are usually sterilized in flasks or test tubes that have been plugged with cotton; the cotton prevents the entry of contaminants after sterilization without restraining the free interchange of gases. Solid media are usually prepared by adding a dehydrated gelling agent to a liquid medium prior to autoclaving; the medium comes out of the autoclave in the liquified state and is poured into sterile test tubes or petri dishes, where it solidifies upon cooling. Sterility of the medium is meaningless, however, unless the person using it is proficient in sterile technique.

B. Sterile technique

Sterile technique is an art that requires some practice to master, but that becomes second nature after a short time; your instructor will demonstrate various aspects of this technique for you at appropriate times during the course. You should *always* use sterile technique when working with microorganisms, even if what you are doing does not actually require it. This is especially important when you are working with conidiating fungi or potentially pathogenic bacteria. Carelessness when making a slide of *Neurospora,* for example, can lead to the release of mil-

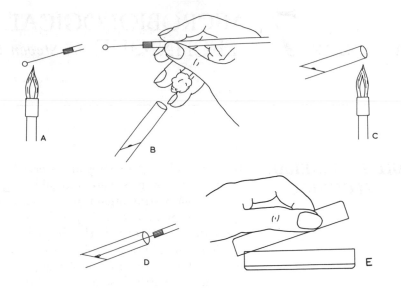

FIG. A7-1. Sequence of steps in a typical sterile transfer. (A) Flame the wire loop with Bunsen or alcohol burner. (B) Remove cotton plug with little finger or between fourth and little fingers. (C) Flame mouth of tube. (D) Remove portion of culture, repeat C, and replace cotton plug. (E) Lift top from petri dish. It is not necessary to flame the lip of the petri dish. Streaking operation is shown in Figure A7-2.

lions of conidia into the laboratory air; these conidia will remain in the air for months and will make contamination by *Neurospora* a serious problem for anyone working in that laboratory. The use of sterile technique minimizes the number of conidia or cells contributed to the laboratory air. This indirectly increases the effectiveness of the direct measures of sterile technique, measures aimed at preventing pure cultures from becoming contaminated during transfer operations. Some of the precautionary measures embodied in sterile technique are described below; many of the techniques involved are illustrated in FIGURE A7-1.

1. Always sponge off your laboratory bench with a germicide solution before making any sterile transfers. Why? Repeat this operation as part of laboratory cleanup.

2. Work quickly, but disturb the air about you as little as possible. Have all the necessary equipment at your desk before you carry out sterile transfers, so that delays are unnecessary.

3. Take special care in opening containers; use of the following methods minimizes the possibility of contamination by organisms in the air or on your fingers.

 a. Hold test tubes or flasks at an angle when removing cotton plugs (or any other bacteriological caps or stoppers) from them (FIGURE A7-1). Making sure that the cotton plug is loose, remove it either by grasping it with your little finger or by holding it between your fourth and little fingers, whichever seems more comfortable to you; keep it in that position in your hand until you replace it. *Never place a cotton plug down on the laboratory bench* (why?).

 b. Lift the lids of petri dishes only enough to permit easy access of transfer instruments (FIGURES A7-1 and A7-3). *Never completely remove the lid of a petri dish.*

4. Get into the habit of *flaming* transfer instruments and the lips of all containers, both before and after making a sterile transfer; this sterilizes the transfer

instrument and prevents organisms on the outside lip of the container from being introduced accidentally into the culture. Flaming consists of heating the object in the flame of either an alcohol burner or, preferably, a Bunsen burner. If you must use an alcohol burner, pull the wick out to make the flame as tall as possible; if a Bunsen burner is available, use the hottest part of the flame, namely the tip of the inner cone.

The transfer instruments most commonly used in microbiology are sterile pipettes and wire inoculating loops. It is a good idea to briefly flame the tips of pipettes, even though they have previously been sterilized; make sure that you let the tip cool, however, before introducing it into a living culture (why?). To sterilize a wire inoculating loop (FIGURE A7-1), heat the entire length of the wire in the flame until the wire is red-hot. Start at the handle and progress toward the tip, so that any moisture at the tip will not be heated suddenly and thus splatter about the laboratory. (This is especially important if the loop has been used to transfer fungal spores or pathogenic bacteria.) Let the loop cool a few seconds in the air, or cool it on sterile agar, before using it to transfer organisms. *Always* remember to flame the loop again before putting it back down on the laboratory bench.

A typical sterile transfer might involve the following sequence of events (FIGURE A7-1):

a. Flame an inoculating loop until it is red-hot, or lightly flame the tip of a sterile pipette.

b. Remove the cotton plug from the tube or flask containing the culture that you are using as a source of inoculum, making sure not to touch the sterile loop or pipette against anything.

c. Flame the mouth of the container.

d. Remove a portion of the culture.

e. Again flame the mouth of the container and replace the cotton plug.

f. Open the container to which you wish to transfer the inoculum and, if this container is a test tube or flask, flame it; it is unnecessary to flame petri dishes, since the dish is constructed in such a way that the lip of the bottom half remains sterile as long as the lid is only partially lifted off.

g. Transfer the inoculum to the medium.

h. Flame the transfer instrument and the lip of the container you have just inoculated.

Further refinements of this general sequence are described in the following section on pure-culture methods.

PART II. PURE-CULTURE METHODS

The microbial community in any particular habitat is made up of many different species. Hence, whenever microorganisms are isolated from nature there is always the problem of separating one organism from another. The small size and paucity of distinct morphological characteristics of microorganisms, especially bacteria, would seem to make the problem of obtaining pure cultures insoluble. Bacteriologists, however, have devised the so-called **pure-culture methods** for separating mixtures of microorganisms into pure cultures of their component species.

The common denominator of all the pure culture methods is the use of a medium solidified with an inert agent like agar. The solidifying agent physically separates the cells of the inoculum, and the ready supply of nutrients in the medium allows the cells to multiply very rapidly. If the medium is incubated under suitable conditions for a day or so, each viable cell grows into a macroscopically visible colony of cells. These colonies appear on or within the agar and are usually about the size of a pinhead; they often have distinct morphological characteristics. In principle, it can be assumed that if measures are taken to break up clumps of cells prior to inoculation and if the colonies that develop after incubation are well-isolated from one another, each colony represents the progeny (often called a clone) of a single parent cell. Theoretically, then, a mixture of morphologically different colonies, each composed of cells belonging to a single species, should result when a mixture of microorganisms is used as an inoculum for a petri dish of an agar medium. (The inoculation of a solid medium is usually referred to as a **plating operation.**) Therefore, isolating a pure culture should be a simple matter of choosing, as a source of inoculum for the culture, any desired colony from the assortment of colonies arising from the primary plating. In practice, however, it is usually a good idea to use cells from an isolated colony to make a second plating; only if all the colonies on this second plate are morphologically homogeneous can it be assumed that the culture is indeed pure. This is necessary because there are various practical situations in which isolated colonies might harbor cells of a contaminant organism. For example, a mixed colony might arise if two morphologically similar cells in the inoculum were deposited very close together, or if cells that remained dormant on the medium employed for the isolation were inadvertently incorporated into a colony. Irrespective of these complications, the basic requirement for the successful isolation of a pure culture is a plating method that deposits the cells of the inoculum far enough apart to yield isolated colonies after growth. Three such methods, each with some advantages and disadvantages, are in common use: the streak-, pour-, and spread-plate methods.

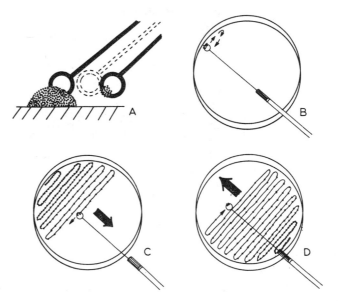

Fig. A7-2. Streaking an agar plate. Black dots represent bacterial cells, highly magnified to show how cells are deposited during movement of the wire loop over the agar surface. A colony will develop wherever a viable cell was deposited. (A) Remove tiny portion of a colony with a sterile wire loop. (B) Touch loop to agar and begin streaking. (C) Streak back and forth toward you, until you reach the center of the plate. (D) Turn the plate halfway round and streak away from you until the entire plate has been streaked.

A. Streak-plate method

For routine isolation of pure cultures the **streak-plate method** (Figure A7-2) is most commonly used. In this method a drop of a liquid cell suspension or a tiny portion of a colony is picked up with a sterile wire loop and inoculated onto agar by streaking the loop back and forth in parallel lines across the agar surface. The loop deposits cells on the agar as it moves across the surface. At first the cells are deposited very close together, but as the number of cells remaining on the loop decreases, the distance between the deposited cells increases. Thus, although the colonies along the first part of the streak may run together, the distance between colony centers increases along the path of the loop, until eventually the colonies are separated from one another. While this method has the advantage of simplicity and speed, it cannot be used to quantify the number of cells in a sample; the spread- and pour-plate methods are used when population counts are desired. The detailed procedure for plating a sample by streaking is described in the steps below, and key points are illustrated in Figure A7-2.

1. Secure a suitable inoculum with your wire loop; be sure to use sterile technique. If the source of your inoculum is a liquid cell suspension, merely dip the loop into the liquid. If it is a colony on a solid medium, just touch the loop to the colony; never pick up an entire loopful of cells, unless you are specifically directed to use a "heavy" inoculum (as in making the cross-streak in the experiments on microbial interactions in Topic 30). (*Note:* If the colony is very hard or is buried in the agar, you may have to dig up a tiny portion of the colony with your loop.)

2. Tilt the lid of the petri dish to be inoculated, and touch the inoculating loop to the agar surface near the point most distant from you.

3. Streak the loop sideways along a horizontal line until you reach the edge of the agar, make a hairpin turn, and streak in the reverse direction along a line parallel to the first. When you reach the opposite edge, again reverse direction. Continue streaking back and forth in a continuous motion, making each successive streak slightly closer to you; keep the streaks as close together as possible. Be careful not to dip the loop into the agar; just let it glide over the surface. When you reach the center of the plate, spin it halfway around and continue streaking along lines parallel to the previous streaks, now making each successive streak farther away from you. By making this reversal, you will avoid the annoyance of having the lip of the petri dish interfere with the motion of the loop sometime before the streaked path has progressed all the way across the agar.

4. Flame the loop and replace the lid of the petri dish.

5. Turn the dish upside down and incubate it under appropriate conditions. If the dish is incubated right-side up, condensation is likely to form on the cover and to drip onto the agar surface, causing the colonies to run together.

B. Pour-Plate method

For obtaining isolated colonies of fungi, the **pour-plate method** (Figure A7-3) is usually chosen. In this method, an appropriate number of vegetative cells or spores is mixed with a tube of plain agar in the liquid state and then poured in a layer over the surface of an agar medium in a petri dish. (Alternatively, the cells may be suspended in a liquified agar medium and poured into an empty petri dish.) Although solid agar must be heated to 100°C in order to melt it, once melted it remains liquid until approximately 40-42°C. The vegetative cells of many microorganisms can withstand a temperature of 45°C for short periods of time; their resting spores (e.g., bacterial endospores and *Neurospora* ascospores) can withstand this and higher temperatures for prolonged periods of time — in fact, such resting spores are often *activated* by heat. This temperature tolerance allows the cells or spores to be evenly distributed throughout the liquid agar, and then fixed in place by letting the agar solidify soon after pouring. The nutrients from the lower layer of agar diffuse upward into the newly poured layer contain-

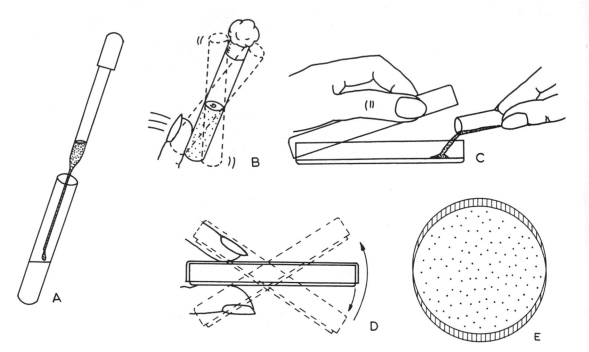

Fig. A7-3. Making a pour plate. The black dots represent individual spores, highly magnified to show how they are suspended in the liquid agar and then fixed in place when the agar solidifies. (A) Inoculate the liquid agar with bacterial cells or fungal spores. (B) Tap the tube briskly to suspend the spores in the agar. (C) Pour the liquid agar into a petri dish of already-solidified agar. (D) Tilt the petri dish back and forth to spread the liquid agar over the surface of the solidified agar. (E) The cells or spores are fixed in place in the agar "overlay."

ing the inoculum; these nutrients allow the cells or spores to develop into colonies. (Whereas a bacterial colony consists of the progeny of a single cell, a fungal "colony" is merely the mycelium resulting from a single propagative unit.) The colonies will develop throughout the upper agar layer, not only on the surface.

Isolated colonies, however, will develop only if the inoculum contained neither too many nor too few cells. If the inoculum was too concentrated, the colonies will be too crowded; if it was too dilute, few if any colonies will develop. In practice, therefore, several plates are usually inoculated, each with a different dilution of the original sample. How a microbiologist might go about securing a proper-sized inoculum by dilution is discussed in C (the spread-plate method). Having obtained isolated colonies on a pour-plate inoculated with a known dilution of a sample, it is possible to estimate the number of vegetative cells or spores in that sample; how this is done is also explained in C.

You will use the pour-plate method in Topic 18 to make it possible to count the number of white and pink colonies developing from a suspension of *Neurospora* ascospores of mixed genotype. You will be given an ascospore suspension that has been diluted to contain an appropriate number of spores for counting. The procedure for preparing a pour-plate, and the way to make fungi with spreading habits grow as isolated colonies, is described in Topic 18, p. 208.

When using the pour-plate method with the vegetative cells of bacteria, the cells are suspended in the liquid agar at 45°C, and the agar is poured immediately; the procedure for mixing and pouring the agar is the same as that described for *Neurospora*. Pure cultures of anaerobic bacteria that must be grown in oxygen-free environments can be obtained by a modification of this method in which the agar suspension of cells is allowed to solidify in a test tube instead of in a petri

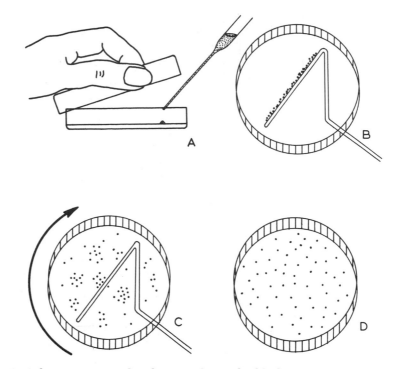

Fɪɢ. A7-4. Inoculating an agar plate by spreading. The black dots represent individual bacterial cells, highly magnified to show how cells are deposited on the agar as the liquid suspending medium is spread over the surface and gradually worked into the agar gel. (A) Place the specified volume of a liquid inoculum on the agar surface. (B) Touch a sterile glass spreader to the drop of liquid and let the drop spread out along the rod. (C) Press the glass rod gently against the agar with one hand and hold the lid in place with the other hand, while a helper turns the plate round and round. (D) Continue turning the plate until all the liquid has been worked into the agar. Replace the lid and invert the plate for incubation.

dish; by sealing the surface with a mixture of sterile vaseline and paraffin, oxygen is prevented from entering the agar.

C. Spread-plate method (optional)

For estimating the number of microorganisms in a sample, the **spread-plate method** (Fɪɢᴜʀᴇ A7-4) is often used. In this method, a small volume of a dilute liquid cell suspension is placed on the center of the agar surface, and the liquid is then spread over the surface with a sterile glass rod, bent to form a handle and a spreading surface. During the process of spreading, slight pressure is applied to the spreading rod to work the liquid gradually into the agar gel. Providing that the inoculum was sufficiently dilute and that the spreading was done correctly, isolated colonies will appear uniformly over the plate. If too many cells were inoculated onto the plate, the colonies will merge to form a confluent mass of cells.

In practice, therefore, a **dilution series** is usually made prior to inoculation, and several plates are inoculated, each with a different dilution of the original sample. Since it is not uncommon for a liquid culture of bacteria to contain 10-100 million (10^7-10^8) cells per milliliter, a series of successive ten-fold dilutions must be made to get into the proper range; each of these dilutions is made by mixing one volume of the previous dilution with nine volumes of a sterile, isosmotic solution. By knowing the total dilution and the volume (usually 0.1 ml) of the final dilution that was plated, the number of viable cells in the original sample can be calculated from the number of colonies appearing on the plate. This, of course, presupposes that each viable cell in the inoculum gives rise to one and only one colony. This assumption is valid as long as all the liquid has been pressed into the agar during spreading. If a liquid film were to remain on the agar for any length of time before evaporating or being absorbed, the cells would be able to move freely during growth; the number of colonies developing would greatly exceed the number of cells inoculated.

If counts are unnecessary, isolated colonies can usually be obtained by an approximate dilution series in which a loop is used instead of a pipette. Where you will have the option to use the spread-plate technique (Topic 30), you should be able to get satisfactory results without making any dilutions; this is because only a very small number of the cells in the inoculum will be able to grow under the conditions you will be using. The mechanics of inoculating a sample by spreading are described below and illustrated in FIGURE A7-4.

1. Place the specified volume of the sample, or an appropriate dilution thereof, on the center of the plate to be inoculated; use sterile technique.

2. Sterilize a glass spreading-rod by dipping it into a beaker of alcohol and passing it through a flame to ignite the alcohol.

3. Tilt the lid of the petri dish and spread the liquid evenly over the surface of the agar by holding the spreader stationary while a helper turns the plate round and round. Press gently on the spreader to work the liquid into the agar; be careful not to press too hard, however, for this will crack the agar. Continue turning the plate until no liquid is visible on the agar when you lift the spreader off the surface.

4. Dip the spreader into alcohol and flame it. Incubate the plate upside down, as in the previous method.

PART III. STAINING TECHNIQUE

Microorganisms are usually colorless, and they differ only slightly from water in the extent to which they bend light rays; therefore, they provide very little contrast to the surrounding medium. This makes microscopic study of their structure very difficult, especially in the case of bacteria whose size is near the limit of resolution of most student microscopes. Although it is possible to increase the contrast between living cells and the surrounding medium by using microscopes equipped with special optical systems, microorganisms are often made visible by staining the cells with colored dyes prior to microscopic examination. Preparing a mount for staining involves smearing a suspension of the cells to be examined onto a clean glass slide and then fixing the cells to the slide by heating. **Fixation** kills the cells and coagulates their protoplasm, thus immobilizing cell structures and causing the cells to adhere to the slide; this is necessary to prevent the cells from being washed off the slide during the staining procedure. The dyes that are used for staining have a higher affinity for the chemical environment of protoplasm than they do for water. When a solution of one of these dyes is placed on a fixed smear of cells, the dye molecules accumulate within the cells. After the excess stain has been washed off with water and the slide has been dried, only the cells retain the dye molecules; the cells are thus clearly visible under the microscope as colored objects against a transparent background. Besides simple stains that uniformly stain any cells on the slide, there are specialized stains that react only with certain classes of organisms or that only stain certain structures of cells. The former so-called **differential stains** are very useful in the classification of microorganisms and the latter **structural stains** reveal structures such as flagella and spores. Mastery of the simple staining technique presented below will aid you immeasurably in your study of microorganisms.

1. Place a drop of cell suspension on a clean glass slide; use your inoculating loop and make sure not to contaminate the culture from which you remove a sample. If you wish to observe cells grown on a solid medium, place a drop of water on a slide and mix a tiny portion of a colony with it. The resulting suspension should appear only slightly cloudy (turbid); avoid making the suspension too heavy. The suspension should look like a glass of water to which several drops of milk have been added.

2. Smear the drop over the surface of the slide, using your wire loop.

3. Dry the smear by passing the slide through a flame several times. Make sure that you do not burn the cells by inadvertently turning the slide upside down in the flame, and do not let the slide get too hot; the slide should feel warm, not hot, to the back of your hand.

5. Flood the fixed smear with several drops of stain, and let the cells absorb the dye for the amount of time specified for that particular dye.

6. Wash off the excess stain by holding the slide vertically in the sink and directing a gentle stream of water at the upper part of the slide on the side *opposite* the stained side. The water will flow around to the stained side and will remove excess stain without washing off the cells, as might happen if the water were directed at the smear.

7. Dry the slides gently between sheets of blotting paper or paper towels. Let the paper absorb the water; do *not* rub the slides to dry them.

8. Observe the stained slides under high power of your microscope; use oil-immersion (Appendix 6) if available.